Prentice Hall Advanced Reference Series

Physical and Life Sciences

Binkley   *The Pineal: Endocrine and Nonendocrine Function*
Eisen   *Mathematical Methods and Models in the Biological Sciences: Linear and One-Dimensional Theory*
Eisen   *Mathematical Methods and Models in the Biological Sciences: Nonlinear and Multidimensional Theory*
Esogbue   *Dynamic Programming for Optimal Water Resource Systems Analysis*
Fraser   *Clastic Depositional Sequences: Processes of Evolution and Principles of Interpretation*
Jeger, Ed.   *Spatial Components of Plant Disease Epidemics*
McLennan   *Introduction to Nonequilibrium Statistical Mechanics*
Plischke and Bergersen   *Equilibrium Statistical Physics*
Sorbian   *Structure of the Atmospheric Boundary Layer*
Valenzuela and Myers   *Adsorption Equilibrium Data Handbook*
Walchen   *Genetic Transformation in Plants* (Open University Press)
Ward   *Fermentation Biotechnology*
Warren   *Evaporite Sedimentology: Importance in Hydrocarbon Accumulation*

CLASTIC DEPOSITIONAL SEQUENCES

Processes of Evolution and Principles of Interpretation

Gordon S. Fraser

*Indiana Geological Survey and
Indiana University
Bloomington, Indiana*

Prentice Hall
Englewood Cliffs, New Jersey 07632

Library of Congress Cataloging-in-Publication Data

Fraser, Gordon S.
  Clastic depositional sequences.

  Bibliography: P.
  Includes index.
  1. Facies (Geology)   2. Sedimentation and deposition.
I. Title.
QE651.F73   1989   551.7   88-32547
ISBN 0-13-851767-3

Cover design: Edsal Enterprises
Manufacturing buyer: Mary Ann Gloriande

 © 1989 by Prentice-Hall, Inc.
A Division of Simon & Schuster
Englewood Cliffs, New Jersey 07632

Printed in the United States of America

10 9 8 7 6 5 4 3 2 1

ISBN 0-13-851767-3

Prentice-Hall International (UK) Limited, *London*
Prentice-Hall of Australia Pty. Limited, *Sydney*
Prentice-Hall Canada Inc., *Toronto*
Prentice-Hall Hispanoamericana, S.A., *Mexico*
Prentice-Hall of India Private Limited, *New Delhi*
Prentice-Hall of Japan, Inc., *Tokyo*
Simon & Schuster Asia Pte. Ltd., *Singapore*
Editora Prentice-Hall do Brasil, Ltda., *Rio de Janeiro*

PSC   6-14-89

Contents

Preface   *xi*

Acknowledgments   *xiii*

Introduction   *1*

PART I   FOUNDATIONS   *3*

Chapter 1   Physical Controls on Sediment Accumulation   *5*

Stream Flow   5
Sediment Entrainment   6
Interaction with the Substrate   6
Regimes of Flow   7
Sediment Movement by Wind   8
Wave-induced Sediment Movement   10
Mass Movement of Sediment   12
Deposition from Suspension   14

Chapter 2   Principles of Sequence Analysis   *15*

Geomorphic Principles   15
    Uniformity   15

Threshold Stress   16
Complex Response   17
Time   18
Stratigraphic Principles   22
Continuity Principle   22
Law of Superposition   22
Facies Concept   22
Walther's Principle   23

Chapter 3   Classification of Environments   24

Discussion: Application of Principles to Analysis of Stratigraphic Sequences   27

PART II   CONTINENTAL ENVIRONMENT   29

Chapter 4   Arid System   31

Introduction   31
Components   32
Alluvial Fan   32
Intermittent Streams   40
Playas and Saline Lakes   48
Ergs   53
The Arid System in the Rock Record   69

Chapter 5   Humid System   76

Introduction   76
Components   77
Alluvial Fans   77
Alluvial Plains   87
Lakes   119
The Humid System in the Rock Record   127
Discussion: Sequence Evolution in Continental Environments   131

PART III   COASTAL TRANSITION ZONE   135

Introduction   135

Chapter 6 Tide-Dominated Coastlines 138

Sedimentary Processes and Products 138
Intrinsic and Extrinsic Controls on Evolution of Tidal Flat Sequences 144
Tidal Flat Deposits in the Rock Record 147

Chapter 7 Muddy Coastlines 151

Sedimentary Processes and Products 151
Intrinsic and Extrinsic Controls on Muddy Shoreline Sequences 153
Muddy Coastline Sequences in the Rock Record 155

Chapter 8 Wave-Dominated Coastlines 157

Sedimentary Processes and Products 157
Depositional Sequences of Wave-Dominated Coasts 164
Wave-Dominated Shoreline Sequences in the Rock Record 166
Systems on Wave-Dominated Coastlines 169
    Introduction 169
    Barrier Island/Lagoon Systems 169
    Strandplain Systems 192

Chapter 9 Deltaic Coastlines 202

Introduction 202
Sedimentary Processes and Products 207
    Delta Plain 207
    Delta Front 210
    Prodelta 212
Intrinsic and Extrinsic Controls on the Evolution of Delta Sequences 213
Delta Sequences in the Rock Record 222

Chapter 10 Estuarine Coasts 231

Introduction 231
Sedimentary Processes and Products 232
    Protected Estuaries 232
    Unprotected Estuaries 235
Intrinsic and Extrinsic Controls on Evolution of Estuarine Sequences 238
Estuarine Sequences in the Rock Record 240
Discussion: Stability of Coastal Zone Systems 244

PART IV   SHALLOW MARINE ENVIRONMENT   251

Introduction   251

Chapter 11   Continental Shelves   253

Introduction   253
Sedimentary Processes and Products   257
    Transport Processes   257
    Sediment Facies   261
Intrinsic and Extrinsic Controls on the Evolution of Shelf Sequences   272
Shelf Sequences in the Rock Record   293
    Evolution of Sand Waves   294
    Transgressive Shelf Sequences   297
    Regressive Shelf Sequences   301

Chapter 12   Epicontinental Seas   306

Introduction   306
Sedimentary Processes and Products   306
Discussion: Interaction of Eustasy and Sediment Influx on Evolving
    Shelf Sequences   316

PART V   CONTINENTAL SLOPE TRANSITION ZONE   321

Introduction   321

Chapter 13   Morphology and Sedimentology of Continental Slopes   323

Morphology   323
Classification   326
Sedimentary Processes   329
    Mass Gravity Movements   329
    Basin-generated Currents   331
    Suspension Settling   332
    Nepheloid Layers   333
Slope Sediment Facies   333
    Undisturbed Sediments   333
    Deformed Sediments   335
    Chaotic Sediments   337

Chapter 14   Evolution of Slope Sequences   339

   Intrinsic and Extrinsic Controls on the Evolution of Slope Sequences   339
      Introduction   339
      Sequences on Passive Margins   342
      Sequences on Active Margins   344
      Sequences in Canyons   347
   Slope Sequences in the Rock Record   351
   Discussion: Controls on Cyclic Sedimentation in Evolving Slope Sequences   359

PART VI   DEEP MARINE ENVIRONMENT   363

   Introduction   363

Chapter 15   Components   365

   Submarine Fans   365
      Sedimentary Processes and Products   365
      Submarine Fan Facies   369
      Intrinsic and Extrinsic Controls on the Evolution of Fan Sequences   371
      Submarine Fan Sequences in the Rock Record   377
   Basin Plains   383
      Sedimentary Processes and Products   383
      Extrinsic Controls on Basin Plain Sequences   386
      Basin Plain Sequences in the Rock Record   388

Chapter 16   Deep Marine Systems   390

   Trench–Fan Systems   390
      Sedimentary Processes and Products   390
      Controls on Evolution of Trench–Fan Systems   394
      Trench–Fan Sequences in the Rock Record   399
   Continental Rise–Abyssal Plain System   402
      Sedimentary Processes and Products   402
      Continental Rise–Abyssal Plain Sequences   406

Conclusion: Event Stratigraphy in Clastic Depositional Sequences   411

References   417

Index   451

Preface

Beginning in the early 1960s, Donald J. P. Swift, John Kraft, and their co-workers began publishing a series of papers on the marginal marine and shallow shelf sediments along the east coast of the United States. Until that time I fully accepted the proposition that depositional models derived from the study of modern coastal environments could only be applied with great care because of the effects of the rapid post-Pleistocene sea-level rise.

Kraft, Swift, and their co-workers, however, were studying these sediments as evolving stratigraphic sequences. They were using the rise in sea level to model the response of coastal sedimentary environments to transgression by determining how sedimentary systems changed in response to altered rates of sediment supply, styles of sedimentation, and patterns of deposition. I felt then, as now, that the principles they derived have a profound effect on stratigraphic concepts in the same way that the facies concept revolutionized the study of rock sequences.

I applied these concepts to my own studies of marine cratonic sedimentation, and later I also utilized them in studies of marine sedimentary environments in other tectonic settings. It appeared to me that intensity of tectonic activity, rate of sediment supply, and rate of sea-level rise or fall were related in a complex manner that simultaneously affected coastal, shallow marine, and deep marine settings. Principles that had been derived from studies done on the east coast of the United States apparently could be applied independently of tectonic settings.

At the same time that studies on the east coast were proceeding, a number of geomorphologists were investigating the effect of intrinsic and extrinsic factors on rates of erosion, transportation, and deposition in geomorphic systems. This body of work has been synthesized in a book and several papers written by Stanley

Schumm over the last ten years. I felt that this work had a direct bearing on stratigraphy in that it demonstrated how geomorphic processes affect the manner in which sediment is delivered to a depositional basin and how it is ultimately distributed. These, of course, are among the main factors involved in the evolution of rock sequences.

In the last several years, numerous researchers have attempted to interpret the dynamic processes involved in the evolution of stratigraphic intervals formed by a variety of depositional systems. These studies have gone beyond identification of depositional systems in rock units by reconstructing the sequence of events that occurred during their deposition, and identifying and analyzing the effects of various intrinsic and extrinsic factors that operated. Sedimentologists are able to do this only because facies analysis has been refined to the point that first principles can now be derived from the study of rocks.

These two trends in geology have advanced to the point where I believe it is now possible to make a first attempt at synthesis. I hope to arrange divergent studies into a concept of how geomorphic and sedimentologic processes act together during the formation of rock sequences.

Gordon S. Fraser
Bloomington, Indiana

Acknowledgments

This book is about the complex interaction of the various processes of sediment accumulation, and it was only through the interaction of numerous people that production of this book was possible. I would like especially to single out the efforts of Richard and Barbara Hill and Roger Purcell, who drafted many of the illustrations in this text, and William Hamm and Claire Kiehle who performed the word processing. Not only did they perform these tasks, but in the manner of true friends they provided encouragement and support throughout. In addition, the staff of the Geology Library at Indiana University, including Louis Heiser, Maureen Balke, and Heidi Detrick, performed the essential service of assembling the literature used in this study with good humor and in the spirit of cooperation, in spite of the often unreasonable demands I made on them.

Special thanks must go to George de V. Klein, who provided the initial impetus to write this book. The ideas presented here benefited greatly through his careful review, as well as that of Donald J. P. Swift. I would also like to express my heartfelt gratitude to Arthur Hagner, professor emeritus at the University of Illinois, without whose unsolicited kindness on one afternoon in 1971 this book might ultimately never have been written.

CLASTIC DEPOSITIONAL SEQUENCES

I wonder whether mankind could not get along without all these names, which keep increasing every day, and hour, and moment; till at last the very air will be full of them; and even in a great plain, men will be breathing each other's breath, owing to the vast multitude of words they use, that consume all the air just as lamp-burners do gas. But people seem to have a great love for names; for to know a great many names, seems to look like knowing a good many things; though I should not be surprised, if there were a great many more names than things in the world.

Herman Melville, *Redburn*, Chapter 13

Introduction

Stratigraphy may be considered the mother science of geology. Early philosophers were intrigued by the very apparent layering in sedimentary rocks, as well as by the occurrence of identifiable organic remains. Some of the earliest inductive studies in geology attempted explanations (often very sophisticated) of the origin of stratified rocks.

Post-Wernerian geologists, realizing the vertical complexity and lateral non-conformity of rock sequences, began the systematic classification and interregional correlation of sedimentary rocks. These studies in time, however, became ends in themselves, and stratigraphy evolved into a science based primarily on petrography and concerned more with nomenclature rather than with interpreting earth history. Even the revolutionary concepts of facies and lateral succession were blunted because the more far-reaching implications of these principles were not vigorously applied everywhere. This certainly must have been on Johannes Walther's mind when he stated that only sedimentology could save us from stratigraphy.

But stratigraphy is a fusion science, and it must be evident to most that a complete understanding of rock sequences cannot be attained until the study is undertaken in that context. All the component parts of the science must be utilized, and efforts to describe, classify, map, and correlate rocks cannot be made at the expense of interpreting their origin. Indeed, the most far-reaching implication of the facies concept is that valid correlations cannot be made without first understanding how the rocks were deposited.

Stratigraphy will be operatively defined in this book as the study of the manner in which rock sequences accumulate. The emphasis in this presentation will be directed, therefore, toward an understanding of how the various processes involved in the production, transportation, deposition, and preservation of sediments interact. These may be broadly classified into sedimentologic and geomorphic processes.

Sedimentologic processes include microprocesses that govern fluid-to-grain interactions, mesoprocesses that describe how fluids interact with the substrate in transporting sediment, and macroprocesses that control the lateral disposition of sediments in various depositional settings. Geomorphic processes determine how sediment is provided to the system and how the system will react to changes in intrinsic and extrinsic variables.

In relative geologic terms, sedimentologic processes act instantaneously, but geomorphic processes, which ultimately control the evolution of depositional landforms, act within a framework of geologic time. Rock sequences evolve as the extrinsic and intrinsic factors that govern the development of depositional landforms change through time.

For this reason, the importance of geomorphic processes as they operate in depositional systems will be stressed. Particular emphasis will be placed on cyclical changes in how sediment is produced in the source area, how it moves through transit systems, and how it accumulates in depositional systems. How the action of feedback mechanisms affects these operations will also be analyzed.

The concept of the depositional system as expressed by the "Texas School" will be modified for use in this book, however. A system as used here will be defined as the collection of all those depositional components that can be expected to interact in a mutually dependent manner and can be expected to succeed one another in a predictable series during the evolution of a rock sequence in a given setting. Use of this definition permits us to emphasize the factors that influence accumulation and preservation of sediments and how these vary through time under different combinations of tectonic setting and climate regime.

The concept of *event stratigraphy* will also be used in the analysis of rock sequences. Attempts will be made to describe how events in one part of a system influence events in other parts and how sediments respond to these events in various parts of the system.

Finally, I will present examples of how systems interact during the accumulation of sedimentary sequences found in some of the great sedimentary basins of the world. The emphasis here will be on the style of accumulation, rather than its timing. Events in the evolution of basins are often marked by gross changes in lithology in response to alterations in basic geomorphic and sedimentologic processes. Tectonic events or changes in climate are capable of acting on a basin-wide scale by triggering waves of complex and interactive responses that sweep through the depositional systems in the basin. An attempt will be made in the last section to show how the style of sediment accumulation responded to these changes.

Part I

FOUNDATIONS

1

Physical Controls on Sediment Accumulation

The action of microprocesses and mesoprocesses will be described in this chapter. Although the basic properties of fluids are not environmentally specific, the ways in which they act to transport and deposit sediment are. An understanding of the way these processes operate is the foundation on which facies analysis is based.

Stream Flow

When a fluid flows over a substrate, friction at the base of the fluid column causes a velocity gradient to form, with velocity greatest near the surface of the flow and least at the base. For the fluid to accommodate itself to this velocity profile, the velocity must change from one fluid layer to another, producing shear between the layers. At slow velocities, water layers shear smoothly over one another. Energy is transmitted between layers on a microscopic scale, and the streamlines in the flow remain parallel to one another in a laminar flow.

At a critical velocity, however, flow becomes turbulent, with secondary velocity vectors superposed on the main flow. The factors that determine the critical velocity are fluid density, fluid viscosity, and water depth. These variables are related to the critical velocity by the Reynold's number. For small Reynold's numbers (less than 500), flow is laminar, and for larger numbers, flow is turbulent.

Most flow in nature is turbulent. When gravity forces acting on a turbulent flow are less than the forces needed to bring the flow to rest over a given distance, the flow is said to be tranquil or subcritical. Such a flow usually occurs in the pool reaches of streams.

When inertial forces exceed gravity forces, the flow is said to be shooting or supercritical. Such flows commonly occur in riffles of streams. The factors that determine the flow state are velocity and depth, which are related by the Froude number. At Froude numbers less than 1, flow is subcritical, and as velocity increases and depth decreases there is a smooth transition into supercritical flow at Froude numbers greater than 1. When deceleration occurs, however, water level rises with the formation of an hydraulic jump, and a great deal of turbulence is created.

Sediment Entrainment

The force necessary to move a grain is a function of the particle size of the grain and its submerged weight, viscosity of the fluid, and the shear stress on the substrate. The critical boundary shear stress at which motion is initiated is dependent on the nature of the bed. Cohesionless beds are formed of loose grains that are not bound together in any way. Cohesive beds, on the other hand, are bound together by adhesive and electrostatic forces, among others, and tend to resist erosion, because the critical shear must overcome these forces as well as gravity forces that hold grains to the bed.

Gravity forces may be overcome by lift forces that act where streamlines converge as they pass over a grain. This causes a decrease in pressure, which tends to lift the particle off the bed. Once the grain leaves the bed, however, pressure forces are equalized, because streamlines must then converge both above and below the grain. At this point, the only forces acting on the grain are vertically directed fluid drag tending to keep the particle in suspension and a horizontally directed drag tending to move the particle in the direction of mean flow. Transfer of momentum to the substrate by grains leaving the flow is also an important mechanism in causing entrainment.

The vertically directed drag force is the result of turbulent eddies. To keep grains in suspension, the velocity of the eddies directed in an upward sense must exceed the downward force of gravity. Because the velocity of turbulent eddies normally is only a small fraction of the mean flow, the velocity of the flow must be much greater than the settling velocity of the grains.

Interaction with the Substrate

Stream power is the rate at which a fluid flow does work. The work performed by a stream, including erosion, transportation, and deposition, is measured by the potential and kinetic energies. Potential energy is simply the difference in elevations between two points on the flow, whereas kinetic energy is a function of the mass and velocity of flow, viscosity of the fluid, and frictional effects of the bed. Frictional effects, including those dependent on channel pattern and cross-sectional profile, as well as elements of bed roughness, greatly affect the ability of the stream to do work. Much of a flow's potential energy is, in fact, lost owing to frictional heat losses.

The relationship between all the factors making up the energy profile of a stream can be expressed by the Chezy equation, which relates flow velocity to channel slope and cross-sectional radius, and a variable coefficient, which is a function, primarily, of boundary flow resistance. The relationship can also be expressed by the Manning equation, which takes nearly the same form, but also includes a roughness factor as a measure of flow resistance.

Regimes of Flow

Only in rare cases is enough information available to rate a stream using the Chezy or Manning relationships. But to understand the processes functioning during the accumulation of sediments, it is important to know, at least inferentially, factors like relative flow velocity. Since the work of Simons and Richardson (1961), sedimentologists have been able to compare a given sedimentary structure to its equivalent bedform and to estimate some of the hydraulic conditions under which these structures were formed.

Stability diagrams have been constructed by a number of workers since then showing the combination of grain size and various hydraulic parameters for which various bedforms are stable (Fig. 1.1). The type of structure present in the rock can then be used to estimate the relative state of the flow (Fig. 1.2).

With tranquil flow at Froude numbers very much less than 1, ripples are the stable bedform as soon as sediment transport is initiated. Resistance to flow is great and grains move individually. With increasing flow velocity, grains are carried beyond the crest of the ripple, and turbulence in the flow erodes the bed in front. Straight-crested ripples become more sinuous, and internal bedding of the ripple changes from tabular with an angular basal contact angle to trough-shaped with a tangential basal contact (Fig. 1.1).

With increasing shear, sand waves and dunes form. They occupy a stability field distinct from ripples, but their shape is similar and their response to increasing shear within their field of stability approximates that of ripples. Together with ripples (and low-shear plane beds in coarse sand), dunes and sand waves characterize the lower flow-regime.

At Froude numbers above about 0.8, flow becomes critical, dunes disappear, and a flat bed is formed, with grain transport occurring as a moving traction carpet (Fig. 1.2). Above a Froude number of 1.0, flow is supercritical or shooting. Antidunes form and surface waves are in phase with antidunes on the bed. Plane beds and antidunes form in the upper flow regime where resistance to flow is low.

Sediment Movement by Wind

Wind may entrain grains by lift or by momentum transfer from grains already in motion. The critical velocity necessary to lift grains is a function of the grain size, density of the grain, and density of the fluid. Impact reduces slightly the velocity necessary to induce motion for all grain sizes.

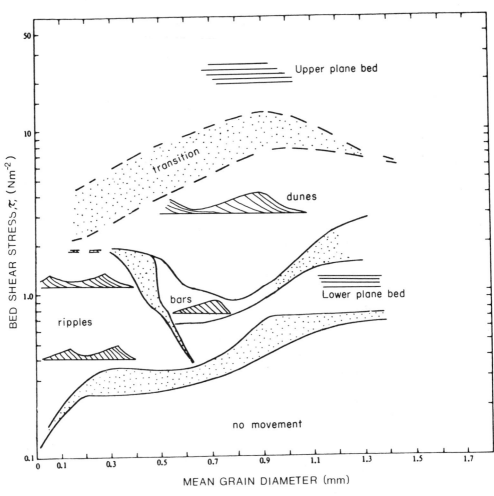

Figure 1.1 The stable fields of various bedforms developed under conditions of uniform flow over a granular bed. Within stable fields, bedforms also change in response to increased bed shear by decreasing slipface angles (after Leeder, 1982).

Once in motion, grains travel in suspension, by saltation, or by surface creep. Suspension occurs only when the vertical velocity component of the mean flow exceeds the settling velocity of the grain. Because of the low relative viscosity of air, grains larger than very fine sand normally cannot be maintained in suspension except near the boundary, where the strongest vertical velocity vectors caused by turbulence occur. The greatest concentration of all grain sizes in suspension occurs near the boundary and decreases upward in relation to the settling velocity of the grains. In addition to turbulence induced by bottom shear, however, other vertical air movements are effective in raising particles upward.

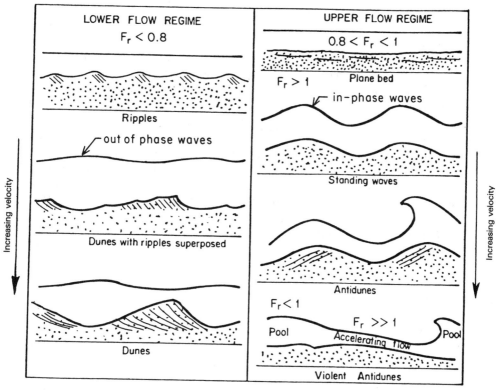

Figure 1.2 Bedforms and expected internal structures observed in flows over sand bed channels (after Simons and Richardson, 1961).

Grain saltation also occurs in water, but it is much more effective in air because downward-directed drag forces are much less than in water and the accelerated particles can rise to greater heights. Grains lifted from the bed are carried forward low to the ground by drag forces along paths that approximate ballistic trajectories. When the particle returns to the bed, it may bounce off again or induce other grains to saltate through momentum transfer. Most of the shear on the bed is imparted indirectly by the air acting through a layer of saltating grains.

Grains too large to rise off the bed through impact and lift forces are driven forward by the impacting grains as a moving carpet at the surface. This movement is termed surface creep, and its magnitude is dependent on the velocity at impact of the saltating grains and, to a lesser extent, on the flow velocity of the air near the bed induced by frictional drag on the saltating grains.

The flow regime theory can be applied after a fashion to airflow. At a critical velocity, grains are set in motion and almost immediately ripples form. These are ballistic ripples rather than hydraulic ripples. They form initially in the lee of obstructions to the flow, where grains at the bed are protected from collision with saltating grains. As ripples grow, they become self-perpetuating as they collect grains coming out of saltation by reducing the angle of impact on their lee surfaces. By this process, ripple trains are formed quickly, with the ripple wavelength dependent on the length of the saltation trajectory (and, therefore, on the wind velocity).

Wind ripples tend to maintain long, straight crests, even during increasing flow velocity, because pronounced turbulence caused by lee eddies is not produced. But, with increasing flow velocity, the amplitude of ripples tends to decrease as the greater impact velocities of saltating grains flattens crests. Eventually, a transition to plane bed develops as velocity is further increased. Bedforms on the scale of subaqueous dunes and sand waves tend not to form in wind flows.

The only true hydraulic bedforms formed by wind are eolian dunes. Flow separation takes place at the crest, but lee eddies are not developed uniformly. At slow flow velocities, sand may move up the windward side of the dunes and then avalanche over the crest, forming leeside laminae. At times, this movement may take place in the form of ripple trains. At faster velocities, some sand may be taken into suspension at the crest and then fall out at different points at the lee of the dune.

Wave-induced Sediment Movement

Wind waves are formed when flowing air exerts a shear stress on a water surface. Energy transmitted to the water moves as a wave that can be described by the trough-to-crest height, crest-to-crest length, and the time period between passage of successive crests past a given point.

In deep water, linear wave theory describes the motion of water particles during passage of a wave. Water particles move in closed orbits that decay exponentially in size downward. Because orbits are closed, no mass transfer is associated with wave motion. In water depths less than about 1:50 of the wavelength, however, orbits are not closed, and a transfer of mass occurs in the direction of wave propagation. In water depths shallower than one-half wavelength, bottom friction becomes sufficient to enhance mass transfer by impeding the seaward return movement under the trough.

Wave-induced bottom shear acts in the direction of wave propagation under the crest and in an opposite direction under the trough. Grains on a cohesionless bed are subjected to these bidirectional forces acting parallel to the bottom. In addition, they are subject to lift forces that are modified by the pressure changes that occur as a wave passes. Near shorelines where bottom slopes are steep, gravity forces may be strong enough to enhance or detract from horizontally directed shear.

Wave base may be defined as the water depth at which significant shear is induced on the bottom by the orbital motion of water particles, normally about one-half wavelength. Initially, this shear affects material such as organic debris and colloidal clays, producing a gentle back and forth motion with passage of the wave. With increasing shear, a similar motion is produced in sand grains, which are organized almost immediately into ripples (Fig. 1.3).

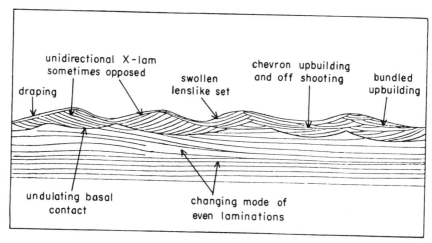

Figure 1.3   Some characteristic wave-formed laminations (after de Raaf, and others, 1977).

True oscillatory ripples are symmetrical and sharp-crested with bimodal dip directions displayed by internal laminae. Most often, however, wave ripples show a preferred orientation of internal structures, reflecting the normal asymmetry of wave-induced shear intensities. Asymmetry in the ripples becomes increasingly pronounced in shallow water.

Ripples form as threshold shear is reached. With increasing shear, a second threshold is reached where ripples disappear and a plane bed state is attained (Fig. 1.4). Larger-scale structures found in zones of wave activity are apparently not related to wave hydraulics but rather to unidirectional flows set up as waves approach shorelines.

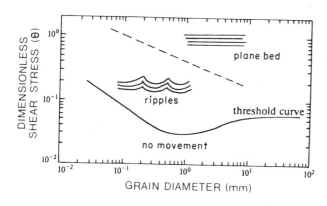

Figure 1.4   Stability diagram for bedforms under conditions of wave-induced oscillatory flow over a granular bed (after Komar, 1976).

Mass Movement of Sediment

Mass sediment movements are those that occur when sediment or, more usually, sediment–water mixtures, move under the influence of gravity. Internal flow is relatively minor in slumps and slides. Movement of the sediment mass occurs along external shear planes, and even shear transmitted into the body of the sediment mass is accommodated by discrete internal shear planes. Slumps and slides occur in both subaerial and subaqueous environments, but even in subaerial environments, water is an important component of the process, acting as a lubricant along basal and internal shear planes.

In the other class of mass sediment movement, fluid or fluidlike flow takes place internally in response to the movement. Water may act as a lubricant in grain-to-grain interactions, but its primary role is in creating excess pore pressure that maintains the fluidity of the flow. The water content determines the nature of the flow, with debris flows, grain flows, fluidized flows, and turbidity currents forming a continuum of flow types in response to increasing fluid content (Fig. 1.5).

The water content of debris flows is sufficiently small that they behave like plastics. Although like fluids in that shear is uniformly distributed through the mass,

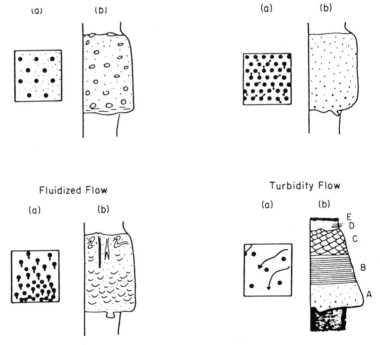

Figure 1.5 Types of gravity-induced mass movements showing mechanisms of sediment support and deposition (a), and expected internal structures and stratification sequences (b) (after Leeder, 1982; Middleton and Hampton, 1973).

debris flows do not occur until a critical shear stress has been exceeded. This critical stress is a function of the internal angle of friction, the viscosity of the sediment–water mixture, the rate of strain, and the cohesion of the material.

In general, flow is laminar in debris flows and grains are supported by buoyancy and resistance to shear of the material, rather than turbulence. Where the density of the material is relatively great, the maximum clast diameter can be quite large. Because of the cohesive nature and great viscosity of the material, debris flows can transport a wide range of grains sizes (Fig. 1.5).

Grain flows are downslope movements of cohesionless sediment that are kept in a fluid state by grain-to-grain interaction. In some types of grain flows, these interactions produce a dispersion through grain collisions, whereas in other types shear is transmitted through the sediment by the fluid when grains approach each other. The fluid in grain flows may be either air or water, but relatively steep slopes are necessary where air is the dispersing medium. On subaerial dunes, for instance, slopes approach the internal angle of friction of the sediment. In subaqueous environments, the critical slope necessary to maintain the flow can be substantially less.

Sedimentation occurs in grain flows when downslope movement slows enough to reduce the frequency or intensity of grain interactions. When this occurs, sedimentation is rapid, even in thick flows, producing a massive, or at best faintly laminated, structure (Fig. 1.5). As water is expelled, discontinuous concave-upward laminae (dish structures) may form as water-escape structures.

Fluid flow through a granular sediment mass may be sufficient in itself to cause grains to disperse and to fluidize the sediments. Pore pressures far in excess of confining hydrostatic pressures may originate in several ways. As water flows over large-scale bedforms in streams, pressure decreases over the crest where streamlines converge, causing an upward flow of pore water to the crest. Cyclic shear or loading on a granular mass may rearrange pore geometries sufficiently to produce excess pore pressures. Lowering of water levels faster than pore fluids can readjust to the new pressure regime can also cause outward flow of pore fluids. Even normal groundwater discharge may suffice to cause a "quick" condition in sands.

The efficacy of fluidization as a transport mechanism of sediment has yet to be demonstrated. Immediately after fluidization occurs, pore pressures begin to equilibrate with the surrounding hydrostatic pressure, and resedimentation begins from the base of the sediment mass upward (Fig. 1.5). Probably its main importance as a sedimentary process is as a trigger for other mass movements, especially grain flows and turbidity flows.

Deformation structures are common responses to fluidization. Injection structures such as pipes or dikes occur where intrastratal liquefaction occurs. Water-escape structures form during the process of resedimentation, and overturned or oversteepened cross-bedding may result from shear acting on fluidized beds. Slump structures may form where fluidized sediments occur on a slope (Fig. 1.5).

Turbidity flows are a particular form of density current. If two fluids of unequal density are placed side by side and allowed to flow together, the heavier will flow under the lighter. No slope is necessary, but the flow will stop when the heavier lies completely under the lighter and a vertical equilibrium profile is attained. Where a slope is present, however, the heavier fluid will flow downslope until shear-induced mixing homogenizes the fluids. The velocity of such a flow is a function of the density difference between the fluids, the slope, and gravitational acceleration, with resistance to flow occurring at both the fluid interface and bed interface.

The density difference in turbidity currents is due to the presence of suspended particles in the flow. Particles are kept in suspension by turbulence produced and maintained by the motion of the flow itself through a process of autosuspension. Absolute density difference between the fluids can also be maintained, given the continued presence of a sufficient slope. Material lost to the flow by turbulent mixing or deposition is added to the body of the flow from the head, where erosion of the substrate takes place.

Flow stops when the density difference between the two fluids is eliminated. This may occur through turbulent mixing or by inability of the head of the flow to add sufficient material to the body and tail to replace that lost through deposition. Once deposition begins, it continues in an accelerating manner. The reduction in the absolute density difference through deposition slows the flow, which, in turn, reduces the ability of the head to erode the substrate and replace the lost material.

The body of sediment deposited by a turbidity flow displays a hierarchical assemblage of characteristics reflecting this rapid deceleration of flow velocity (Fig. 1.5) (Bouma, 1962). Its basal units consist of deposits accumulating under conditions of very high shear stress at the bed. Overlying these are units deposited under conditions of rapidly decelerating flow. The turbidity current loses its capacity to maintain relatively coarse material in suspension, while retaining some capacity for traction transport. The uppermost unit of the sequence is deposited entirely from suspension.

Deposition from Suspension

Once sediment is taken into suspension or is introduced into the fluid in suspension, it will remain in that state as long as the upward-directed forces exceed the gravity force. Upward-directed shear caused by turbulence is the most important of these forces; but even if turbulence stops, two other forces resist the downward motion of particles. Inertial forces resist the lateral displacement of the fluid under the particle as it settles downward, and viscous forces are the result of the resistance of the fluid to shear. For larger particles, inertial forces are more important than viscous forces, whereas the reverse is true for smaller particles. The settling velocity of a particle is a function of the particle's density and diameter, gravitational acceleration, and viscosity of the fluid, and it is this velocity that must be at least balanced by turbulence to maintain the particle in suspension.

2

Principles of
Sequence Analysis

Geomorphic Principles

Geomorphology is the study of landforms, and that branch of geomorphology that studies landforms as physical systems with histories must be considered one of the bases of stratigraphy. The way in which erosional landscapes evolve, for instance, determines the manner in which sediment is provided to points of deposition, and stratigraphic sequences form as depositional landscapes evolve. Schumm (1977) has defined three principles basic to the study of landform evolution. These are uniformity, threshold stress, and complex response. It is through these that the processes affecting evolution of source areas, zones of transport, and depositional areas can be understood.

Uniformity

The modified concept of uniformity does not say that events occurred in the past the same as in the present. Rather, it states that the same processes operated, because the laws of chemistry and physics that operate today also operated in the past. This modification thus allows for progressive changes in the earth's chemistry, which certainly occurred, and also allows changes in the magnitude of physical processes and the rate at which they acted.

The implication of this modified concept is that simple observations of the way things change are not enough to lead to an understanding of stratigraphic sequences; even considering the historical record, our length of observation is entirely too short. Rather, the laws governing the way things change must be understood.

This is especially important considering the possibility that catastrophic events have been important in shaping the stratigraphic succession. Understanding of the great flood that formed the Channeled Scabland (Bretz and others, 1956), for instance, came about not because a modern counterpart existed for an event of that magnitude but because counterparts acting on a much smaller scale existed for the processes.

It must also be remembered that uniformity does not imply continuity. Breaks in the stratigraphic record, from profound unconformities to bedding planes, are a testament to the fact that processes do not act in the same way for very long periods of time.

Threshold Stress

Not only does the way in which physical processes operate vary over even short intervals, but the way in which systems respond is discontinuous even if the stress is applied continuously. In many cases, as stress is gradually increased there is no change in the system until a dramatic response occurs. For instance, shear stress may be applied gradually to a granular bed until a critical point is reached at which sediment movement is initiated. Normally, however, the critical shear is in excess of that necessary to keep the grain in motion, and for a time, therefore, the bed is in a metastable condition. In cases such as this, and others where the stress being applied is external to the system, the point of critical response is known as an extrinsic threshold.

Intrinsic thresholds are those that are responses to internal stress applied as a natural product of the evolution of the system. Channel avulsion occurs, for instance, when the meander belt of a river aggrades to the point of instability in its floodplain. Meander cutoffs occur when a river must steepen its longitudinal profile to account for elongation of its channel as its delta progrades.

In some cases, it is a combination of intrinsic and extrinsic stresses that causes thresholds to be exceeded. In source areas, for instance, valleys may aggrade to the point of instability during dry periods. A metastable condition will be maintained, however, until a flood event of sufficient magnitude occurs that is able to flush the valley of sediment (Fig. 2.1).

The most important sources of extrinsic stress on systems are tectonism, climate, and eustatic base-level changes. A change in stream regimen, for instance, may occur as a result of uplift in the source area that imposes an increased sediment load on the system. Base-level changes cause coastline migrations that may change the way sediment is delivered to offshore areas, and they may also change the way rivers transport sediment.

The effects of intrinsic stress are superimposed on those produced by extrinsic stress. When extrinsic thresholds are exceeded, major changes may occur in the system, but when intrinsic thresholds are exceeded, series of cyclical responses may occur. A major uplift in a source area can cause increased coarse sediment yield to reach an alluvial fan. Progradation of fan facies will occur, which will be evident

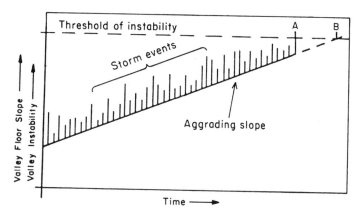

Figure 2.1 Changing stability of an aggrading valley floor through time. As the valley floor slope increases, it approaches a threshold of instability. The valley floor will begin to erode if the threshold is crossed, but it may also erode if an extreme storm event induces instability (after Schumm, 1968).

in the stratigraphic sequence, but aggradation will also occur, causing the fan slope to steepen beyond an intrinsic threshold of stability. The instability is relieved periodically by fanhead erosion and reworking, resulting in cyclic pulses of increased sediment yield to mid- and distal fan environments. Thus a major sedimentary event is produced by extrinsic stress with superposed minor cycles produced by intrinsic stress. When thresholds are exceeded, response in the system may be felt far beyond the point of stress; but in these areas the system may respond in different ways, and the magnitude of the response may progressively be dampened with distance.

Complex Response

The third important geomorphic principle is that of complex response. There are three characteristics of a process–response system (Chorley and Kennedy, 1971):

1. The operation is controlled by the magnitude and frequency of the inputs (that is, factors involved in the system's operation).
2. Progressive changes in the morphology and operation of the system can occur if input changes or if there is degradation of the system. (This is modified by thresholds inherent in the system.)
3. Self-regulation or negative feedback occurs to create an equilibrium state between the morphological and cascading components of the process–response system.

The third characteristic describes the way in which a system responds in a complex manner to external stresses. When extrinsic thresholds are exceeded, feedback mechanisms are triggered within the system, causing complex responses as internal thresholds are approached and exceeded. Lowering of the base level of a trunk stream, for instance, will result in an initial phase of downcutting in the main valley, remobilizing stored sediments and delivering them as a major pulse of sedimentation. Initial lowering of base level is an extrinsic stress, but as the trunk valley is lowered it will place intrinsic stress on the tributaries by lowering their local base level and causing them to degrade their valleys. They, in turn, will lower the base levels of their tributaries, and the increased sediment yield will choke the main valley until it aggrades beyond the point of stability, and a second phase of downcutting will result in another major pulse of sedimentation. This will continue until the system regains its former state of equilibrium. In this way, a single external change may cause cyclic response in the system through a complex series of feedback responses (Schumm, 1977).

Another way in which systems may respond complexly occurs when more than one response may be elicited from the same external stress. In arid climates, for instance, sediments tend to collect in valleys because of the inability of streams to maintain them in transport. If the climate changes to a more humid one, however, this sediment may be flushed from the valleys, resulting in a pulse of sediment transport out of the system. This response would be reinforced by increased slope erosion (Fig. 2.2).

But this condition would not last indefinitely, because more humid conditions would promote increased vegetative cover on the slopes, ultimately causing a reduction in the amount of sediment being provided. Thus, two divergent responses may result from a single external stress being placed on the system.

Time

Many sedimentary and geomorphic processes act instantaneously in geologic terms. But it is through their repeated action over time, acting in the broader context of landscape evolution, that rock sequences are built. Even single stratigraphic units do not represent a period of time, but rather a progression of events that occurred through time.

Sedimentary and geomorphic processes occur at various rates, and they are represented in the rock record at different scales. One process may act so slowly that, to our limited observational time base, the system is not responding measurably. Another may occur so rapidly that the change in the system may be measured visually. Often, however, it is the long-acting processes that are the most represented in the rock record.

Schumm (1977) defined three time scales in which various geomorphic processes act. These are steady time, graded time, and cyclic time, and to these I would add "instantaneous" time. Relative to geologic time, many sedimentary processes act instantaneously. Particles are entrained and deposited, laminae form, bedforms appear and are destroyed. These processes are initiated over very short periods of time, and the individual responses normally show relatively little chance of being

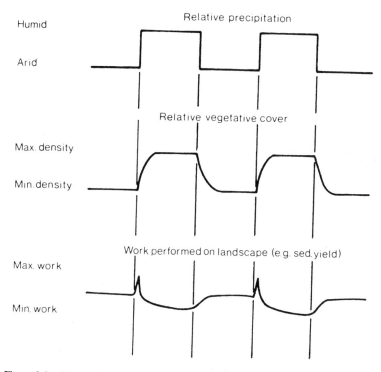

Figure 2.2 Changes in sediment yield produced in response to changes in climate and vegetative cover. An increase in precipitation may cause a sharp increase in sediment yield that soon begins to decline as the vegetative cover increases in response to greater availability of water. Sediment yield increases slowly in response to an abrupt decline in precipitation as vegetative cover slowly decreases (after Knox, 1972).

preserved. The time interval represented by bedding planes may be magnitudes longer than the bedding units they enclose, and during that time interval many unrecorded events may have occurred. It is to the process-oriented sedimentologist that these instantaneous responses in the system are of the most interest.

During steady time, a system may be in equilibrium even though short-term variations of some chemical and physical processes may be occurring. Once these are integrated, no perceptible changes occur in the system during steady time. River levels rise and fall, and storm cycles may affect beaches, but either the changes wrought by these variations are neither additive nor subtractive (the system varies about a mean), or the changes are incremental but slight (the system moves slightly in one direction from the mean) and the system remains essentially unchanged. It is within this time frame that hydraulics engineers and sedimentologists studying equilibrium sediment distribution patterns work.

Graded time is an interval during which rhythmic changes in a system may occur, but the basic system remains the same. Major changes in the landscape may occur, but they occur within the context of a complex feedback mechanism that acts to maintain the system. Fan head accretion and trenching occur rhythmically, but always within the context of an alluvial fan. Stream avulsion occurs on floodplains, delta lobes form and are abandoned, and shorelines prograde or retreat, but in each case, even though major landscape changes occur, the processes are all intrinsic to a system that remains essentially unaltered.

Rhythmic time is the basic time framework of interest to geomorphologists studying the history of a system. It is also the realm of the stratigrapher using Walther's principle, because it is within this time framework that facies migrations take place and vertical sequences are built.

Cyclic time encompasses a major interval of geologic time during which systems change in response to major changes in the environment. External stresses placed on the system by tectonism or climatic changes induce profound changes in the way sediment enters the system and is transported, and this results in major alterations in depositional patterns. Renewed tectonism in the source area or increased aridity may change stream patterns from meandering to braided. A eustatic fall in sea level may affect the same changes, whereas transgression will affect not only coastal sedimentation patterns, but will also affect the sediment yield to the shelf and to ocean basins.

The degree to which such large-scale changes in geomorphic processes affect the rock record is studied by basin analysts, and it is in this time framework that the greatest difficulties occur. The changes that are wrought bring systems so far out of equilibrium that partial or complete destruction of previous deposits often occurs, leaving the stratigrapher only disparate fragments. To understand the incomplete record, it is necessary to understand the processes involved in the succession of landforms during the evolution of the landscape.

W. M. Davis (1899) proposed a model of landscape evolution in which steady progressive erosion caused drainage divides and valley slopes to converge in elevation on base level (Fig. 2.3a). But geomorphic systems respond in a complex manner and at different rates to stress. A conceptual model, therefore, must be capable of explaining events of varying duration and magnitude.

Schumm (1977) proposed a model in which the time framework adjusted itself to the rate at which various events occurred during the evolution of a landscape. The Davisian cycle may adequately explain the way in which denudation of a landscape occurs following an initial phase of uplift, but even on a gross scale this is an oversimplification of the process, because isostatic readjustment may periodically raise elevations of divides and valley floors (Fig. 2.3b).

Expansion of the time frame in which isostatic readjustment occurs permits a more detailed examination of the way valley slopes respond to these periodic changes in elevation. In this time frame, valley slopes are not reduced progressively during periods of tectonic inactivity. Rather, alternating periods of storage and flushing of sediments in the valleys result in a stepped profile of slope readjustment controlled by complex response to intrinsic stress (Fig. 2.3c).

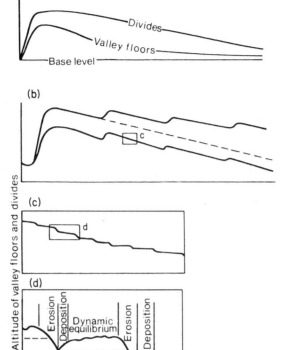

Figure 2.3 Response of landscapes to uplift followed by erosion. (a) Erosional cycle of Davis shows progressive decrease in slopes of divides and valley floors following uplift. (b) The erosional cycle would be modified by subsequent episodes of isostatic readjustment (dashed line shows Davisian cycle). (c) Superimposed on the response caused by isostatic readjustment are episodic slope changes caused by periodic flushing of stored sediment during cyclic time. (d) During graded time, even shorter-term changes in valley floor slope caused by episodes of erosion and aggradation can be accommodated (after Schumm, 1977).

Further expansion of the time frame may reveal that, even between periods of rapid readjustment, valley slope reduction is occurring within a context of dynamic equilibrium. Other intrinsic stresses cause episodes of major erosion and rapid aggradation that are separated by intervals during which the valley slope is constantly making smaller readjustments of its profile as it aggrades to a threshold of instability (Fig. 2.3d).

Basinal deposits will reflect the varying responses that landscapes make to intrinsic and extrinsic stresses in the source areas. Stratigraphers must be able to differentiate major extrinsic stresses operating continuously over long periods from intrinsic ones that cause shorter-term fluctuations of the system around steady-state or graded equilibrium conditions. The first will certainly have basin-wide impact, and the recognition of such stresses as they are reflected in the deposits of one area may be used in predicting responses in other parts of the basin. The record may only show local impact as each system within a basin responds in a different way and at different times to internal stresses.

Stratigraphic Principles

The basis of stratigraphic analysis can be found in three concepts: the continuity principle, the law of superposition, and the facies concept. These, in turn, are combined in a fourth concept, Walther's principle.

Continuity Principle

The continuity principle simply states that when an equilibrium state is attained the amount of sediment leaving an area is equal to the amount entering. When an unbalanced stress is placed on a system, however, the quantities of sediment leaving and entering an area are not equal, and either net deposition or net erosion occurs as the system attempts to reestablish equilibrium. It is during periods of nonequilibrium that stratigraphic systems evolve, and by analyzing the factors that produce disequilibrium, it is possible to predict the way in which sediments are distributed vertically and laterally.

It was not until studies of modern surficial sediments were combined with studies of their subsurface distribution that the importance of the continuity principle became apparent. For deposition to occur, disequilibrium must be produced, but as systems respond to stress, facies not only shift laterally, but they may also change their relative distribution as well, and deposition in one area may induce erosion in another. The added complexity, of course, is reflected in the architecture of the resulting deposits.

Law of Superposition

The law of superposition states that, in the absence of overturned sequences, underlying strata are older than the strata overlying them. The concept is simple in its application to rocks in a given area, but difficulties arise with interregional correlations of rocks that may not be the same age everywhere or show the same appearance.

Facies Concept

The first step in resolving these problems came with the observation by Gressly in 1838 that correlatable strata in the Jura Mountains changed laterally in appearance. The facies concept implies that similarity of physical and paleontological characteristics is not the only criterion to use in correlating rock units. Rigid use of the criterion may, in fact, lead to error. Besides easing many of the problems associated with stratigraphic classification and correlation, facies definitions are the basis of environmental reconstructions. In fact, Gressly identified two different facies in his study and gave them paleonenvironmental interpretations (Middleton, 1978).

Modern application of the facies concept is based on the recognition that lateral variations may occur within units, but the more common use of the concept is

to make informal divisions among different rock units along lithologic or paleonto-
logic lines. Divisions may occur within formal rock units, may cross unit boundaries, or may occur repeatedly within one or more formal units.

Walther's Principle

The second step came with Walther's enunciation of his principle of the correlation
of facies (1893–1894). The principle states that "Only those facies . . . can be superimposed which can be observed beside each other at the present time" (translation
by Middleton, 1973, p. 982). It should be observed that the inverse of this principle,
which is, in fact, the way it is most often stated, is not true. With the statement of
this principle, Walther showed that the law of superposition must be applied with
care because time lines cross facies boundaries.

Two other concepts are derived from Walther's principle. First, the principle
makes an actualistic approach to genetic interpretation of stratigraphic sequences
mandatory, because it emphasizes the need to compare rock sequences to lateral
distributions of sediments in modern environments. Walther himself clearly stated
this concept by saying that "only the ontological method [the study of present natural phenomena] will save us from stratigraphy" (translation by Middleton, 1973,
p. 983).

Second, the principle is the foundation of basin analysis and facies prediction.
It states that rock sequences are built by the migration of facies through time.
Walther stated that "Every facies is related to other contemporaneous facies and
when we want to interpret a fossil deposit we must compare it with the sediments it
is connected with at the present time. We must research in the surroundings of the
fossil deposit and complete the missing facies according to the laws of correlation"
(translation by Middleton, 1973, p. 982). These correlations would not be made on
the basis of similar lithologies, but on the expected transitions of facies known to
be laterally equivalent in modern facies tracts. If the facies of a given rock sequence
can be interpreted, and if an actualistic model of facies distribution can be applied,
then the lateral equivalents of the facies can be predicted.

It must be stressed that the inverse of the law is not necessarily true. The law
does not state that the vertical sequence always reproduces the horizontal facies
distribution (Middleton, 1973), although the literature contains many synthetic
stratigraphic sequences produced by assuming laterally migrating facies will all be
represented vertically.

3

Classification
of Environments

In the broadest sense, the term sedimentary environment consists of the physical, chemical, and biological processes affecting sediment. The scale of the environment is entirely without restriction, and the boundaries between environments may be set arbitrarily. Some geochemists, for instance, may study the chemical environment in a single pore space, whereas others may study the chemical balance of the oceans. Some sedimentologists refer to the fluvial environment as composed of channel and floodplain facies, but others refer to a fluvial system composed of channel and floodplain environments, with the floodplain further divided into levee, splay, and swamp or marsh facies. Because of previous different uses of the term sedimentary environment, it is necessary first to propose a classification that will establish a common nomenclature.

The classification proposed here divides all depositional settings into three environments that are characterized by the dominance of a single sedimentary process (Table 3.1). In the continental environment, most sediment transport is accomplished by flows that move in response to the gravity field from positions of greater to lesser potential energy. These flows may vary considerably in magnitude, but, except in the case of ephemeral streams in arid climates, flows vary over periods long enough for equilibrium conditions at the bed to be attained. Wind flows are episodic, however, and flow paths are determined by the local barometric gradient rather than the earth's gravity field. The local barometric gradient changes in response to transient atmospheric conditions, and sediment transport directions may, therefore, vary.

Fluid flows in the shallow marine environment are not only normally episodic, but some are oscillatory as well. Tidal flows and wave-induced currents vary in

TABLE 3.1   CLASSIFICATION OF DEPOSITIONAL ENVIRONMENTS

Environment	Transition zone	System	Component
Continental		Arid	Alluvial fan Ephemeral stream Playa/lake Erg
		Humid	Alluvial fan Alluvial plain Lake
Shallow marine	Coastal	Tide dominated	Supratidal Intertidal Subtidal
		Muddy coastline	Chenier plain Tidal flat
		Wave dominated	Strand plain Barrier island Lagoon
		Deltaic	Delta plain Delta margin Prodelta
		Estuarine	Tidal flat Tidal channel
		Marginal marine	Inner shelf Outer shelf
		Epeiric sea/epicontinental seaway	
Deep marine	Continental slope	Canyon	Canyon wall Canyon axis
		Slope	Interslope basin shape
		Active margin	Trench/basin plain Submarine fan
		Passive margin	Continental rise Abyssal plain

intensity and direction over very short periods (seconds to hours), and even wind-forced flows may have recurrence intervals measured in days. Currents flow without regard to the gravity field. Initially, they respond to directions imposed on them by the propagating force, but after flow is established, rectifying forces, acting in the context of the topographic character of the substrate, may force currents into shore-wise flows.

Suspension settling and unidirectional flow produced by density currents are the dominant processes in the deep marine environment. Deep oceanic currents may modify sediments deposited by these other mechanisms, and mass movements are common near continental margins.

The three environments are separated by transition zones. The transition zones may act as barriers, conduits, or filters by modifying dominant processes where they mutually interact. The coastal transition zone occurs between the continental and shallow marine environments. Oscillatory motion is modified to translational motion along shoaling coastlines, unidirectional stream flow changes to jet flow at stream mouths, and tidal and fluvial processes interact and are mutually modified in estuaries.

The coastal zone also modifies the sediment provided to it. Coarse-grained sediments are stored preferentially along sandy coastlines, at delta mouths, and at estuary mouths. Fine-grained sediments may be trapped on tidal flats or delta plains, and often it is only the sediments most easily remobilized by flood flow in distributaries and by storm waves that are passed on to the shallow marine environment.

The shelf edge–uppper slope transition zone occurs between the shallow and deep marine environments. Very large amplitude waves may entrain sediments at the shelf edge. They progressively leapfrog sediments to the slope break, where they build to instability that is relieved by periodic slope failure, transporting this sediment downslope into the basin. Oceanic currents may impinge seasonally on the outer shelf, where they are capable of transporting sediment or remolding previously deposited sediments. Upwelling deep ocean waters can modify the chemical environment locally by causing blooms of marine organisms that sink to the bottom on death, causing sediment deoxygenation and denitrification.

Within environments are systems that can dominate deposition in an area for geologically long periods. These systems are the products of a combination of tectonic setting and climate, and they consist of mutually interactive components that may be considered landforms. In common usage today, components are considered to be depositional systems that are composed of interacting environments or facies. The classification proposed here is intended for use primarily by stratigraphers, and the term system is retained for the combination of depositional components that interact in the modern environment to form deposits that succeed each other in a predictable manner during the evolution of the system through time. This is a somewhat broader definition than commonly used today because it does not set arbitrary boundaries between mutually interdependent depositional landforms that interact in unique tectonic and climatic settings.

Humid and arid systems occur within the continental environment. Components of the humid system include humid region alluvial fans, alluvial plains, and lakes. The arid system also contains alluvial fan and alluvial plain components, but these differ considerably from their humid counterparts because of the climatic overprint on sedimentary and geomorphic processes. The arid system also includes eolian plain (erg) and playa components.

Five systems may occur in the coastal transition zone. Deltas occur where sediment yield overwhelms marine processes, whereas nondeltaic clastic shorelines, including tide-dominated, wave-dominated, and muddy coasts, occur where these processes redistribute incoming sediment extensively. Estuarine systems are transitory features that occur on transgressive coasts. All these systems may contain elements that are common to all of them, including barrier island, tidal flat, and brackish lagoon or marsh, as well as components that are unique to one or two, such as distributary mouth bar and delta plain.

The shallow marine environment consists of inner shelf and outer shelf systems. These systems do interact contemporaneously, but because of the dramatic impact caused by eustatic sea-level rises on shelf sedimentation patterns, the systems are also divided into allochthonous and autochthonous components that succeed each other in time.

The continental margin transition zone consists of a slope system and a canyon system. On the slope, upper and lower slope components occur, and in canyons, axial and flank components are found. Some continental margins may also possess intraslope basin systems.

Where the deep marine environment is bounded by passive continental margins, it consists normally of continental rise and abyssal plain components. Where actively subducting continental margins occur, the main components are trenches and submarine fans. Transform margins are complex and consist of marginal fans and small basins connected by narrow channel systems that cut through sediments.

In the absence of major changes in their tectonic or climatic setting, systems are the basic units of stratigraphic sequences. Components of the various systems occur together in vertical sequence in response to facies migrations as the systems evolve through time. Systems may succeed one another in response to major environmental changes, and in the broad context of basin development, environments are often found succeeding one another in vertical sequence.

It is important to remember that even environments do not exist in isolation. Processes within a given environment affect events in others, even though the processes are modified as they pass through the transition zones. The concept of event stratigraphy may help to analyze the progression of events that occurs from original point of stress to ultimate depositional sink.

Discussion: Application of Principles to Analysis of Stratigraphic Sequences

In this section the concepts essential to understanding the evolution of rock sequences have been introduced. When the three sets of principles are combined, they form a powerful tool to be used to conceptualize the three-dimensional aspects of sedimentary deposits.

The principles of sedimentology explain the processes involved in erosion, transportation, and deposition of sediments. By considering depositional environments to be assemblages of processes, we can construct conceptual models that show universal applicability in explaining the almost infinite diversity of sediment responses that can occur even within simple systems. Sedimentological principles, however, cannot be used out of the context of their operation within geomorphic systems. Most studies of modern depositional systems consider only the surface sediments, which only represent the lateral facies distribution at a particular time and under a unique set of circumstances. If the stresses on the system change, not only do the facies shift, but their positions relative to one another may change in a manner not entirely predicted by Walther's principle.

The principles of geomorphology explain the processes operating during the evolution of landscapes. The way in which landscapes respond to stresses controls the way in which sediment is delivered to a depositional site and how it accumulates there. An understanding of how these processes operate in various systems is essential in interpreting the sequence of events occurring during the deposition of stratigraphic sequences. The principles of stratigraphy are used to establish the position of rock sequences in space and time. It is these principles that enable stratigraphers to recreate through modeling the events that occurred during the filling of depositional basins by extrapolating the responses to stress operating in one system to other related systems.

To predict how basinal sequences evolve, it is useful to consider depositional systems in a framework or classification that will establish the hierarchical ability of the various divisions to interact and thus produce a set of commonly associated responses in the rock record. Such a classification will be of particular use to stratigraphers in producing models of sedimentary sequences, because it will group those depositional systems that can be expected to succeed one another during the evolution of a basin.

Part **II**

CONTINENTAL ENVIRONMENT

4

Arid System

Introduction

Deserts occur where water loss through evaporation exceeds precipitation. Arid regions are defined as those areas where annual rainfall does not exceed 25 cm. But potential evaporation varies considerably with temperature, and no fixed value for precipitation can be used to predict aridity. Arid regions are commonly bordered by semiarid regions or steppes, and together these dry regions occupy about 25% of the world's land surface.

Deserts may occur in low, middle, or high latitudes. Low-latitude deserts are characterized by diurnal temperature extremes and summer daytime temperatures in excess of 50°C. Mid-latitude deserts may record very high summer temperatures, but winter temperatures are cold, with some of the annual precipitation falling as snow. Their lower evaporation rates also increase residency times of soil moisture that is available for plants. Polar regions may also experience low precipitation. In these climates, low temperatures may severely limit rates of evaporation; but because of sparse vegetation, desertlike conditions may prevail despite relatively large amounts of available water.

Both low- and mid-latitude deserts are characterized by constant winds as a result of severe instability in the air column caused by rapid daytime heating. Lack of plant cover promotes the effectiveness of wind as a transport agent because of reduced frictional drag and the poor baffling and binding capacity of widely spaced plants.

Lack of vegetative cover is the result of reduced availability of water and high salt concentration in the soil, or cold temperatures in high-latitude deserts. In either case, the result is a land surface subject to severe erosion by flowing water as well as wind.

In deserts, the mechanical weathering of rocks may produce abundant coarse clastic debris. Extreme diurnal temperature ranges alternately expand and contract surface skins of rocks, causing exfoliation. In high-latitude deserts, a similar process results from frost action.

Rainfall is so sparse in some deserts as to be a negligible factor in sedimentation. Most deserts, however, do experience seasonal rains, and these rains may be extremely effective transporters of sediment because of sparse plant cover and the episodic, often catastrophic, nature of runoff. The episodic runoff coupled with the rapid rise to flood stage and equally fast decline produce some fluvial processes unique to desert areas.

Components

The main components of the arid system are alluvial fans where deserts are bordered by uplands, wadis or ephemeral streams (some of which impinge on fan surfaces), sabkhas or playas, which may occupy structural depressions or shallow basins of interior drainage, and eolian plains. Minor components include hamadas, which are areas of debris-covered bedrock, and serirs, which are deflationary surfaces of coarse debris that armor more easily eroded material.

Alluvial Fan

Sedimentary processes and products.   Alluvial fans are cone-shaped deposits of clastic debris that form against upland scarps where sediment yield is abundant. They form where upland streams emerging from highlands experience flow expansion and deceleration and a corresponding decrease in competence and capacity. The best-studied alluvial fans occur in arid and semiarid regions, but alluvial fans also occur in other areas, especially in proglacial settings, and these are receiving increased attention.

Most fans show a concave-upward profile in radial cross section that is composed of segments that can be related to successive growth periods (Bull, 1964). Growth periods may result from readjustment to renewed uplift in source areas or complex responses to intrinsic geomorphic thresholds when fans aggrade to instability and then are eroded. Fans are convex-upward in transverse cross section, and along tectonically active fronts, fans may coalesce into fringing landforms called bajadas.

Fans range in size from a few to hundreds of square kilometers, with size apparently being controlled by the size of the drainage basin and the relative resistance of source areas to erosion (Bull, 1972). Dip angles of as much as 10° may occur on fan surfaces, but normally surface gradients do not exceed 6°. Fans built of fine-grained materials tend to possess steeper slopes than coarse-grained fans.

The surface features of fans consist of a fan apex, fan head channel that is incised into the apex, mid-fan, where numerous shallow braid channels occur, and a distal fan that coalesces with adjacent fans and grades into basinal deposits. Within these zones, a variety of sedimentary processes occur, which include sheet-flood, stream channel flow, debris flows, and sieve processes (Fig. 4.1).

Debris flows or mudflows have long been known to be major components of fan sedimentation (see Church, 1978, for reviews). Debris flows occur preferentially where source rocks produce sufficient fine materials for their formation. The intense nature of rainfall in desert regions, susceptibility of substrates to erosion, and rapid runoff all contribute to the formation of debris flows.

Flows begin to move only after a critical shear threshold has been exceeded. Movement ceases if shear drops below the threshold owing to loss of water, increase in viscosity, or decrease in slope. Because of the great viscosity, sediment is carried without sorting, and when movement ceases, the poorly sorted, often extremely coarse, load accumulates in place. Distribution of coarse materials may not be uniform, however, with coarsest clasts often occurring at the front of the flow or on

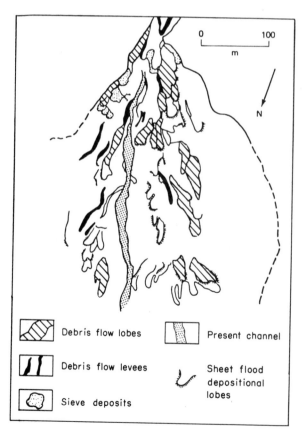

Debris flow lobes   Present channel

Debris flow levees   Sheet flood
                           depositional
Sieve deposits   lobes

Figure 4.1 Depositional processes on the Trollheim Fan, Death Valley, California (after Hooke, 1967; Galloway and Hobday, 1983).

its upper surface. Stream flow commonly follows deposition of debris flows, resulting in an intimate association of coarse, poorly sorted debris-flow deposits and water-laid sand and gravel (Fig. 4.2).

Floods frequently percolate into fan surfaces before reaching the toe of the fan. Loads carried by such floods may be rapidly deposited when this occurs, leading to the formation of lobes of coarse, relatively well sorted debris termed sieve deposits by Hooke (1967). A sieve deposit may also show a very coarse terminus that acts as a strainer, allowing water to percolate through, but trapping fines in interstices (Fig. 4.3). Sieve deposits may occur, therefore, where water percolates rapidly into the fan surface or at slope changes where the coarse load of a sheetflood is deposited.

Stream channels on alluvial fans include major channels entrenched in fan heads that extend downfan to the point where the channel bottom intersects the fan surface, minor channels that commonly occur as distributaries from the main channel, and washes that head on generally inactive portions of the fan. Channels shift frequently because of backfilling, aggradation of channels to the point of avulsion (intrinsic stress), and headward erosion of washes resulting in stream capture. Although deposition may be localized on a portion of the fan surface during steady-

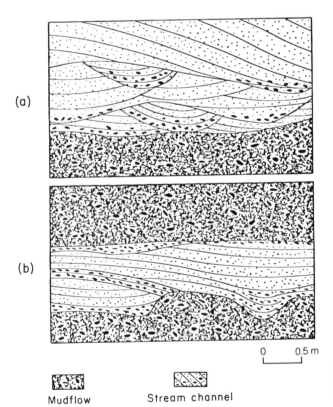

(a)

(b)

0   0.5 m

Mudflow   Stream channel

Figure 4.2 Interbedded mudflow and stream channel deposits on an alluvial fan (after Blissenbach, 1954).

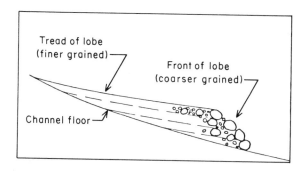

Figure 4.3 Diagrammatic cross section of a sieve deposit (after Hooke, 1967).

state time, channel shifting over longer periods of time results in uniform deposition.

Stream channel deposits are continuous in an axial direction, but they are lenticular in transverse section and commonly display cut-and-fill relationships with enclosing sediments (Fig. 4.2). Pebble imbrication and upper flow regime bedforms such as plane beds are common, and sands are often crossbedded with clay drapes on cross sets.

Sheetfloods occur when floods leave the mouth of a channel at the lower portion of the midfan and spread out in a broad sheet, giving rise to a very low relief, braided stream pattern. Shallow flow depths give rise to a series of low bars separated by shallow channels. Bars may appear to be primary sites of initial deposition, whereas channels tend to be active longer, sometimes forming small-scale subsidiary braid patterns within the channel itself as flow wanes. Flows tend to be in the upper regime, except during late stages, both in response to high flow velocities and shallow water depths. Massive gravels and plane bedded sands predominate in sheetflood deposits, although crossbedding in sands does occur.

Facies distributions.   Facies distributions on modern arid-region alluvial fans correspond closely to the geomorphic divisions and can thus be divided into proximal, medial, and distal facies (Fig. 4.4).

The proximal facies is characteristic of the fan head, which is dominated by a fan head trench that acts as a conduit for debris flows and sediment-laden flood flows. Sediments in this facies consist of debris flows and sieve deposits in the channel, debris flow levees in interchannel areas, and occasional sheetflood deposits laid down during extreme flows that exceed channel capacity. The proximal facies consists of the coarsest, most poorly sorted and angular of all fan sediments except for scree deposits, which are limited to a small area adjacent to the scarp.

Medial facies are deposited in depositional lobes in the central portion of the fan downslope from the intersection point (Fig. 4.5). Sediments consist predominantly of sieve and channel flow deposits, with minor debris flows forming depositional lobes heading at intersection points. The mid-fan is traversed by channels containing deposits that accumulate during cyclic flood conditions characterized by rapid flow decay. The deposits reflect this regime where they consist of upper flow

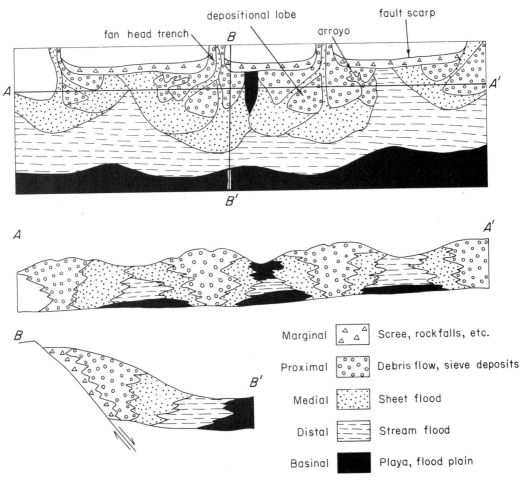

Figure 4.4 Distribution of alluvial fan facies shown in plan view (upper), transverse or strike section (middle), and longitudinal or dip section (lower).

regime structures at the base that grade rapidly upward into sediments that settled out of suspension.

Active deposition does not occur over the entire fan, but is concentrated in lobes that head at intersection points. Interlobe areas may receive no sediments for long periods of time, resulting in formation of soils and caliche on the accumulated debris.

Deposits of the distal facies are generally much finer grained than those farther upfan. Sheetflood deposition is an important process in the distal facies, but washes or gullies frequently head in the distal fan where infiltrated water from upfan emerges onto the surface at hydrologic discontinuities. Sands in plane beds and

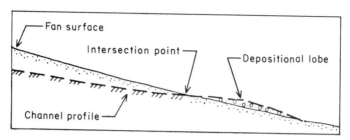

Figure 4.5 Diagrammatic representation of the relationship between the intersection point and the position of the mid-fan depositional lobe of an alluvial fan (after Hooke, 1967).

low-angle crossbeds predominate in the facies with subordinate occurrence of ripple crossbedded sands. The distal facies forms a transitional zone between fan and low-land deposits.

Intrinsic and extrinsic controls on alluvial fan sequences.   The downfan change in process and products should be reflected in the vertical sequence of a prograding fan. Playa or floodplain deposits should pass upward into sheetflood deposits of the mid- and distal fan, which in turn would be overlain by channeled debris flow and sieve deposits of the upper fan and apex. The most proximal scree breccias probably would not be represented in vertical sequence, but might be laterally equivalent to all the other deposits because their depositional setting is restricted to the fault scarp (Fig. 4.4).

Proximal and distal relationships, however, are complicated somewhat by the conical shape of fans. Lateral deposits of a fan even near the upland margin are not qualitatively different from distal deposits along the fan axis. Thus, aggrading fans may produce similar coarsening upward sequences both down dip and along strike (Fig. 4.4).

A simple coarsening upward sequence represents an initial response to uplift, but fan sequences commonly represent variously scaled cyclic responses to a number of variables (Fraser and Suttner, 1986). First-order cycles begin with fan construction following an initial phase of tectonism. Breccias and scree deposits accumulate next to the fault scarp at the same time streams flowing from reentrants in the fault scarp initiate fan deposition. As the drainage basin enlarges, the fan surface grows and proximal facies begin to migrate over more distal ones. Maximum fan extension occurs when the drainage basin is well-developed, with narrow interfluves and steep valley walls. Fan progradation at this point has produced a coarsening upward sequence.

From the point of maximum extension, however, fan growth declines as slopes of streams and interfluves decrease and valley floors widen as they become temporary depositories of large quantities of sediment. This period of fan growth is termed the abstraction phase (Schumm, 1977).

Denudation of the source area, however, would trigger isostatic rebound. The expected response would be a fan sequence similar to the initial one, but isostatic rebound may produce a series of complex responses in the existing drainage basin. The first response would be headward erosion of the main trunk stream that would release large amounts of stored sediment out onto the fan. The initial flood of sediment would decrease with time as sediment yields declined. However, erosion in the trunk stream would lower base levels of major tributaries, causing erosion of their floodplains. This pattern would be repeated as succesively higher order tributaries released the sediments stored in their floodplains, resulting in a sediment yield of cyclical peaks with progressively dampened amplitudes (Parker, 1976). Each episode of isostatic readjustment might result, therefore, in deposition of a major fining-upward, thinning-upward sequence consisting of smaller-scale fining-upward, thinning-upward cycles.

Progradation of depositional lobes forms thickening-upward, coarsening-upward sequences. Commonly, growth of depositional lobes occurs when stream avulsion or fan head entrenchment takes place (Fig. 4.5). Such cycles might be on the same scale as one of the minor cycles caused by isostatic readjustment. Whereas climatic changes or tectonism might cause progressive changes in fan sedimentation, stream avulsion or fan head entrenchment causes abrupt changes that form sharp hiatal boundaries between depositional units. Such boundaries may involve only small portions of the fan where mid-lobe avulsions occur, but if entrenchment of the apex is involved, large portions of the fan surface may be affected (Fig. 4.6). Thin, fining-upward cycles record storm events, and these cycles are superimposed on all other cycles deposited during fan evolution.

Alluvial fans in the rock record.   Two of the best documented sequences of arid-region alluvial fan deposits occur in the New Red Sandstone (Steel, 1974) and Old Red Sandstone (Bluck, 1967) of northern Britain.

Four distinct conglomerate types occur in the Old Red Sandstone in the Clyde area of Scotland (Fig. 4.7). Mudflow conglomerates consist of poorly sorted, unoriented, pebble- to cobble-sized clasts in a sand and mud matrix. The upper portions of these deposits consist of thin sandstones that probably are the product of reworking of the upper surfaces of flows during the waning stage.

Stream flood deposits consist of crossbedded conglomerate overlain by crossbedded standstones occasionally separated by a thin lag deposit. Basal contacts of the conglomerates are erosional. The coarse-grained basal unit probably accumulated during maximum flow, and the crossbedded sandstones represent late-stage reworking during waning flow. Typically these deposits are associated with mudflow conglomerates.

Braided stream conglomerates consist of crossbedded sand and gravel and occasional beds of plane bedded sandstone and intraformational conglomerates. These beds appear to have formed in a setting characterized by rapid lateral channel shifts, and their local association with stream deposits suggests a depositional process not unlike sheetfloods. Finer-grained units are commonly associated with playa depos-

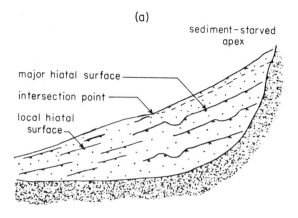

(a)

sediment–starved
apex

major hiatal surface

intersection point

local hiatal
surface

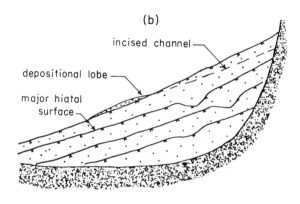

(b)

incised channel

depositional lobe

major hiatal
surface

Figure 4.6 Diagrammatic cross section of an alluvial fan showing the internal structure of the fan. Major hiatal surfaces are caused by extrinsic factors and are fan-wide, whereas minor hiatal surfaces are caused by intrinsic factors and affect only portions of the fan surface (after Talbot and Williams, 1979).

its, suggesting a distal fan or basinal setting. Floodplain conglomerates occur as thin lenticular units enclosed in sandstone and probably represent lag deposits in channels on the floodplain.

In comparison with modern fans, mudflow and stream deposits are thought to represent more proximal facies, and sheetflood and floodplain deposits were deposited with increasing distance from the source. The vertical sequence fines upward, with proximal deposits at the base and distal deposits at the top, suggesting that the sequence was deposited during progressive reduction of the source area after a single rapid uplift.

Similar sediments occur in the New Red Sandstone of the Hebrides (Steel, 1974). Conglomerates with a silt and clay matrix are interpreted as mudflow deposits (Fig. 4.8). Discontinuous beds of well-sorted, clean conglomerate, often in large-scale crossbeds, are associated frequently with the mudflows and apparently formed as flood-stage channel deposits. Braided stream deposits are well sorted, but are characterized by rapid lateral variations in texture. They are finer-grained than stream flood conglomerates and may have formed in channels heading on the fan

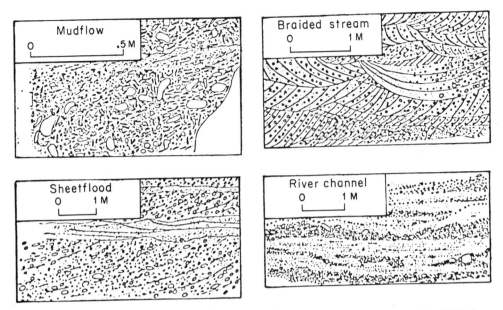

Figure 4.7 Structures in conglomerates of alluvial fan and floodplain deposits of the Old Red Sandstone in the Clyde area, Scotland (after Bluck, 1967).

surface (Fig. 4.8), where they assume a lenticular shape. Other similar deposits are more laterally continuous and may represent sheetflood deposits of depositional lobes.

Subordinate facies include playa deposits commonly associated with distal components of mudflows and stream floods, and caliche crusts found in sandstones capping braided stream deposits. The occurrence of both facies is suggestive of an arid climate.

In addition to alluvial fan sediments, floodplain deposits of both bedload and suspended-load streams also occur in the New Red Sandstone. They may occur at the base of a sequence where they are in sharp contact with overlying fining-upward alluvial fan sequences. In other sequences, alluvial fan sediments fine upward and are capped by one or more floodplain sequences (Fig. 4.8).

Intermittent Streams

Sedimentary processes and products.   Commonly called wadis or arroyos, intermittent streams are characterized by extremely episodic flow. The lack of vegetation and often close association with upland areas produces a hydraulic situation where floods peak soon after storm peaks. Floods may abate quickly as well, where waters percolate rapidly into porous alluvium, or may continue for several days, as falling flows are augmented by subsequent storms.

Alluvial fans are often considered to be products of intermittent streams. But

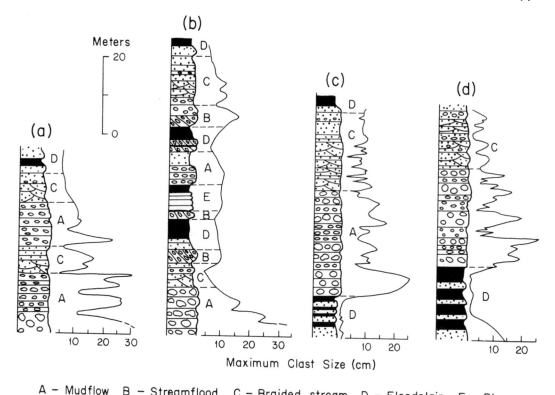

A – Mudflow   B – Streamflood   C – Braided stream   D – Floodplain   E – Playa

Figure 4.8 Stratification sequences in alluvial fan deposits in the New Red Sandstone of Scotland. The sequences fine upward from erosional bases and represent changes from proximal to distal depositional settings (after Steel, 1974).

except for the fan head trench and for arroyos or washes being cut in inactive fan surfaces, flow on fans is at best poorly channelized. The products of stream floods and sheetfloods reflect the unconfined nature of the flow, and their geometries are quite different from those produced by the channelized flow of intermittent streams.

Intermittent streams that cross basin plains of internal drainage end in sabkhas or playas. In block-faulted basins, drainage from fringing fans may coalesce into a single longitudinal trunk stream carrying drainage from a large upland area. Other streams may cross large desert basins if precipitation in bordering or central massifs is sufficient. The scale of such streams varies greatly, and they may be sinuous to meandering or braided in form. Commonly, the stream courses are inherited from earlier periods of different climate. Such stream courses are not necessarily stable with current factors of slope, discharge, and load.

Sediments in wadis are also highly variable. The erosive capability of flowing water is magnified by the noncohesive, unvegetated soils of deserts. Under such conditions, large amounts of sediment may be incorporated in flood waters and collected in channels, forming turbid flows that nearly approximate debris flows. The resulting deposits may be very poorly sorted and matrix-rich, with little or no internal organization.

Other deposits are more typical of normal stream sedimentation. Sediments are better sorted and internally well-organized into a variety of sedimentary structures. The bedforms, however, are not well-organized in a predictable manner in response to a hierarchical structure of flow (Karcz, 1972). Rather, they respond to rapidly changing conditions, where the flow following flood peak is modified in a complex manner by bed relief.

Erosion of the channel and the formation of lag deposits appear to dominate peak flow, but during waning flow, upper flow regime bedforms are gradually succeeded by lower form regime bedforms with a wide range of scale. Characteristically, larger-scale bedforms act to modify waning flow by imposing some transverse elements on it, and the bedforms themselves are, in turn, reshaped by the flow.

Some floods may exceed bank-full stage, producing a sheet flow over the adjacent floodplain. Local irregularities in the surface control sedimentation to a large degree, with plane beds forming on highs on the surface and climbing ripples and tabular cross-bedded microdeltas filling lows (McKee and others, 1967).

Although hierarchical assemblages of bedforms have not been recognized in ephemeral streams, characteristic associations do occur, and they result in a variety of vertical sequences. Karcz (1972), for instance, divided ephemeral stream beds into those characterized by either homogeneous or heterogeneous associations. He found that long reaches of streams were covered by ripples or erosional structures, including harrow marks and obstacle marks. Heterogeneous associations include large-scale transverse and lateral bars, both often covered by ripples, large- and small-scale bedforms modified by erosion during rapid flow changes, and small-scale forms deposited by transverse flows that had been redirected by primary bed relief. The deposits of the heterogeneous associations would, therefore, display a great degree of lateral variability.

Picard and High (1973) identified a variety of stratification types that could be grouped into three typical sedimentation units (Fig. 4.9). Point-bar sequences were deposited during waning flow and record the resulting progressive decline in velocity and flow depth. The sequence fines upward and consists of low-angle cross-stratification at the base, overlain in turn by micro cross-stratification, climbing ripples, and horizontal laminations with a mud film at the surface.

Both longitudinal and transverse bars occurred in braided reaches of streams. They produced sequences with large-scale planar crossbeds overlain by discontinuous plane beds that may, in turn, be overlain by trough cross-stratification, micro cross-stratification, or continuous plane bed. Mud films may also accumulate on bar surfaces where flow has ceased.

Channel deposits accumulate during waning flow by backfilling from distal portions of the channel toward more proximal portions. Basal channel lags may be overlain by upper flow regime plane beds formed just after flood peak. Festoon crossbedding formed by migrating dunes characteristically makes up the bulk of

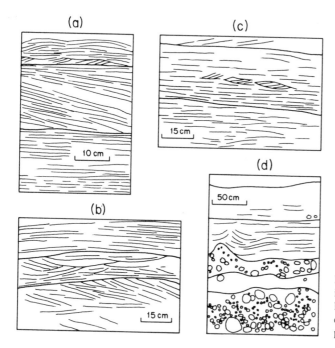

Figure 4.9 (a) Stratification sequences in point bar; (b) transverse bar; (c) longitudinal bar; (d) channel deposits of ephemeral streams (drawn from photographs in Picard and High, 1973).

channel sequences, and these may be overlain by micro cross-lamination or plane beds formed in shallow water near the end of the flow.

Other channel deposits are the result of turbid flows, which because of their great viscosity can carry coarse bedloads. Gravelly basal deposits are overlain by poorly sorted, fine-grained sediments that accumulate rapidly and maintain high pore pressures so that they are easily deformed.

Intrinsic and extrinsic controls on ephemeral stream sequences.   Three elements are involved in cyclic deposition by ephemeral streams. Individual flood events produce relatively thin packages of sediments bounded at the base by lag deposits and at the top by an upper surface displaying, where preserved, indications of subaerial exposure. Individual sequences fine upward with pronounced coarse and fine-tail grading. Internal structures record the rapid onset of flood, the somewhat less rapid waning, and eventual cessation of flow. In some settings, individual storm deposits may be interbedded with eolian sands.

Sneh (1983) was able to differentiate three types of desert stream sequences based primarily on the relative degree of confinement of the flow that deposited them. Flows in confined floodplains produced sequences consisting of repeated, flood-originated sedimentation units, each about 50 cm thick. Peak flow in these units is represented by sandy, pebble-bearing horizontal beds overlain by sandy cross-beds, and ripple beds deposited during waning flow.

Sequences deposited in open floodplains are also responses to flood-flow cycles. However, they are thinner than those of confined floodplains, horizontal stratification is less well-developed, and ripple bedding, especially climbing ripples, are characteristic of these units. Mud layers are minor constituents of confined floodplain sequences, but are more numerous and thicker in open floodplain ones.

Thin units of mud and ripple-bedded fine sand are deposited by low-intensity, lower flow regime flows on terminal floodplains where flood flows rapidly dissipate. Thicker mud layers in the sequences represent episodes when water was ponded for short periods.

The floodplain configuration is controlled mainly by channel gradient, with flows occurring in confined channels near stream heads, in unconfined channels along the middle portions of stream courses, and as sheetfloods in terminal floodplains. It is conceivable that a stacking of sequence types might occur during an aggradational phase where steeper gradients were progressively translated away from a mountain front (Fig. 4.10).

Flood cycles are superposed on climatic cycles of various lengths. In arid regions, a slight increase in precipitation in the source area may cause a stream to

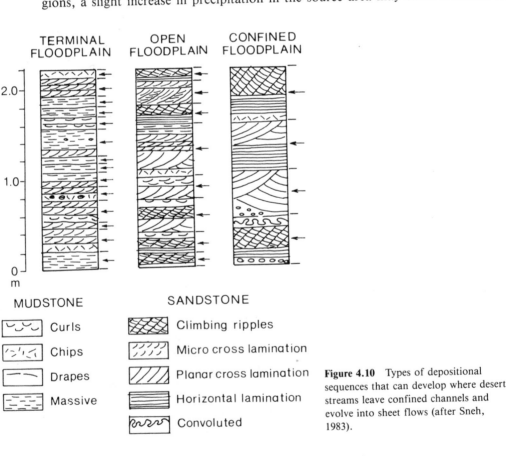

Figure 4.10 Types of depositional sequences that can develop where desert streams leave confined channels and evolve into sheet flows (after Sneh, 1983).

aggrade, whereas a decrease may cause it to entrench its floodplain. Some cycles of entrenchment and aggradation may last for decades and effect minor or even ephemeral changes, but others may last for hundreds or thousands of years, effecting major alterations on the pattern and distribution of sediment facies (Love, 1979).

Channel morphologies also change in response to hydrographic variations. During aggradational phases, channels tend to be relatively straight, stream discharge less flashy, and channel position relatively stable (Shepard, 1978). Rapid aggradation combined with channel stability produces abrupt lateral contacts between vertically continuous depositional elements. Narrow thalweg deposits bounded by overbank sediments may build vertical sequences of considerable thickness with very little lateral facies shifting. Degradational sequences, on the other hand, are characterized by sinuous channel patterns that produce laterally continuous accretionary deposits with considerable interfingering of channel and overbank sediments.

Sequences produced during a degradational phase are not commonly preserved. A package of sediments produced during a climatic cycle would contain only a basal sequence representing the last phase of deposition during the degradational phase. The bulk of the sequence would consist of sediments deposited during aggradation (Shepard, 1978). The nature of the deposits laid down during this phase depends on the climate and the relationship of source area to site of deposition (Love, 1979).

Longer-period cycles would form in response to uplift in the source area or subsidence in the basin. The complex pattern of responses to the initial and subsequent phases of tectonism would not be unlike those occurring on alluvial fans, which have already been discussed at length.

Ephemeral streams in the rock record.   Numerous examples of ephemeral streams have been described in the literature, and many are closely associated with eolian sediments. Ephemeral stream deposits in the Permian Rotliegendes of northern Europe, for instance, commonly consist of basal conglomerates, often containing clay pebbles, overlain by micro cross-laminated sandstones capped by clays with shrinkage cracks and sandstone. These sequences invariably are interbedded with eolian sandstones (Glennie, 1972).

Fluvial sediments are also interbedded with eolian sands in the Arikaree Group of the Great Plains, USA (Bart, 1977). The fluvial sediments consist of ripple- and parallel-laminated sands and massive units of silty clay that laterally interfinger with crossbedded eolian sands. Facies are vertically persistent but laterally variable, suggesting that rapid accumulation took place during an aggradational phase.

Sneh (1983) also reported the presence of eolian sands interbedded with ephemeral stream sediments in Pleistocene deposits of the Sinai Peninsula. However, he also noted the ability of streams in nearby modern wadis to remove eolian sediments, and he postulated that only significant changes in climate from the present, very arid conditions would permit coexistence of eolian and ephemeral stream sediments in vertical sequence.

Two types of ephemeral stream sequences occur in the New Red Sandstone in Scotland (Steel, 1974). One type probably originated in washes eroding inactive portions of alluvial fans. Basal conglomerates are overlain either by coarse sand in horizontal or cross-laminations or by fine-grained sandstone or siltstone with ripple drift lamination. Caliche commonly occurs at the top of these sequences (Fig. 4.11).

Floodplains were also the site of ephemeral stream flow. Fining-upward sequences consist of a thin, impersistent lag conglomerate overlain by coarse-grained, flat-bedded sandstones with parting lineation or by crossbedded sandstones. Coarse members of the sequences rarely exceed 1 m in thickness and are overlain by microtrough cross-laminated sandstone with a caliche profile developed on top.

Tunbridge (1984) identified three types of ephemeral stream sequences in the Middle Devonian Trentishoe Formation of Great Britain that were controlled by distance from source area. Proximal sequences consist of multistory, cross-cutting channel sandstones containing only subordinate amounts of mudstone. Sandstones are crossbedded at the base of these sequences and parallel-laminated at the top and were apparently deposited by high-stage shallow flows in rapidly migrating, low-sinuosity channels (Fig. 4.12).

Medial sequences consist of a variety of sediments deposited by sheetfloods on sandflats and confined flows in channels. Sheetfloods in proximal settings produced parallel-laminated sheet sandstones up to 2.5 m thick. Thinner bedded parallel- and cross-laminated sandstones interbedded with mudstones were deposited by sheetfloods farther from the source area. Confined flows in broad, shallow washes deposited intraformational conglomerates overlain by parallel-laminated sandstone and siltstone, but flows in deeper channels produced similar deposits organized in lateral accretionary beds.

Facies sequences individually fine upward, but they are broadly organized in coarsening-upward sequences representing an upward change from distal sheetflood deposition to proximal sheetflood deposition. Proximal deposits are cut by channel-

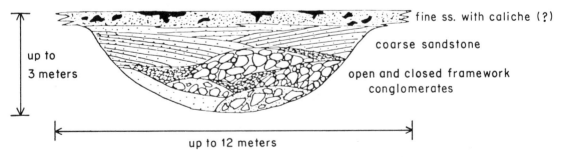

Figure 4.11 Characteristic channel fill sequence in ephemeral braided stream deposits of the New Red Sandstone of Scotland (after Steel, 1974).

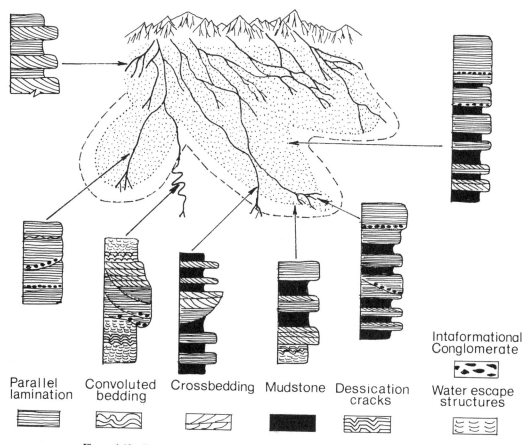

Figure 4.12 Types of sequences deposited by ephemeral stream systems in the Middle Devonian Trentishoe Formation of Great Britain (after Tunbridge, 1984).

fill sandstones that are probably counterparts of the confined floodplain sequences of Sneh (1983).

Distal sequences were deposited on a clay playa and consist of alternating layers of fine-grained sandstone and laminated mudstone in 2-cm-thick beds. Mudstones are marked by desiccation features, and sandstones are locally wave-ripple bedded. Similar deposits occur in the sandy facies of the East Berlin Formation (Triassic) in the Hartford Basin of the eastern United States. These deposits occur as single, decimeter-thick bedding units that fine upward or as stacked units up to 1 m thick that thin upward from basal units that are locally crossbedded. These sediments were deposited by sheetfloods flowing over dried lake beds under conditions of semiaridity that occurred periodically in the basin (De Micco and Kordesch, 1986). Both the distal facies of the Trentishoe Formation and the sandy facies of the East Berlin Formation are probably counterparts of Sneh's (1983) terminal floodplain deposits.

Playas and Saline Lakes

Sedimentary processes and products.   Playa lakes possess salinities in excess of 5000 ppm dissolved solutes, which is the upper limit of tolerance of most nonmarine aquatic organisms (Hardie and others, 1978). Such lakes are common in desert regions where evaporation exceeds precipitation and there is a closed drainage system.

The causes of a closed drainage system are varied. Fault-block basins normally possess interior drainage, and local orographic effects may add to the high temperatures and aridity of the climate. Such basins may hold perennial saline lakes if streams rising in adjacent highlands flow year-round. Deflation hollows may be sufficiently deep to intersect the groundwater table. In such cases, long-standing but shallow lakes may form where the surface is at least kept perpetually damp. Migrating dunes may derange the drainage by blocking stream courses, or stream avulsion may cause alluvial fan lobes to prograde across lowland areas laterally adjacent to the fan apex. Playas commonly form at the toes of fans where groundwater from the fan alluvium emerges as springs at hydrographic discontinuities. Because of the great variability of cause and occurrence, playa lakes vary greatly in size and depth.

The chemical composition of the rocks of the watersheds surrounding playas controls, to a large measure, the chemical composition of the water in the lakes and the end products of the evaporative concentration of brines (Hardie and others, 1978). The first deposits are relatively less soluble alkaline earth carbonates, followed by gypsum and then soluble salts. Final products include minerals precipitated directly from the brine, such as halite and trona, as well as authigenic clays and zeolites, where brines interact with detrital sediments.

Although they are not environments of great organic productivity, neither are saline lakes sterile. There are salt-tolerant fauna, such as brine shrimp, and halophytic flora that may produce a particularly important effect on sedimentation. Blue-green algal mats, for instance, may decompose in oxygen-starved bottom waters through bacterial action and produce deposits of kerogen-rich sediments, such as those of the Green River Formation (Bradley, 1973). Higher-order plants develop a characteristic zonation around playas, and their rooting activity may entirely homogenize substrates in areas of small sediment yield, producing a characteristic texture called dikaka (Glennie, 1972). The central areas of the playa may be devoid of flowering plants, but fringing areas may have salt-tolerant phreatophytes. Freshwater phreatophytes and xerophytes are normally confined to areas beyond the fringe (Neal, 1965).

Hardie and others (1978) differentiate 10 major facies of saline lakes. Of these, alluvial fans and associated sand flats, dune fields, floodplains, and shoreline features are discussed elsewhere in this book. The remainder, including mudflats, ephemeral and perennial saline lakes, and springs are directly related to playas.

Mudflats fringe the central playa (Fig. 4.13). They derive their water from floods en route from alluvial fan to playa or from expansion of the playa over the fringing mud. A variety of sedimentary structures occur on mudflats, including finely laminated muds, graded beds of sand and silt, climbing ripples, and adhesion ripples of windblown sand. After deposition, these structures commonly are modified by desiccation or shrinkage cracks, sand dikes, crystal growth, and disruption by roots of halophytic plants.

The surface of the mudflat is normally covered by an efflorescent crust that may consist of micritic carbonates or more soluble salts such as halite. A lateral zonation is produced when more soluble salts are brought to the surface of the flat by evaporative pumping and then dissolved by surface waters and carried farther basinward. Less soluble salts such as magnesium and calcium carbonates are left on the outer zone of the mudflat (Hardie and Eugster, 1970).

Ephemeral saline lakes undergo cyclical periods when they exist as standing bodies of water, followed by progressive evaporative concentrations of brines, leading eventually to total evaporation. The evaporated lake consists of an outer fringe of muddy sediments intergrown with salt crystals and a central area or salt pan defined by the extent of the lake at the time the salts began to precipitate.

Fine-grained sediments are brought to the center of the basin during flooding. They settle from suspension and form the initial deposits of the flood cycle. Algal blooms may add organic material to bottom sediments, and progressively more soluble salts are precipitated as evaporation takes place. The resultant deposits in the pan may then consist of mud–salt couplets (Hunt and Mabey, 1966).

Perennial saline lakes occur where inflow provided by perennial streams, springs, or episodic stream flow is sufficient to maintain water in the basin. Nor-

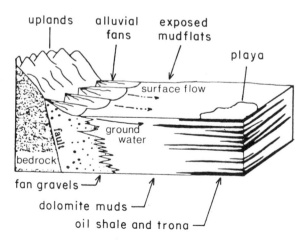

Figure 4.13 Relationships among source area, alluvial fans, mudflats, and playas on a broad, nearly flat floored basin (after Eugster and Hardie, 1975).

mally, the inflow is fresh, although some springs may be strongly saline, and it may flow into a highly stratified lake, with brine concentration increasing downward. Surface waters eventually mix with underlying waters as evaporation increases their salinity.

Sedimentary structures in perennial lakes consist of laminated salt deposits of varied composition separated by thin clastic laminae that record episodes of terrigenous influx. Algal blooms in perennial lakes may also add organic matter to bottom sediments, where complex interactions between the organic matter, anaerobic bacteria, and sulfate minerals occur.

Three basic types of facies associations occur in the playa component (Hardie and others, 1978). In narrow fault-block basins, alluvial fans fringing upthrown blocks grade basinward into sandflats, mudflats, and salt pans (Hunt and others, 1966). Extreme floods may cross the sandflat, but for the most part water is added to the playa by fan-edge springs.

In wide basins, ephemeral saline lakes are not influenced by alluvial fans, but instead are bordered by dune fields and ephemeral stream floodplains. Some lakes may be dry for longer periods than they are wet (Twidale, 1972), but nonetheless can be very large when flooded. Smaller lakes may be the result of wind deflation down to the water table and will remain wet as long as the water table remains high. The sediments of both types of lakes may be intimately associated with deposits of eolian dunes and ephemeral streams.

Some perennial saline lakes are associated with perennial stream floodplains. Some, as in the case of the Great Salt Lake, can be quite shallow when they are situated in the middle of a wide, nearly level basin plain. Others may be deep, as in the case of the Dead Sea, which occurs in a narrow fault-block basin. In the first case, the lake will be associated with mudflats of previous lake-level stands, floodplains, and shoreline features (Fig. 4.13), whereas, in the second, alluvial fans fringing nearby fault scarps prograde directly into the lake (Fig. 4.14).

Intrinsic and extrinsic controls on lacustrine sequences.   The most important factor in sedimentation in saline lakes and playas is climate. As climate changes from wetter to drier, lake levels rise and fall, margins expand and contract, clastic sediment yield increases and decreases, and water chemistry changes.

Storm events provide the impetus for minor sedimentary cycles. In ephemeral lakes, a storm will result in the formation of a standing body of often turbid water. Laminated silts and clays will settle from suspension, and temporary blooms of blue-green algae can add organic matter to the sediments. As the lake dries, brines will become progressively more concentrated and a series of salts will be precipitated. After the lake dries, windblown sand may be trapped on the surface as adhesion ripples.

Storm events on the mudflats surrounding large lakes are recorded by thin cycles of graded beds in response to rapid deceleration of flow over the fringing flats and into the lake (Hubert and Hyde, 1982). Storm floods flowing over terrigenous mudflats may deliver large amounts of fine sediment that can temporarily over-

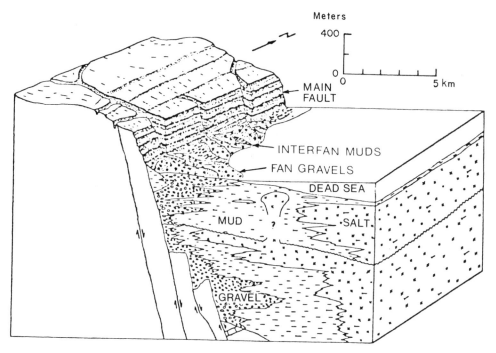

Figure 4.14 Relationships among source area, alluvial fans, and saline lakes in a fault-bound basin, Dead Sea of Israel (after Langozky and Sneh, 1966; Friedman and Sanders, 1978).

whelm chemical sedimentation, resulting in clastic–evaporite couplets. Where storm floods flow over carbonate mudflats, however, they can carry carbonates into the lake as peloids (not as solutes), where they form deposits that can alternate with evaporite or organic-rich beds (Hardie and others, 1978).

Storm events on the transitional sandflats between playas and alluvial fans are represented by thin beds, commonly centimeters thick, of sand in plane beds or wavy horizontal beds. As lake level expands, ponds may form on the very low sloping sand plain, resulting in wave reworking of the surface and the formation of shallow-water wave ripples (Hardie and others, 1978).

Longer-period climatic cycles cause different effects on different types of playa lakes. During prolonged periods of drought, groundwater tables might be lowered to such an extent that ephemeral lakes would dry completely and allow dunes to encroach on the lake basin. Shallow perennial lakes might evolve into ephemeral lakes, resulting in the precipitation of thicker and more abundant salt layers (G. I. Smith, 1962) and the deposition of fewer mud layers or organic layers (Bradley and Eugster, 1969). In deep lakes, thick, massive, soluble-salt deposits with few mud partings might result from prolonged dry periods.

It might be expected that playas associated with fans would complete a cycle with encroachment of alluvial fan sediments into the central basin. There is evidence to suggest, however, that this does not happen. Playa sediments commonly are interbedded with fan sediments only at their margins, and even a completely dry lake is only the site of sheet flow during extreme floods.

The areas occupied by fans and playas in closed basins are in delicate equilibrium (Denny, 1967). Uplift at the source may bring additional coarse debris to the fan, causing an initial enlargement, but this is soon compensated for by the additional delivery of fine sediments to the playa. This allows the playa to aggrade and reestablish its equilibrium area. The same factors that cause fan growth also cause playa growth, resulting in vertically continuous facies showing relatively little lateral shifting of facies boundaries.

The culminating phase of playa deposition occurs when through-flowing drainage is established, either because the basin floor has aggraded above enclosing barriers or because the barrier has been breached by tectonism or erosion. At that point the lake must drain, and fluvial deposition will dominate the central basin.

Playas and saline lakes in the rock record.   There are numerous examples in the rock record of lacustrine sediments interbedded with eolian sandstones that probably represent dried interdune playas. These consist mainly of mudstones, carbonates, and evaporite minerals, with internal structures consisting of laminae, adhesion ripples, root traces and crawling traces, and salt crystal molds.

The lacustrine sediments occur commonly in small, concave depressions that have been interpreted as blowouts or as small lenses lying on widespread truncation surfaces. Where they are laterally extensive, interdune playa sediments may form vertical impediments to migration of fluids in what are otherwise excellent reservoir rocks (Ahlbrandt and Fryberger, 1983).

Playa sediments in the Lyons Formation (Permian) of Colorado overlie truncation surfaces developed on thick, well-sorted, crossbedded sandstones of probable eolian origin (Adams and Patton, 1979). Lacustrine deposits consist of poorly sorted sandstones interbedded with calcareous silty sandstones. Laminae in the sediments are wavy and contorted, often displaying diapirlike structures, and the upper surface is commonly deformed by load and flame structures where it is in contact with eolian sandstones. Polygonal shrinkage structures are evident on bedding surfaces, and the sediments may be bioturbated with burrows and trails.

Deposits of large, perennial saline lakes are commonly the initial deposits of pull-apart basins where they occupy narrow grabens. A vertically persistent facies distribution is often the result of sedimentation in these settings, with alluvial fan sediments lining the trough and separated from lacustrine sediments in the center of the basin by a narrow band of alluvial muds (Langozky and Sneh, 1966) (Fig. 4.14).

Large perennial lakes may also occupy broad flexural basins. Perhaps the best known of these are the lakes in which the Tertiary Green River Formation was deposited. The Green River Formation was deposited in large basins occupying a nearly featureless plain (Surdam and Wolfbauer, 1975), bounded on the west by the overthrust belt and on the east by the Rawlins Uplift (Surdam and Stanley, 1979).

The lake stood at its highest levels early and late in its development when it covered an area of up to 50,000 square miles. Paleobotanical evidence suggests a subhumid climate during these times, and evaporite facies are not well-developed. The climate must have been more extreme, however, during deposition of the middle Wilkins Peak Member when the lake was at its smallest, because evaporite minerals make up a significant portion of the sediments.

The Wilkins Peak Member was deposited in a shallow lake that was subject to rapid expansions and contractions caused by climatic variations (Eugster and Hardie, 1975; Smoot, 1983). During low stands of the lake, when the rate of evaporation was accelerated, trona and halite were deposited alternately. Occasional storms washed mud into the lake, giving rise to terrigenous partings in the salts, or they washed in carbonate peloids from the surrounding mudflats, forming micritic carbonate beds.

The lake expanded during pluvial periods. Flat-pebble carbonate conglomerates accumulated when the transgression was rapid, but eroded mud chips were more completely broken down during slow transgressions, producing carbonate muds that were deposited as beds of relatively clean carbonates as the lake expanded. Oil shale accumulated in the relatively fresh water where blue-green algae thrived. Organic muds alternate with dolomitic laminae formed when dolomitic peloids were washed into the lake from the surrounding mudflats during storms (Fig. 4.15).

The alternate expansions and contractions were a response of the lake to cyclic climatic fluctuations lasting 20,000 to 50,000 years. Cycles began as the lake transgressed over the fringing mudflats during expansion. Basal beds consist of flat-pebble conglomerates or lime mudstone, depending on the rate of transgression (Fig. 4.16). During the maximum extent of the lake, oil shales were the dominant sediment; but as the lake evaporated slowly, salinites increased and dolomite became increasingly abundant at the expense of the organic material. Cyclic deposits are somewhat different in the center of the basin, away from the fringing mudflats. There the cycles consist solely of oil shale, deposited during maximum extension, and trona–halite, which accumulated during lowest lake levels.

Ergs

Sedimentary processes and products.   An erg is an area that is covered by more than 20% windblown sand and is large enough to contain very large sand ridges (draas) (Wilson, 1973). Active ergs develop in areas where rainfall is insufficient to sustain an appreciable plant cover that can interfere with dune movement. When dunes become inactive the erg is relict.

Ergs tend to occur in basin plains with sand deposition ceasing at pronounced slope breaks (Wilson, 1973), possibly because wind patterns are altered by the presence of highlands, and an abundance of fine-grained alluvium tends to be found in basins rather than uplands. The close association of ergs with structural or physiographic basins enhances the preservation of these sediments.

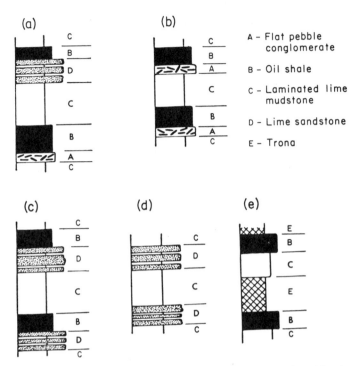

Figure 4.15 Various stratigraphic sequences developed during depositional cycles of the Wilkins Peak Member of the Green River Formation. Sequences (a) through (d) develop near the lake margin, whereas sequence (e) is restricted to the central portion of the playa (after Eugster and Hardie, 1975).

Single ergs range in size up to 560,000 km², and continuous sand sheets containing several ergs may cover areas up to 1 million km². The size of some large ergs is on the order of some of the very large ancient eolian sand bodies, such as the Botucatu Sandstone of Brazil (Salamuni and Bigarella, 1967).

Although ergs may cover large areas, the sand cover when averaged over the entire area is thin. The average sand thickness in some ergs may be as little as 1 m (Wilson, 1973), and only in core areas do sand thicknesses reach the order of magnitude of some ancient eolian sand bodies.

The three major facies of the erg component include dunes and draas, interdune, and eolian plain. The proportion of the erg area that these facies occupy is extremely variable and dependent to a large degree on the availability of sand.

Dunes or draas. Sand accumulations in ergs range in scale from ripples with heights measured in millimeters to draas with heights of 100 m or more and wave lengths up to kilometers long. Dunes are intermediate in scale with heights to tens of meters and wave lengths up to hundreds of meters (Wilson, 1972). Ripples are unrestricted in their occurrence, and dunes may also occur in areas not totally cov-

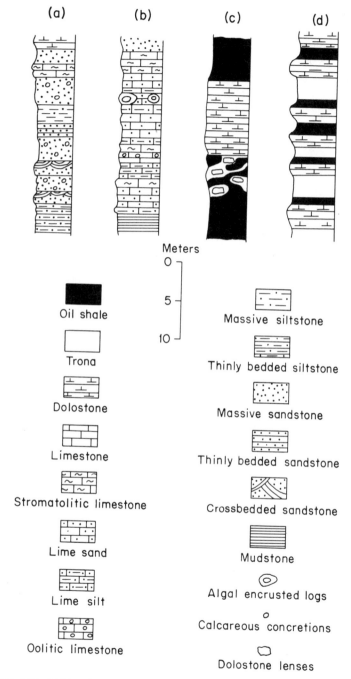

Figure 4.16 Lithic columns of typical marginal (a), inner mudflat (b), outer mud-flat (c), and lacustrine (d) facies of the Green River Formation (after Surdam and Wolfbauer, 1975).

ered by sand. Draas, however, can only occur in those areas of the erg that are saturated with sand.

There are two main types of ripples, neither of which is a true bedform. Ballistic ripples form when sand collects in the lee of obstructions to the wind flow, where grains are protected from further impact. Adhesion ripples form when sand adheres to a damp surface in a regular or irregular pattern. They occur commonly in interdune areas or playas where the groundwater table intersects the surface (Kocurek and Fiedler, 1982).

Dunes represent the only true eolian bedforms and they may be divided into two basic types: transverse dunes and longitudinal dunes. Transverse dunes are those oriented more or less perpendicular to the resultant of airflow. They are characterized by a gently sloping windward face and a steep slipface sloping at or very near the internal angle of friction of the sand. The slipface of barchan dunes is concave downwind, with the maximum crest elevation at the center and decreasing along the downwind projections. Parabolic dunes possess relatively shallow slipfaces that are convex in a downwind direction because the center of the dune moves faster than the sides, which may be partially stabilized by vegetation.

Long-crested dunes with more or less continuous crests tend to form in areas with a complete sand cover. Isolated dunes, such as parabolic dunes or barchans, form where the airflow is unsaturated with sand.

Crests of longitudinal dunes or seif dunes are oriented parallel to windflow. Crests are continuous and often extend for 10 km along wind flow. The form of the dune may be the result of the repeated occurrence of occasional strong winds blowing at an angle to the general sand drift (Bagnold, 1954). Longitudinal dunes, however, commonly possess sinuous crests with alternating slipfaces, suggesting the presence of spiral vortices or a bidirectional wind regime (Bagnold, 1941; Folk, 1971; McKee, 1982). Sideslopes of large longitudinal dunes are often marked by the presence of superimposed smaller-scale dunes.

A third type of dune appears to show no preferred orientation to resultant wind flow. It includes rhourds, star dunes, and dome-shaped dunes that apparently form when more than one effective wind resultant affects sand transport. Crestlines are sometimes present, but crestlines more often are broken up into isolated peaks.

The laminae within dunes result from three depositional modes (Hunter, 1977). Accretion bedding results from the transport of sand up windward slopes, principally by saltation and often in the form of wind ripples. Foreset laminae may develop from movement of sand down lee slopes through grain flowage or mass movement. They may also develop from grainfall of sand from suspension as it is carried over the crest. Within a single dune there may be a characteristic distribution of sediments deposited by these various modes of sedimentation (Fig. 4.17).

Some structures appear to be common to most dunes. Horizontal laminae form on windward slopes, flanks, and bottomsets. Large-scale sets of crossbeds form on lee slopes that dip at steep angles (30° to 34°), very near the internal angle of friction of dry sand, and groups of cross sets may be separated by horizontal or near horizontal bounding surfaces. Deformation structures, such as contorted

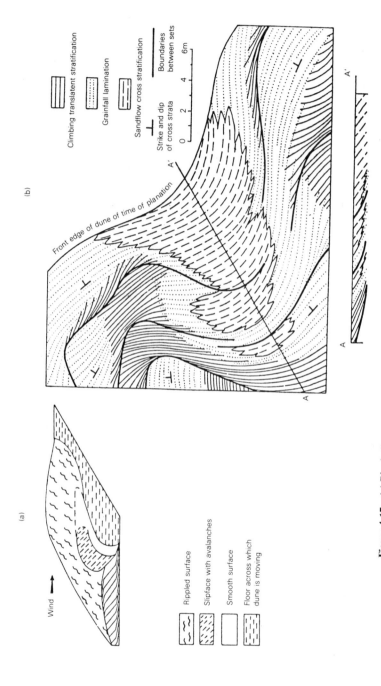

(a)

Wind

Rippled surface

Slipface with avalanches

Smooth surface

Floor across which
dune is moving

(b)

Front edge of dune of time of planation

Climbing translatent stratification

Grainfall lamination

Sandflow cross stratification

Strike and dip
of cross strata

Boundaries
between sets

0   2   4   6m

A'

A'

A

A

Figure 4.17 (a) Block diagram showing distribution of surface features on a barchanoid dune;
(b) map with cross section showing distribution of the various types of internal structures in
a planed-off transverse dune (after Hunter, 1977).

bedding, folding, faulting, and blurring of laminations are ubiquitous in slipface beds. Trough crossbeds are not common structures, although they develop in many settings. Most are apparently related to infilling of blowouts, although large-scale festoon bedding can develop during migration of sinuous, crested, longitudinal dunes (Brookfield, 1977).

Brookfield (1977) has defined a hierarchy of bounding surfaces in eolian deposits (Fig. 4.18). Third-order surfaces separate bundles of cross strata within major sets of crossbeds and are probably produced by short-term fluctuations in wind strength or flow direction. Second-order surfaces separate major crossbed sets and represent aggradational cycles during the growth of individual dunes. The cycles probably occur in response to seasonal fluctuations in flow direction and/or strength (Hunter and Rubin, 1983). Minor changes in the wind regime generally vary widely in duration and strength, but seasonal cycles tend to exhibit relatively small variations in time-averaged transport vectors. They are, thus, more likely to generate long series of similarly styled cycles separated by second-order surfaces.

Sequences of major crossbed sets may be separated by horizontal or near-horizontal first-order bounding surfaces whose origin has been the focus of long-standing controversy. Stokes (1968) attributed them to periodic deflation of sand down to a groundwater table. Some first-order surfaces may be related to major events in eolian plain evolution, such as profound climatic changes (Talbot, 1985), or marine transgression, long-term change in wind direction, and cutoff of sand supply (Kocurek, 1984). Such events might produce Stokes-type surfaces that would be recognized in ancient deposits as extensive, nonclimbing, irregularly spaced surfaces associated with marked changes in the character of dune and interdune sediments.

More regulary spaced surfaces separating dune deposits of mean constant thickness or systematically changing thickness, however, are more probably related to less substantive changes in erg dynamics. Rubin and Hunter (1984), for example, have shown that the time required to deflate and then reestablish an eolian plain is probably too long to permit most sequences of eolian sediments displaying repetitive first-order bounding surfaces to accumulate.

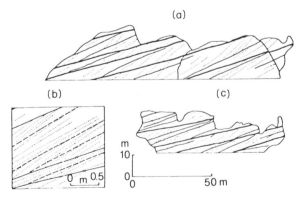

Figure 4.18 Outcrop drawings illustrating the hierarchical arrangement of bounding surfaces in ancient eolian sandstones. First-order surfaces are shown in thick lines; second-order surfaces are shown with thin lines; third-order surfaces are shown with dashed lines; and surfaces of crossbeds are shown with dotted lines (after Brookfield, 1977).

McKee and Moiola (1975) attribute first-order surfaces to migration of dunes or draas over planed-off remnants of preceding deposits. Unless substantial deflation takes place, however, succeeding deposits must climb over preserved remnants of the underlying deposits (Kocurek, 1981b; 1984). This may easily be accomplished repetitively within the context of a single evolving erg by altering the angle of climb (Rubin and Hunter, 1984). When such alterations take place, nearly isochronous surfaces are produced between cosets.

Other structures are characteristic of specific dune types. Barchans develop sets of high-angle foresets 1 to 2 m high separated by gently dipping bounding surfaces (Fig. 4.17). Dip angles are steepest at the axis of the dune and decrease toward the horns. Dip directions are downwind at the axis, but in the horns they become transverse to wind flow (McKee, 1966; Ahlbrandt, 1975).

Transverse dunes consist of high-angle crossbeds in stacked sets that tend to thin upward in the dune (Fig. 4.19). In sections oriented perpendicular to wind flow, laminae show great lateral extent (McKee, 1978).

Slipfaces of seif dunes occur on either side of the crest and develop foreset bedding dipping in nearly opposing directions (Fig. 4.20) (McKee and Tibbitts, 1964). Wedge-planar cross sets, rare in most dunes, may be common in seif dunes where alternating periods of deposition and erosion occur in a regime of reversing wind direction.

The ability of dome dunes to grow upward is limited by large wind velocities, and they do not develop high-angle slip faces except in an upwind core, suggesting that dome dunes have been modified from other dunes (Fig. 4.21) (McKee, 1978). They are characterized by abundant scour and fill structures and well-developed topsets that grade into gently dipping leeside foresets.

The internal structures of star dunes reflects the complex wind regime in which they form (McKee, 1966). Each arm that extends from the central crest may have a separate slipface that contributes high-angle cross strata to the composite structure.

Reversing dunes form as the result of reversing wind patterns that produce two principal dip orientations of the cross strata (Merk, 1960). Truncation surfaces are common because the reversing winds tend to level previous deposits, leading Sharp (1966) to suggest that the lack of high-angle foresets, typical of fossil dunes, may be caused by the poor preservation potential of the upper parts of dunes where high-angle dips occur.

Draas are giant landforms that accumulate in areas of complete sand saturation. They are composite features, often covered with superimposed dunes, and they may form massiflike structures that can have a major effect on subsequent wind patterns. Draas are the eolian sand structure with the greatest potential to be included in the rock record.

Because of their size, we can only make inferences about internal structures of draas. The presence of variously oriented slipfaces tens of meters high suggests that draas might have a complex association of mega cross strata. Sharp (1966), however, argued that the internal structures may be consistent with the main wind resultant.

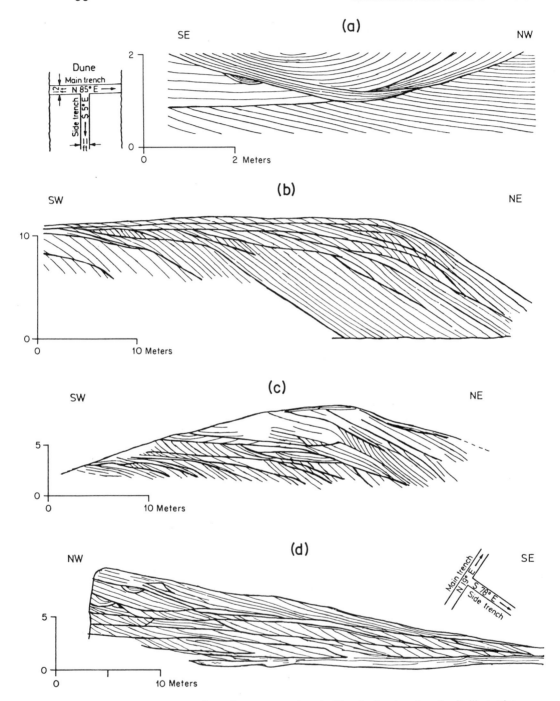

Figure 4.19 Cross sections of a transverse dune (a, b) and a barchan dune (c, d), illustrating the internal organization of crossbeds (thin lines) and bounding surfaces (thick lines) (after McKee, 1966).

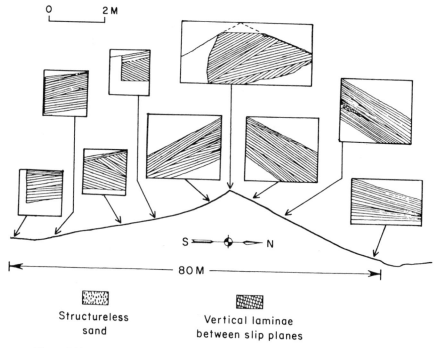

Figure 4.20 Internal structures seen in shallow trenches cut in various locations on a seif dune (after McKee and Tibbitts, 1964).

Interdunes.   The dominant process in interdune areas is deflation, although some desert interdunes may be covered by thin sand sheets. Wind may erode down to bedrock in some areas, but in others it may concentrate a lag armor called a serir that may protect the surface from further erosion. Erosion may remove sand down to the water table. In these cases the damp sand is less susceptible to erosion, and

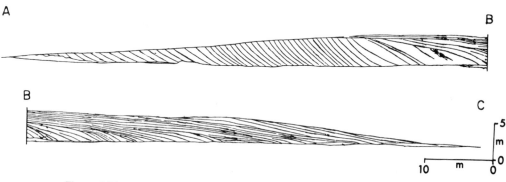

Figure 4.21 Cross section showing the internal structure of a dome-shaped dune (after McKee, 1966).

there may be sufficient water to support plant growth. Interdune sediments may include clay and silt that settle out of suspension, windblown sand that may occur as scattered grains in a mud matrix or as adhesion ripples, lag gravels, salt crusts, and locally abundant organic matter.

Interdunes may be classified into dry, damp, wet, and evaporitic types (Ahlbrandt and Fryberger, 1979; Kocurek, 1981b). Dry interdunes consist of sand sheets of poorly sorted sand in scour and fill structures, ballistic ripples, small eolian dunes, and grainfall deposits over lag grain surfaces (Figs. 4.22 and 4.23).

Wet interdunes contain more fine-grained sediment and much more plant-derived organic matter. Fine-grained sediments occur in thin laminae that may be flat, hummocky, or contorted by liquefaction, salt crystallization, or loading by dune emplacement. In perennially wet interdunes, plant growth may be extensive enough to obliterate bedding structures. Infaunal structures consist mainly of arthropod burrows (Ahlbrandt and others, 1978). Sand layers may display both wave and current ripple bedding.

Fine-grained sediments are also abundant in evaporitic interdunes, but evaporite mineralization is the dominant process. Silt and clay form laminae as they settle out of suspension, and sand may accumulate as adhesion ripples. The surface of the interdune, however, can be encrusted with salt (Glennie, 1970).

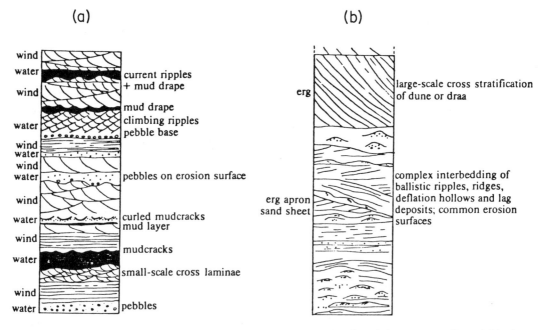

Figure 4.22 (a) Stratigraphic sequences characteristic of sand sheets near wadis, and (b) of sand sheets in areas near erg margins not traversed by wadis (after Fryberger and others, 1979; Leeder, 1982).

DEPOSITIONAL CONDITIONS

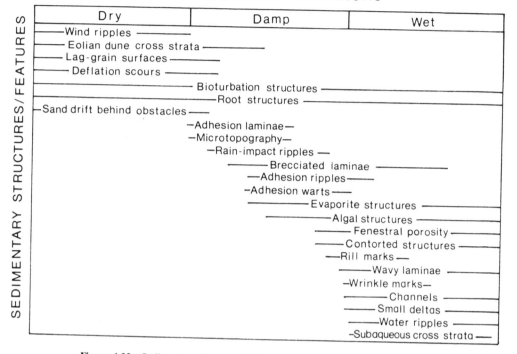

Figure 4.23 Sedimentary features in interdune areas shown as a function of soil moisture conditions (after Kocurek, 1981b).

Damp interdunes are characterized by adhesion structures, including ripples, laminae, and warts, as well as rain impact structures and brecciated laminae composed of broken crusts and algal mats, and flakes of cohesive mud.

Interdune deposits show a variety of sequences (Kocurek, 1981b). Some sequences consist of sediments deposited as the interdune dried, reflecting its aggradation or its encroachment by migrating dunes. Sequences that show evidence of wetter conditions with time reflect a rising water table but are less common, suggesting the ability of interdune areas to trap sediment and aggrade. Many sequences are very complex, reflecting the wide variety of processes that can occur both laterally and temporally in an interdune (Fig. 4.23).

Eolian plains.   Sand sheets are deposits of eolian sand with low-relief, irregular topography and lack of slip surfaces (Fryberger and others, 1979; Kocurek and Nielson, 1986). Bagnold (1954) suggested that they develop under lag gravels, whereas Glennie (1970) believed the low-angle to horizontal bedding to be the result of "upper flow regime" bedforms. They tend to form in areas marginal to dune fields where air, metasaturated in sand, is decelerated by topographic irregularities, baffling effect of vegetation, or complex wind patterns (Fryberger and others,

1979). Conditions within these areas, however, must be outside the range within which dunes form or where some factor such as vegetation, surface cementation or binding, periodic flooding, or significant coarse sediment population interferes with dune formation. Topography is provided by depressions undergoing erosion or deposition, small sand mounds or degrading relict dunes, and sand tongues and sand strips, which are migrating sand bodies with no slipfaces.

Eolian sand sheets near dune fields contain thicker sand deposits, with hummocky surfaces consisting of low dune features, blowouts, and degraded dunes (Kocurek and Nielson, 1986). Farther away from dune fields, sand deposits are thinner, and water-laid deposits accumulate in ephemeral alkaline lakes, marshes, creeks, and hardpans where surfaces are at or near the water table. Topography in distal sand sheets is also provided by blowouts and degraded dunes.

Internal structures of proximal eolian sand sheets consist primarily of planar horizontal laminae, thin sets of tabular, planar cross laminae, and wedge-shaped sets of cut-and-fill structures. Eolian deposits of distal sand sheets are characteristically interbedded with water-laid deposits consisting of ripple drift sands, current ripples with mud drapes, and cut-and-fill structures (Fig. 4.22) (Fryberger and others, 1979).

Facies distribution.   Ergs tend to be convex-upward lenses of sand, thicker in the center where high draas occur and tapering to very thin sand accumulations on the margins (Wilson, 1973). Bedform growth apparently responds to the length of time wind has to entrain sand and transport it. Where deflation rates are large, such as in the lee of alluvial fans, airflow may become saturated quickly and deposition may be initiated at the source. In such cases the erg deposit can become interbedded with its source, but for the most part the area of ergs most likely to be preserved in the rock record is the center of topographic and/or structural basins downwind of the source area (Wilson, 1971).

The resultant facies distribution may show an asymmetric zonation of dune types (Norris and Norris, 1961; Mainguet and Callot, 1974). Isolated barchans tend to occur on the margins of ergs and grade into coalescing barchans toward the center. The interior zone, where the airflow is saturated, contains seif dunes and longitudinal dunes, with a core of draas occurring on the upwind side of the erg (Fig. 4.24), and McKee (1983) reported the tendency of dome dunes to occur on upwind margins of sand seas. The margins of ergs may be fringed by an eolian sand plain that grades outward into an extra dune area (Lupe and Ahlbrandt, 1979).

The distribution and relative development of interdune areas are a function of their position within the erg and the nature of the adjacent dunes. Interdune deposits are small and thin where barchan dunes occur (McKee and Moiola, 1975), suggesting that on the margins of ergs, where mobile barchans predominate, interdune deposits will be thin and of limited areal extent. In areas of slowly migrating star dunes, on the other hand, interdune deposits up to 5 m thick have been recorded (Ahlbrandt and Fryberger, 1983). Thicker deposits, therefore, might be expected in the core areas of ergs where dunes migrate at slow rates.

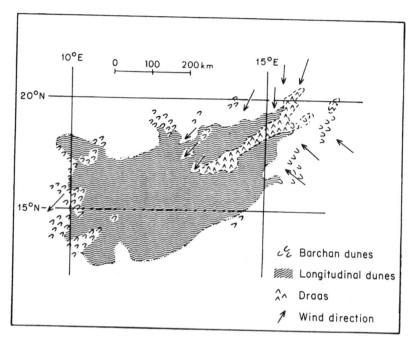

Figure 4.24 Dune facies of the Fachi-Bilma Erg, southeastern Sahara (after Mainguet and Callot, 1974; Walker and Middleton, 1979).

Instrinsic and extrinsic controls on eolian sequences.   Three zones may be defined in a given sand flow. Where the flow is unsaturated, the wind is transporting less sand than its capacity, so deflation takes place. In a metasaturated zone, the wind is also transporting less sand than its capacity, but deposition nonetheless takes place because of local flow divergence or deceleration. Erosion also occurs, however, so there is no net deposition. In saturated flow, the maximum amount of sand is moving, the sand blanket is continuous, and net deposition takes place (Wilson, 1971).

If sufficient sand is available, wind flow becomes quickly saturated downwind from the source and deposition takes place, with dune sands locally interdigitated with their source materials (Wilson, 1973). As sand is progressively deposited, however, the wind flow soon becomes metasaturated farther downwind from the source and eventually it becomes unsaturated. The zonation of dune types in an erg and the areal distribution of sand thicknesses reflect this process (Mainguet and Chemin, 1983). Where the sand budget is positive (input > output), dunes of deposition, such as barchans or transverse chains, occur. Where the budget is negative, however, deflation corridors form, and chains evolve into sand ridges with wide, sand-free, interdune deflation surfaces.

A prograding erg will move downwind from its source, gradually spreading the zone of saturation as the erg itself becomes the source for downwind areas.

Eventually, the erg will reach an obstacle, such as a basin margin or a zone where flow divergence occurs, and migration ceases. The central area of draas will also aggrade until it becomes large enough to modify wind patterns to the extent that aggradation can no longer occur. Even in this phase of erg growth, however, downwind extension is still possible.

The ideal erg sequence, therefore, would consist of basal eolian plain sands interbedded with extra dune deposits, such as playas or serirs, of probable large lateral extent. These would be overlain by deposits of barchan dunes interbedded with diachronous interdune deposits of limited lateral extent. Barchan dune deposits would be overlain by longitudinal dune sands, accompanied by thicker and more laterally continuous interdune sediments. The sequence would be capped by draa deposits associated with relatively few, but thick, interdune deposits.

Like other sedimentary deposits, ergs respond to external stresses by forming cycles. They are the largest sand bodies currently forming, and the rather limited size grade of the sand makes them more sensitive to cyclic changes in sediment yield than other sand deposits. Sandy residuum, for instance, can only be considered a minor source because of the long time it takes to form, and dune fields can only be self-sourcing for a limited time before undersaturation occurs. In most cases, sandy alluvium is the only potential source sufficient to sustain erg growth.

Given even moderate winds, sand flow rates will quickly outstrip most sources except for unconsolidated alluvium. Near mountains, the availability of sand is cyclic to a degree, but given the time necessary to form a dune the yield is relatively constant. Erg formation under these conditions proceeds until the ideal progradational sequence evolves to the point where an intrinsic threshold is reached and growth is stopped.

In the center of large basins, however, thick alluvium is deposited only during relatively wet periods, whereas the deflation of the alluviated surface and the subsequent growth of dunes can only take place during dry periods. However, deflation can proceed only to the point where a protective lag is formed. At that point, the airflow becomes metasaturated or undersaturated and the erg itself is deflated; in extreme cases, deflation may proceed down to the water table. Sand deposition on an erg can proceed only after another wet period delivers alluvium to the basin. A more steady growth of ergs may be achieved in semiarid climates where wet periods may occur with seasonal cyclicity and provide alluvium to a basin more or less continuously (Mainguet and Chemin, 1983).

First-order cycles are related to migration of dune forms. Presumably, in a subsiding basin a condition with a rising water table will apply. As dunes migrate, they will preserve interdune sediments and will themselves be preserved, leading to a cyclic interbedding of dune and interdune facies. Such sequences have been observed in unconsolidated sediments of modern dune fields (McKee and Moiola, 1975) and in fossil dunes as well (Lupe and Ahlbrandt, 1979).

Changes in wind regime would produce the second- and third-order cyclicity of Brookfield (1977) by forming bedding planes between sets of cross strata (Hunter and Rubin, 1983). Longer-term fluctuations in sand flow might lead to more profound alteration and degradation of dune forms. Barchan dunes, for instance, might be altered to dome-shaped dunes; but for the most part, these cycles would not have a great impact on the stratigraphic sequence.

Eolian sequences in the rock record.   The recognition of fossil eolian deposits is based mainly on the occurrence of large-scale crossbeds of well-sorted sand (see McKee, 1978, for a review), and many examples of such deposits can be cited in the literature. Only a few researchers have attempted to make areal reconstructions of ergs or even to interpret vertical variations in dune sequences.

The Lyons Sandstone (Pennsylvanian) is a well-documented fossil dune deposit (Walker and Harms, 1972; Adams and Patton, 1979). The deposit consists of very large sets of planitabular cross strata dipping at angles up to 28°. Low-amplitude ripples were observed crossing foresets diagonally on exhumed faces, and other structures suggestive of subaerial deposition, such as animal tracks and raindrop impressions, were also observed. Minor structures found in the Lyons Sandstone that characteristically occur in dunes included avalanche structures and gravel lags.

Kocurek (1981a) made a regional study of the Entrada Sandstone (Jurassic) of Colorado and northern Utah. The unit was deposited in a cratonic basin between uplands to the east and shallow marine environments to the west (Fig. 4.25). Medium- to large-scale sets of dune cross strata were deposited in the central part of the erg, but near the coast, the sets are smaller and less regular and are interbedded with marine deposits (Fig. 4.25).

On the margins of the erg in inland areas, the deposits consist almost entirely of ripple cross sets of both aqueous and eolian origin, pebble lags, and scattered beds of thin, dune cross sets. These deposits probably accumulated in an area where sand sheets were formed by eolian reworking of fluvial deposits.

A hierarchy of bounding surfaces occurs in the deposits of the central erg facies. It indicates that the dunes there consisted of crescentic draas with small, enclosed interdune areas. Outside the central erg, however, only two orders of bounding surfaces occur, suggesting that these sets were deposited by simple crescentic dunes.

Relatively few eolian deposits of Precambrian age have been confidently identified. One of these is the Bigbear Erg of Northwest Canada, which is one of three clastic units filling an intermontane/intracratonic basin (Ross, 1983). Facies within the unit show consistent regional variations that suggest significant changes within the system, both laterally and through time. The oldest deposits consist of fanglomerates and fluvial sandstones at the upwind margin of the basin and eolian sandstones elsewhere. Dune deposits at the upwind margin consist of thin, tabular and wedge-planar cross sets probably deposited by small, straight-crested dunes (Fig. 4.26). Downwind from the margin, however, the dune cross sets are larger and have curved bounding surfaces, suggesting that they were deposited by bigger dunes with sinuous crestlines (Fig. 4.27) that were evolving into draas.

During deposition of the younger unit, river systems extended farther into the

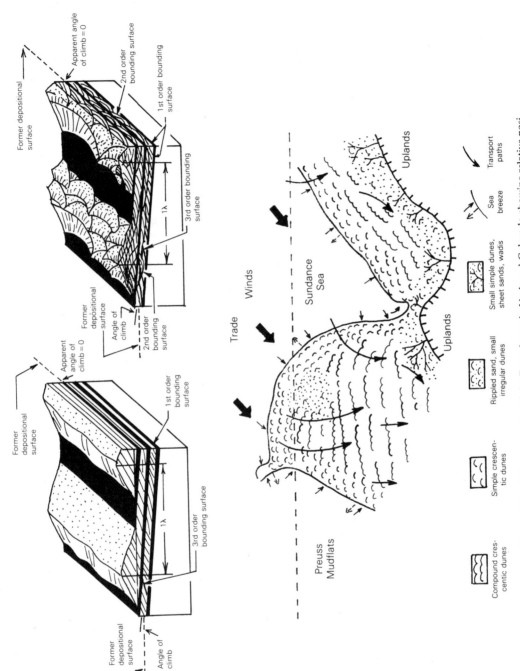

Figure 4.25 Paleogeography of the Entrada erg in Utah and Colorado showing relative positions of marine and nonmarine facies and the distribution of the various dune types represented in the formation (after Kocurek, 1981a).

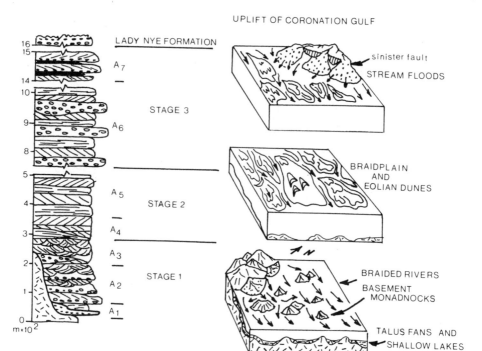

Figure 4.26 Stages in the development of the Big Bear System showing the initial development of a braid plain in the basin followed by the establishment of an eolian plain and ending with uplift at the margins of the basin and development of a mixed eolian and fluvial system (after Ross, 1983).

basin where floodplain deposits alternate with deposits of eolian sand sheets and solitary crescentic dunes. Surficial water did not penetrate to the central part of the basin because fluvial deposits are absent, and the form of the eolian deposits suggests deposition by barchanoid dune ridges (Fig. 4.27). The downwind change from tabular to trough megasets again probably reflects an increase in bedform size and crestline sinuosity.

The Arid System in the Rock Record

Arid to semiarid conditions existed over much of northern Europe during the Permian and early Triassic, and the rocks deposited during this time are among the most well-known arid-region sediments. Perhaps the best known of these rocks is the Permian Rotliegendes of northwestern Europe (Glennie, 1972, 1983).

Two basic sequences occur in the Rotliegendes (Fig. 4.28). Mixed eolian and wadi sediments occur at the base of the first and are replaced by eolian sand and interdune sabkhas upward. The wadi sediments consist of crossbedded or massive

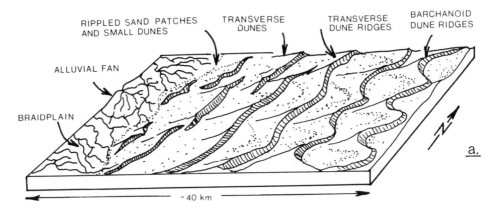

RIPPLED SAND PATCHES AND SMALL DUNES

TRANSVERSE DUNES

TRANSVERSE DUNE RIDGES

BARCHANOID DUNE RIDGES

ALLUVIAL FAN

BRAIDPLAIN

~ 40 km

a.

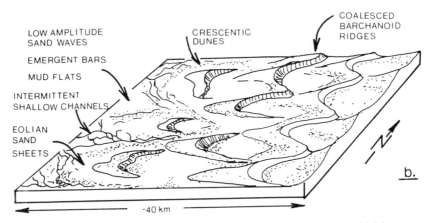

LOW AMPLITUDE SAND WAVES

EMERGENT BARS

MUD FLATS

INTERMITTENT SHALLOW CHANNELS

EOLIAN SAND SHEETS

CRESCENTIC DUNES

COALESCED BARCHANOID RIDGES

~40 km

b.

Figure 4.27 Stages in the development of the Big Bear Erg showing the initial dominance of straight-crested transverse ridges and barchanoid ridges (a) followed by a transition to single crescentic dunes and coalescing barchanoid ridges (b). Streams initially were restricted to the basin margins, but later they extended into the basin interior (after Ross, 1983).

sandstones, clay pebble conglomerates, and mudcracked clays with sandstone dikes (Fig. 4.28). Eolian sands are well-sorted and display planar crossbedding and some horizontal laminations in sequences separated by intraformational unconformities. Adhesion ripples occur occasionally at the base of crossbedded intervals, suggesting that interdune areas were moist. Although interdunes may have been moist, interdune sabkhas are not well-developed in this sequence. Where present, they consist of horizontally bedded clays that commonly are mudcracked and contain minor amounts of nodular anhydrite.

Wadi sediments are better developed at the base of the second basic sequence, and they also occur higher in the sequence. Thick sabkha deposits occur near the

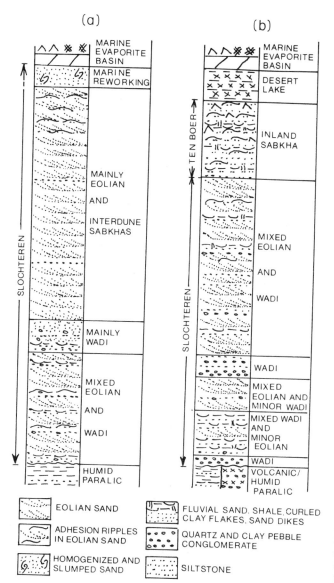

(a)

(b)

Figure 4.28 (a) Stratigraphic sequences in the upper Rotliegendes in the southern Rotliegendes Basin, and (b) the southcentral part of the Rotliegendes Basin (after Glennie, 1972).

top of the sequence, which is capped with lacustrine sediments consisting of interbedded halite and red clay (Fig. 4.28).

The first sequence is typical of the western part of the basin. Wadis originating in the Variscan Highlands to the south apparently terminated in dune fields without reaching the central basin, which was occupied by a lake covering much of the area of the present Caspian Sea (Fig. 4.29). Wadis crossing the eastern side of the basin, however, terminated at the lake and were probably its principal source of water.

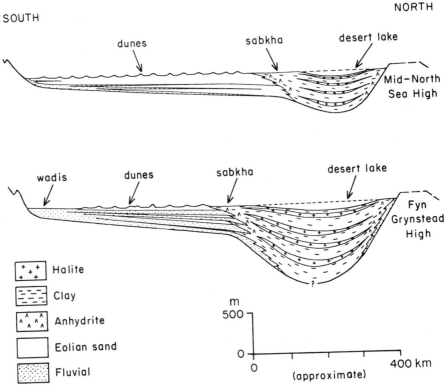

Figure 4.29 Diagrammatic cross sections through the Rotliegendes Basin in the southern North Sea and Eastern Netherlands prior to the Zechstein transgression (after Glennie, 1972).

Other arid-region sediments of Permian age occur in the northern part of Scotland (Clemmensen and Abrahamsen, 1983). The sediments consist of fringing alluvial fans grading basinward into marginal sequences of interbedded eolian sands and fluvial conglomerates (Fig. 4.30). Basinal deposits consist almost entirely of eolian sands with only rare interdune playa deposits (Fig. 4.31).

Eolian sands are well-sorted and occur in large-scale cross sets that apparently were formed by grainfall, sand flow, and grain flow processes. Interdune deposits consist of sand and granules in ripple beds and horizontal laminations and may have been, in part, deflationary.

Eolian sands deposited on draas in the center of the basin are preserved in sets of crossbeds up to 8 m high. Most drainage into the fault-bounded basin apparently percolated into porous sediments before reaching the center, because playa and wadi sediments do not occur.

Dune height decreases toward the margin of the basin and there is evidence of the occurrence of isolated crescentic dunes. Dune sands become interbedded with

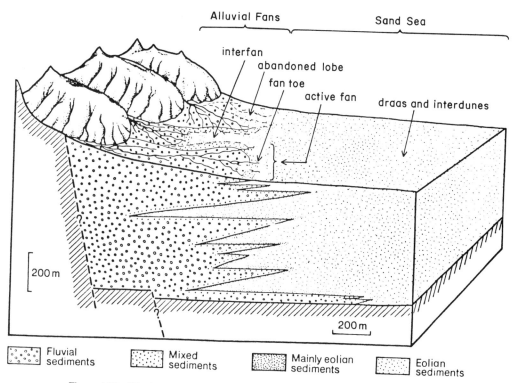

Figure 4.30 Distribution of the various components of a desert system in the Permian sediments of the Arran Basin, Scotland (after Clemmensen and Abrahamsen, 1983).

wadi and playa deposits, and at the margin of the basin, alluvial fan sediments consisting of debris flow, stream flood, and sheetflow deposits occur (Fig. 4.30).

The sequence near the margin can be divided into three major coarsening-upward cycles that were apparently tectonically induced. Sequences consist of eolian sands at the base and fluvial conglomerates at the top that probably formed in response to episodes of fan progradation.

Smaller-scale cycles are superimposed on the larger-scale ones. They consist of water-laid sediments at the base overlain by dune and interdune deposits, which are in turn overlain by fluvial conglomerates. The cycles may have formed in response to climatic variations or to episodes of lobe building and lobe abandonment on the fan.

The examples given thus far are those that developed in basins of internal drainage. While arid systems are often associated with closed basins, they are not restricted to them. The Old Red Sandstone of North Wales, for instance, apparently was deposited in an arid region with through drainage (Allen, 1965b) (Fig. 4.32).

The Old Red Sandstone consists of four major facies. The basal facies consists of sheetflood and stream flood gravels that accumulated on alluvial fans that were

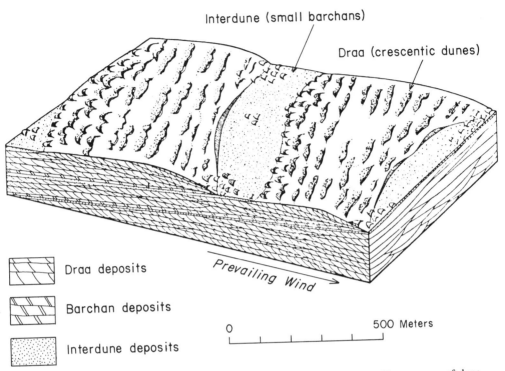

Interdune (small barchans)

Draa (crescentic dunes)

Draa deposits

Barchan deposits

Interdune deposits

Prevailing Wind

0   500 Meters

Figure 4.31 Block diagram illustrating the development of stratigraphic sequences of dune and interdune sediments in the Permian sediments of the Arran Basin, Scotland (after Clemmensen and Abrahamsen, 1983).

prograding basinward from the fault scarp forming the southwest margin of the basin. Pebbly sandstones of this facies are crossbedded, but coarser conglomerates are generally massive.

Conglomerates interfinger with limestones, dolomites, and concretionary siltstones. These beds are up to 2 m thick and consist of red siltstones at the base containing small calcareous concretions that increase in size and abundance upward until coalescence occurs, forming a massive, impure limestone. Beds containing dolomite concretions are similar except that they are reddish in hue and highly brecciated in appearance. These beds probably accumulated in playas, and the crossbedded sandstones that occasionally occur with them may represent ephemeral stream deposits.

At a relatively early stage in the development of the basin, major throughflowing drainage was established. While some lateral flow was derived from the fringing alluvial fans, the major drainage was longitudinal flow to the southwest. Cyclic alluvial deposits of this drainage system consist of a basal coarse member and an upper fine member. The coarse member consists of pebbly sandstones in large-scale crossbeds that overlie an erosional base. The pebbly sands grade upward

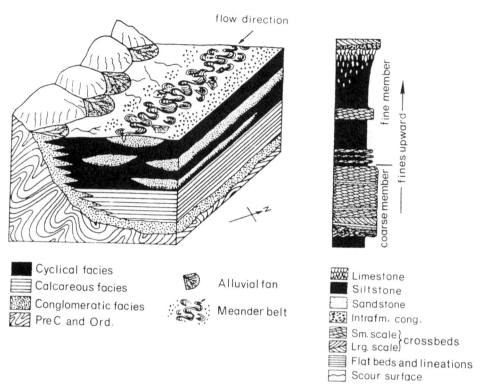

flow direction

fine member

fines upward

coarse member

Cyclical facies
Calcareous facies
Conglomeratic facies
Pre C and Ord.

Alluvial fan

Meander belt

Limestone
Siltstone
Sandstone
Intrafm. cong.
Sm. scale }
Lrg. scale } crossbeds
Flat beds and lineations
Scour surface

Figure 4.32 Model for the deposition of the Old Red Sandstone in Wales showing the development of a through-flowing river in an arid system. The meandering channel was bounded by floodbasins and playas that periodically encroached on fringing alluvial fans (after Allen, 1965b).

to sandstones in small-scale crossbeds or plane beds with parting lineation. The fine member consists of siltstone with occasional thin interbeds of crossbedded fine sand. The sequence is commonly capped by concretionary carbonates.

Coarse members apparently accumulated in alluvial channels, whereas fine members were deposited on an alluvial plain. Concretionary siltstones probably represent arid-region soil profiles where caliche developed. Cycles formed in response to major shifts in channel location, with each cycle representing the deposits of a single stream subject to stream avulsion.

5

Humid System

Introduction

The humid system is characterized by an excess of available moisture produced by relatively abundant precipitation and reduced rates of evaporation. Evaporation may exceed precipitation temporarily during seasons of reduced rainfall, but yearly average precipitation exceeds evaporation and excess water results.

Greater availability of moisture in the environment produces a number of different effects. A greater degree of chemical weathering promotes the production of stable mineral suites and aids mechanical erosion in reducing slopes in source areas. Lower slopes and relatively complete vegetative covers combine to reduce the effectiveness of sheetwash and enhance channelized flow. They also work together to decrease the amplitude and increase the period of flood cycles.

Through-flowing drainage is almost always achieved because topographic basins are filled to overflowing and channels are established through outlets. Although the chemistry may vary radically from lake to lake, the water in them tends to be fresh. Chemical stratification, therefore, does not normally occur in humid-region lakes, although temperature stratification does. Because through-flowing drainage is normally the case in humid regions, lakes do not develop the sedimentological importance that they show in arid regions. Rivers are far more important in humid regions, where they develop better drainage networks and generally carry flow throughout the year.

As in arid regions, temperatures may vary considerably among the humid regions of the world, affecting in turn the way in which processes act in source areas, transport systems, and depositional basins. Abundant moisture and elevated temperatures combine to accelerate chemical weathering in source areas. Rivers tend to be suspended load streams because of the accelerated production of clay, and even single-cycle mineral suites tend to be enriched in quartz. Temperatures are seasonally uniform in low latitudes as well. This leads to persistent temperature stratification in low-latitude lakes where oxygenated surface waters mix only slightly with deoxygenated bottom waters.

Precipitation in mid-latitudes tends to show a distinct seasonal cycle as atmospheric circulation patterns shift with the tracking of the sun. Seasonal temperature ranges may be extreme, producing overturning and mixing of the water column in all but the deepest lakes, and seasonal variations in flow are clearly manifested in the patterns of river sedimentation. Chemical weathering remains an important process, and soil formation occurs characteristically in both source areas and depocenters.

Components

Many components of the arid system are also present in the humid system, but the processes that act on these components act in a far different manner than their arid-region counterparts. The major components of the humid system include alluvial fans along the margins of uplands, alluvial plains, and lakes.

Alluvial Fans

Sedimentary processes and products.   As in arid regions, humid-region alluvial fans form where streams leave confining valleys of upland areas and deposit their load at the slope break. Because alluvial fans are often considered characteristic of arid climates, most studies of fans have been in arid regions, and the models that have been generated from these studies are applicable only to fans that have been built under the unique hydraulic conditions imposed by an arid climate.

The greatest divergence that occurs between the two fan types lies in the relative importance of channel processes. A single channel usually occurs at the head of arid-region fans, but in the mid-fan area, flow leaves the channel and unconfined flooding occurs. Only during low-flow periods is water confined to the shallow channels passing between low-amplitude bar forms that were built during flood stage. It is only at the toe of the fan that waters are again collected into significant channels, suggesting that braided alluvial channels are distinct from alluvial fan facies (Brown and others, 1973).

Channel processes are much more important in humid fans (Boothroyd and Nummedal, 1978). At the head of a fan, several channels may be incised into coarse alluvium. Lower on the fan, these channels bifurcate into smaller channels that develop around large bar forms, and further downfan, additional channel splitting takes place around smaller bars. Unlike arid-region fans, flow may be maintained in these channels between flood episodes.

Most studies of humid-region fans have been of those associated with glaciers. The models developed, therefore, are based on outwash fans (Boothroyd and Nummedal, 1978); but, although some processes may act similarly in both glacial and nonglacial settings, other factors do not. Streams issuing from beneath glaciers onto outwash fans, for instance, possess tremendous erosive capacity owing to the excess hydrostatic head produced by the weight of the glaciers (Fraser and Cobb, 1982). This hydrostatic head may also contribute to excess pore pressures in proglacial sediments, leading to unstable slopes and increasing the importance of gravity-induced mass movement. Extreme diurnal flow fluctuations can occur in some climatic settings, affecting formation, migration, and modification of bar forms; and even though yearly flood cycles may occur in both settings, the amplitude of the flood cycles in glacial settings may be very much greater than in nonglacial ones. Nevertheless, when approached with these reservations in mind, glacial fan models can be used to predict sedimentation patterns on nonglacial fans.

The core of the fan consists of the proximal and mid-fan areas (Fig. 5.1). The coarsest sediments of the fan are deposited here on longitudinal bars separated by channels. The bars consist mainly of gravelly sediments in ill-defined horizontal bedding, although finer-grained foresets may develop on downstream ends of bars during waning flow. The processes involved in the origin and preservation of braid bars will be discussed more fully in the section on alluvial plains. Bars are separated by channels, where flow may be maintained even during low stage. Channel deposits usually consist of a coarse basal lag overlain by finer-grained sediments in trough sets of varying scale. Channels are unstable, and where a shift in channel takes place or a cutoff occurs, small ponds may develop, where silt and clay settle out of suspension, forming drapes over sandy bedforms. Occasionally, deposits up to several centimeters thick can develop where the pond receives washover during successive flood periods.

These sediment types possess a poor preservation potential. The overall appearance of proximal and mid-fan deposit is that of an irregular arrangement of erosional remnants of bar and channel deposits, and only rarely are complete bar forms or channel fills preserved in a regime characterized by rapid lateral migration of bars and channels (Fraser and Cobb, 1982).

In the lower mid-fan, a noticeable tendency exists toward development of transverse bars, and these are the dominant form in distal areas of the fan (Fig. 5.1). Transverse bars build tabular sets of planar crossbeds, although transverse bar deposits may consist of a very complex internal arrangement of bedding types due to modifications of the bar form during waning flow. As in proximal areas, channel deposits consist of relatively fine-grained sediment in trough cross sets.

Unlike arid-region fans, the inactive portions of humid fans commonly are more or less covered completely by vegetation. When overbank flooding occurs, this vegetation acts as a sediment baffle that traps fine-grained sediments, forming significant overbank deposits. These deposits consist of laminated muds and peat interbedded with sandy crevasse splays or sheetflood deposits. Where inactive por-

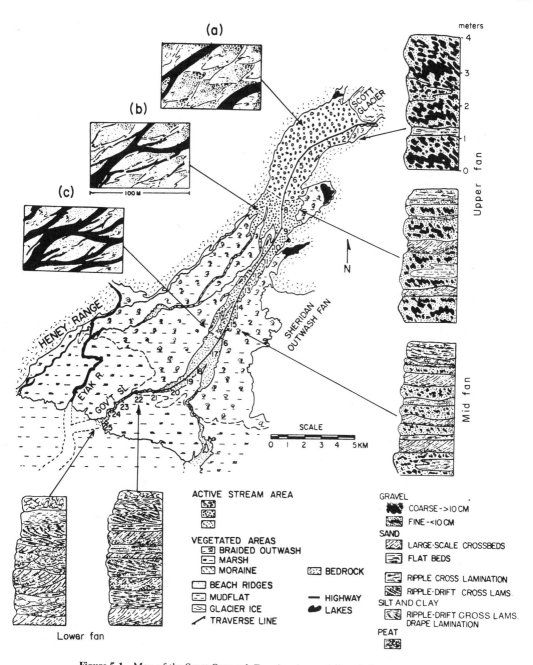

Figure 5.1 Map of the Scott Outwash Fan showing variations in bar morphology and channel pattern (upper left), areal distribution of facies (center), and typical stratigraphic sequences developed in various portions of the fan (right and lower margins) (after Boothroyd and Ashley, 1975).

tions of fans are unvegetated, however, such as in those areas abandoned only recently, eolian dunes may develop.

Extrinsic and intrinsic controls on humid fan sequences. The shortest period cyclicity found in humid-region fans is that caused by diurnal fluctuation in meltwater flow from glaciers or perennial snowfields. Sediment response may be significant but short-lived, so those changes in bar and channel position that can occur are not normally preserved. Smaller-scale responses include modification of bar shapes, channeling of bar surfaces, infiltration of fine-grained sediment into open-framework gravels, and deformation of high-angle slopes on bars, all of which occur during waning flow (Smith, 1974). A diurnal flood cycle, however, would not be characteristic of nonglacial alluvial fans.

Storm cycles are relatively short-term phenomena that can affect fan sedimentation considerably. Under such a regime, sediments respond to gradually strengthening flow that peaks at flood strength and then gradually declines. Sediments might be expected to form coarsening-upward and then fining-upward sequences as they respond to gradually changing flow; but because of rapid and radical shifts in positions of channels and bars during the cycle, sediment responses may be abrupt rather than gradational. In addition, sediments deposited during rising stage may be eroded during peak flow so that flood cycles may show evidence only of waning flow.

A typical response to a storm flood cycle might be initiated by the formation and migration of bars or the reactivation of preexisting bars during flood. Localization of flow would follow during waning flow, and finer-grained sediments would be deposited in lateral or bar-top channels. The cycle would be completed with a second period of bar movement during the next succeeding flood. Fining-upward sequences would normally result from such a cycle, with coarse bar sediment overlying coarser basal lags and, in turn, being overlain by finer-grained channel deposits.

A second type of sequence might develop with the reactivation of an abandoned channel. This process might be related to a flood cycle, but it might also be the result of random processes. Where channel reactivation occurs, coarse channel or bar sediments are deposited on top of muds that settled out of suspension during the time flow was not being maintained in the channel. The contact between the two deposits might be sharp, or it might be gradational if flow was only established gradually.

Seasonal meltwater cycles profoundly affect glacial fans. A similar, although nonglacial, seasonal flood cycle occurs in mid-latitudes when spring storms coincide with the release of winter precipitation stored in watersheds as ice and snow. The effects of this type of hydrologic cycle are not well-documented, but they presumably would not be unlike the response of proglacial fans to seasonal melting cycles when major changes in the position of bars and channels take place.

Longer-term cycles affect fan evolution. The effects of extrinsic factors, such as changes in elevation of source area or base level, on progradational or retrogradational cycles have already been discussed for arid-region fans, and the reader is referred to that section. Intrinsic factors, such as channel instability, probably affect both arid- and humid-region fans in the same way.

The effects of climatic changes, however, would be different. In arid-region fans, sediment production from slopes increases during dry periods, but the sediment is stored in source area valleys (Garner, 1959). During wetter periods, vegetation may stabilize slopes, but valley floors are entrenched, and sediment yield from the source area is increased (Schumm, 1977).

In humid-region fans, an increase in precipitation would increase sediment yield briefly as channels adjust to the increased runoff (Schumm, 1977), but sediment production from slopes would not be affected greatly. A shift toward drier climates, however, might bring the climate into a sensitive zone (Bond, 1967), where major adjustments in sediment concentration and yield would occur with relatively slight changes in precipitation. Sediment production might increase dramatically, with a shift toward a drier climate (see Schumm, 1977, Fig. 3.1). This material would be stored in valleys until a subsequent increase in precipitation would flush it onto the fan.

Humid-system alluvial fans in the rock record.   One of the best-known examples of a humid-region fan in the rock record is the Van Horn Sandstone of West Texas (McGowen and Groat, 1971). The size of this fan was on the order of the present-day Kosi Fan of India and it can be divided into three facies. The proximal facies was deposited in canyons leading out onto the fan surface (Fig. 5.2). Flow was confined and during flood was capable of moving boulders. Horizontally bedded gravels were deposited on longitudinal bars, and thin crossbedded sands accumulated in bar-top channels. Alternating bar and channel deposits were deposited during flood cycles caused by storm events.

The mid-fan area was characterized by rapid lateral shifts of braid channels that resulted in the preservation of complete channel fill sequences. Longitudinal bars were deposited in the main channel during flood, and crossbedded sands were deposited on transverse bars that migrated in interbar channels during low-flow stages. Aggradation was rapid in the mid-fan area, and avulsion of main stream channels occurred when thresholds of instability were exceeded. Cyclic deposition caused by storm events was superimposed on major cycles caused by avulsion.

Discharges of lesser magnitude were accommodated in numerous shallow braid channels in distal portions of the fan. Deposition during flood was on transverse bars and on lunate dunes in interbar channels during low flow. Sequences of crossbedded bar and channel deposits overlying muddy sands probably formed in response to storm events when flow was reestablished in abandoned channels.

The entire facies tract shifted northward toward the source during deposition so that the vertical succession in northern exposures consists of proximal, mid-, and distal fan facies in ascending order. This may represent gradual filling of the feeder canyons, but it also may have occurred in response to gradual slope reduction in source areas.

Other well-documented examples of ancient alluvial fan deposition occur in

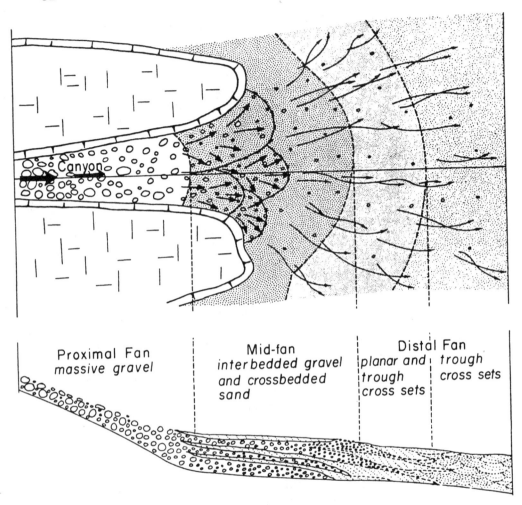

Figure 5.2 Distribution of alluvial fan facies in the Van Horn Sandstone, Texas (after Mc-Gowen and Groat, 1971).

the Devonian of western Norway where breccias, conglomerates, and sandstones were deposited on alluvial fans prograding into grabens or half-grabens (Nilsen, 1969; Steel, 1976; Steel and others, 1977). Although interpreted as semiarid fans, a more humid climate is suggested by abundant floral remains, including Psilophyton (Dilcher, oral comm.) in the Solund Basin, and the presence of widespread, nonevaporitic lacustrine sediments in both the Solund Basin and the Hornelen Basin.

Breccias consist of angular fragments in a matrix of arkosic sand and clay. They accumulated on a highly irregular surface developed on basement rock, and they probably represent talus deposits and debris flow deposited very near the fault scarp (Fig. 5.3). Breccias interfinger with and are overlain by conglomerate consisting of clasts to boulder size in thick, massive beds separated by thin discontinuous beds of sand in planar laminae or very low angle crossbeds. The conglomerates probably were deposited on longitudinal bars or as diffuse gravel sheets that were in motion during flood periods. The thin sandstone lenses probably represent deposition in lows on bar surfaces during waning flow.

Conglomerates are overlain by sandstone in trough and tabular crossbeds or horizontal planar laminae. Beds of conglomerate are common near the base, and thin lag deposits occur elsewhere in the section. Gravel lenses probably accumulated on longitudinal bars, whereas crossbedded sandstones were deposited as transverse bars or in interbar channels.

The relative thickness of the conglomerate suggests that uplift was continuous in the source area. Nilsen (1969) ascribes the gradual overlap of the conglomerate by the sandstone facies as a response to erosion of the source area and retreat of the fault scarp. It is more likely, however, that fan retrogradation occurred in response to evolution of the drainage system in the source area. Erosion of the interfluves reduced the quantity and caliber of the sediment being delivered to the streams, and widening of valleys increased the storage capacity of the flood plains, especially for coarse fractions.

Alluvial fan sediments also occur along the margin of the Hornelen basin to the north, where they display marked cyclicity consisting of 100-m-thick, coarsening-upward subcycles that, in turn, are composed of coarsening-upward subcycles 10 to

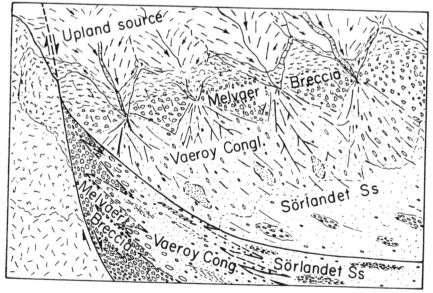

Figure 5.3 Paleogeography and internal facies distribution of the Vaeroy Conglomerate and Sorlandet Sandstone (after Nilsen, 1969).

25 m thick (Steel, 1976; Steel and others, 1977). Fan sediments were deposited by both debris flow and stream flow processes, and the cycles are, in turn, grouped together into sequences thousands of meters thick, consisting of sediments deposited primarily by one or the other of these mechanisms (Fig. 5.4).

The fine-grained unit at the base of the cycles consists most commonly of

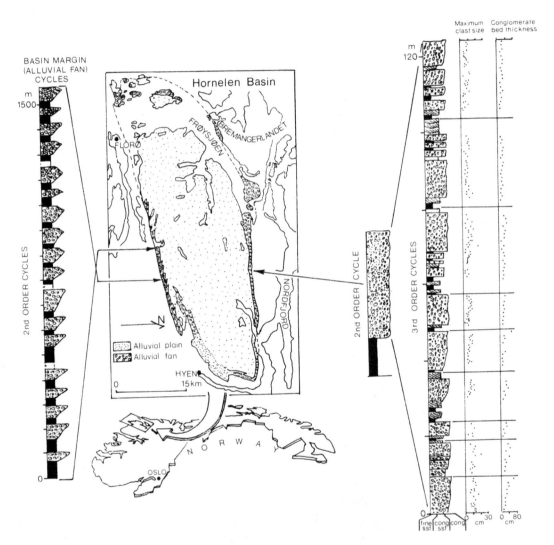

Figure 5.4 Alluvial fan and alluvial basin sequences in the Devonian of the Hornelen Basin of Norway showing stream flow and debris flow-dominated alluvial fans on the margins of the basin, and braided stream, floodbasin, and lacustrine facies in the interior of the basin (after Steel, 1976; and Steel and others, 1977).

sandstones that were probably deposited on alluvial plains of the basin floor. Fan progradation into the basin produced coarsening-upward cycles by a decrease in the percentage of sandstone beds and an increase in the clast size of the conglomerates. Each cycle displays a sharp and apparently planar upper boundary where coarse fanglomerates are abruptly succeeded by sandstones of the next overlying cycle (Fig. 5.4).

The thickest sequences may represent a response to climatic change. Sequences consisting predominantly of debris flow facies probably were deposited during relatively dry periods, whereas those deposited by stream flow presumably accumulated under wetter conditions.

The two smaller-scale cycles, however, are attributed by Steel (1976) and Steel and others (1977) to subsidence of the basin along fault scarps at the margin. Cycles 10 to 25 m thick formed in response to discrete episodes of faulting, and the 100-m-thick cycles accumulated over extended periods of frequent movement. Intermediate cycles began with a severe tectonic event that resulted in a rapid migration of basinal facies over fan facies. As equilibrium profiles were gradually reestablished, fan facies migrated more slowly basinward, creating asymmetric, coarsening-upward cycles. Fan growth was apparently interrupted periodically by tectonic events of lesser magnitude, which resulted in a less radical shift of facies and the creation of the minor coarsening-upward cycles.

Steel and others (1977) reject climatic variations as a viable mechanism because both intermediate and minor cycles occur in megasequences dominated by both debris flow and stream flow deposits. They likewise reject an intrinsic control, such as stream avulsion, because the cyclicity can be traced not only across fan bodies, but into the basin as well.

It might be argued that the minor cycles represent the response of the drainage system to changes in base level modeled by Parker (1976). Steel and others (1977), however, observed that minor cycles thicken upward, and rejuvenation of progressively higher-order tributaries in the source, triggered by a drop in base level of the trunk stream, would most likely produce thinning-upward cycles.

Heward (1978) analyzed the factors involved in the development of alluvial fan sequences in the Stephanian coalfields of northern Spain, in which he could differentiate six facies. Valley and fanhead canyon sediments consist of coarse, unstratified conglomerates, occasional thin coals, trough crossbedded sandstone, and root-mottled mudstone (Fig. 5.5). They were deposited by a variety of processes, including mass movement by debris flows and colluviation, stream flow, and suspension settling in lacustrine environments.

These sediments are organized into fining- and thinning-upward cycles, 15 to 65 m thick, which are superimposed on thinning- and fining-upward megacycles (Fig. 5.5). The smaller cycles may reflect episodes of channel filling and abandonment, whereas the megasequences may represent cycles of uplift, followed by slope reduction in the source area and scarp retreat.

Mid-fan depositional lobes consist of stratified and unstratified conglomerates and crossbedded sandstone. Conglomerates were probably deposited during major

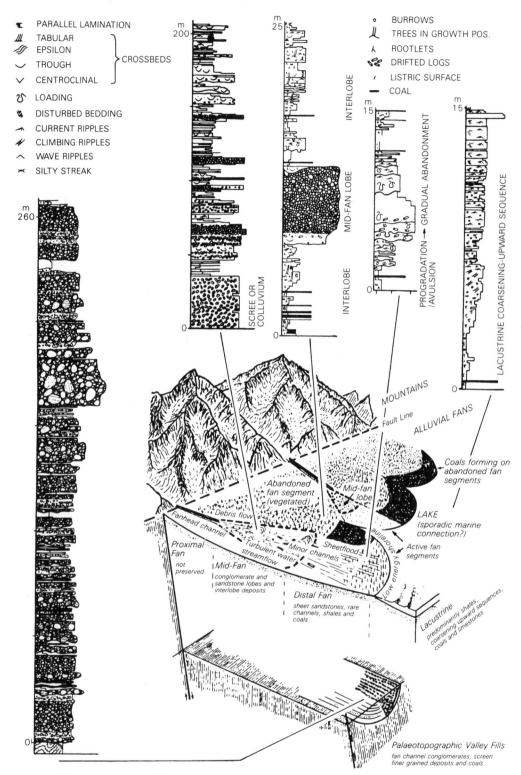

Figure 5.5 Types of stratigraphic sequences developed in various parts of an alluvial fan in the Stephanian A and B sediments of the La Magdalena, Matallana, and Sabers coalfields, Spain (after Heward, 1978).

floods, either as debris flows or stream flood deposits, with crossbedded sands accumulating during waning flow.

These sediments are organized into both fining- and thinning-upward cycles and thickening- and coarsening-upward cycles. Thinning- and fining-upward cycles may record episodes of tectonism, followed by slope reduction and scarp retreat, whereas thickening and coarsening-upward sequences accumulated during periods of lobe progradation caused by tectonism or avulsion. Minor storm cycles consisting of alternating coarser and finer layers are superimposed on these longer-period cycles.

Distal fan sediments consist of mudstones and fine sandstones. Sandstones appear to have been deposited by sheetfloods during peak flow, and they are interbedded with thin mudstones and coals deposited during intervening periods. Sandstones occur in packets up to 10 m thick that may thicken and coarsen upward or thin and fine upward. Sequences that thicken and coarsen upward are probably the distal equivalent of mid-fan lobe progradation, whereas thinning- and fining-upward sequences represent lobe abandonment or periods of progressively reduced flood magnitude. The alluvial fans prograded into lakes where coarsening-upward sequences, up to 140 m thick, were deposited. Large-scale sequences, 15 to 140 m thick, coarsen upward from black shales through gray shales with thin sandstones into fine-grained sandstones with evidence of plant colonization. Thinner sequences 5 to 20 m thick are less extensive and usually lack the basal black shale component. In addition, distal fan sediments commonly occur in the upper part of lacustrine sequences.

Lacustrine sequences probably reflect gradual infilling of large basinal lakes by prograding alluvial shorelines. Fine-grained sediment settled out in the deeper portion of the lake, whereas coarser sediments were deposited by downslope traction currents near the shore. Sequences observed in the lacustrine sediments occur on three scales. Alternating coarse and fine layers might represent storm episodes equivalent to sheetfloods in the distal fan environment, and thinner coarsening-upward sequences represent infilling of small basins and coastal lagoons. Thicker sequences may represent infilling of large basins, perhaps triggered by abandonment and subsidence of depositional fan segments or tectonic readjustments in the basin.

These sequences are superimposed on three basin-filling sequences up to 2500 m thick. Overall, these sequences fine upward, suggesting that the source retreated from the depositional basin. In addition, each succession is marked by the occurrence of fine-grained distal sediment abruptly overlying proximal fan deposits. The change occurs at varying levels and probably reflects back-faulting at the basin margin. Each basin-filling succession is composed of smaller-scale coarsening-upward sequences, reflecting repeated basin margin faulting and source area rejuvenation.

Alluvial Plains

Sedimentary processes and products.   Although river channels may assume a wide variety of shapes, they may be broadly classified into four basic types:

braided, meandering, anastamosing, and straight. Divisions among channel types are somewhat arbitrary, however, because gradation exists among them. Braid bars, for instance, may become emergent in the channel of a meandering stream at reduced flow, and thalwegs of even straight reaches actually meander between alternate side bars. In addition, channels of anastamosing streams may be braided or meandering, and all four basic channel patterns may occur along various reaches of the same stream.

The underlying cause of variation in channel pattern appears to be a complex and poorly understood interrelationship among climate, slope, and quantity and caliber of the sediment load. These are the factors controlling the way a river performs its task of transporting water and sediment, and the way in which a river responds to these factors determines not only channel pattern but also the character of the deposits that are the ultimate product of the river.

By carefully noting details of bedding and vertical sequence, the deposits of braided and meandering streams may be confidently differentiated, and lateral facies patterns appear to offer a criterion for recognition of anastamosing stream deposits. These are the three types of streams that will be discussed; few streams maintain straight reaches for very long distances, suggesting that they probably are unstable and incapable of producing significant deposits in the rock record.

Meandering channels.   Meandering streams display a sinuous channel pattern that traverses floodplains in broad loops. Channels of meandering streams tend to be relatively deep and narrow, making them efficient conduits of water and sediment. The small width-to-depth ratio results in a small wetted perimeter, reducing energy loss caused by friction. Thus, for a given discharge and sediment load, a meandering stream requires a lower slope to accomplish the task than does a braided stream with a large width-to-depth ratio.

The ratio of suspended load to bedload tends to be large in meandering streams (Schumm, 1960). Fine sediments deposited outside the channel during overbank flooding form cohesive banks that are less susceptible to lateral widening. To accommodate flood flow, therefore, meandering streams first deepen their channels and then overstep their banks.

These characteristics of meandering streams do not necessarily offer insight into why meandering occurs. The ultimate cause must lie in the way a stream expends energy, that is, how it adjusts its longitudinal and cross-sectional profile in order to accommodate its imposed load. Schumm and Beathard (1976), for instance, showed that, although braided streams tend to possess steep channel slopes, on those rivers consisting of both meandering and braided reaches, meandering reaches occurred along those portions traversing steeper valley slopes. This relationship suggested that rivers adjust their slope by lengthening their courses when steep valley slopes are encountered or by shortening their courses by straightening it on gentle valley slopes.

Whatever the ultimate reason, once a river begins to meander, a characteristic flow pattern is established and the channel profile is modified in response. As the stream flows around the bend, a weak transverse secondary flow is initiated toward the outside bank at the surface and toward the inside bank at the bottom. Resultant flow vectors diverge slightly from the downstream direction, and a helical flow pattern is established. The stream adjusts its transverse profile in response to helical flow by building a point bar. The slope of the point bar into the thalweg is adjusted until the shear forces driving sediment up onto the bar are balanced by gravity forces acting in a downslope direction.

Predictable changes in bed shear occur from the thalweg up onto the point bar, producing a hierarchical succession of stratification types in a generally fining-upward sequence (Fig. 5.6). The channel floor usually is covered by a coarse lag consisting of the coarsest fraction of the bedload, waterlogged organic debris, or eroded portions of the cutbank. Channel deposits consisting of bedload material in some form of large-scale cross stratification overlie the basal lag. The form of the cross stratification depends on the available stream power, and the scale depends on channel depth.

Point-bar deposits overlie channel sediments. Basal point-bar sediments consist of trough or wedge-planar cross stratification that decreases in scale upward, eventually grading to ripple beds in the upper part of the sequence. Plane beds may occur anywhere in the sequence where large velocities or shallow depth occur, but they are particularly common at the top of the sequence.

The ideal sequence is not found where a helical flow pattern is not fully established or where other hydraulic parameters modify the process. A helical flow pat-

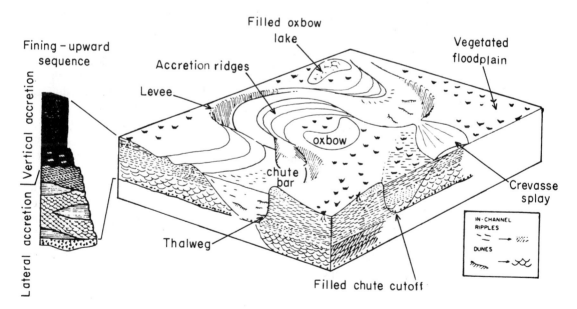

Figure 5.6 Facies distribution and sedimentary sequences of a meandering stream (after Walker and Cant, 1979, reprinted with permission of the Geological Association of Canada).

tern, for instance, is not fully established everywhere around a point bar (Jackson, 1976). A transitional zone at the upstream end of a bar inherits the flow pattern from the preceding bar, and no transverse flow occurs where the thalweg crosses from one bank to the other near the middle of the bar. Not until the flow is well into the meander is a fully developed helical flow pattern established. Each position on the bar, therefore, experiences bed shear of different magnitude and direction, and each develops a different sequence.

Extreme fluctuations in discharge may form stepped profiles in point-bar surfaces (McGowen and Garner, 1970). Under such conditions, no apparent hierarchical sequence of bedforms develops, nor are grain-size variations predictable. Large-scale bedforms may develop under the surface of both lower and upper steps, and coarse sediments of chute bars may be laterally juxtaposed with fine sediments that fill the chute of the lower step (Fig. 5.7).

The sequence is also modified where scroll bars and transverse bars occur on the point bar. Scroll bars are elongate ridges on the upper point-bar surface that parallel the river bank. They form apparently in response to transverse flow during flood, and they introduce bankward-oriented, large-scale foresets high in the point-bar sequence. Commonly, series of scroll bars are separated by swales where fine-grained, often highly organic sediments accumulate.

Transverse bars are sand waves that sweep across point-bar surfaces during flood. Crests are oriented generally transverse to flow, with dips on foresets diverging slightly from mean downstream flow. Transverse bars also introduce large-scale foresets into the upper part of point-bar sequences.

The dynamics of meandering streams of low sinuosity provide additional complications. Such channels are occupied by alternate bars and point bars, which accrete laterally as channel bends migrate by bend expansion and downstream translation (Bridge and others, 1986). This basic pattern of channel evolution is altered, however, by a series of complex responses triggered when chutes cut across bars, detach them from adjacent banks, and re-form them into mid-channel islands. Lateral accretionary surfaces are built into newly formed channels from the flanks of the mid-channel islands in association with migration of the new channels (Fig. 5.8). Chute bars and scroll bars are built at the mouths of the newly formed channels, and they, in turn, modify flow in downstream reaches, causing changes in local patterns of erosion and deposition.

Channel bottoms are covered by sinuous-crested dunes during most flow stages, but shallow areas near channel margins are covered by ripples. Simple lateral accretionary sequences, therefore, consist of large-scale crossbeds overlain by small-scale crossbeds. Chute bars and scroll bars are covered by dunes during high flow stage and ripples during low flow. Bars, therefore, deposit composite sequences composed of stage-controlled alternations of large- and small-scale cross-stratified sand.

Chutes eventually capture most of the flow, causing abandonment of the original channels. Abandoned channels fill progressively from upstream ends, where thick deposits of crossbedded sand in fining- and thinning-upward sequences accu-

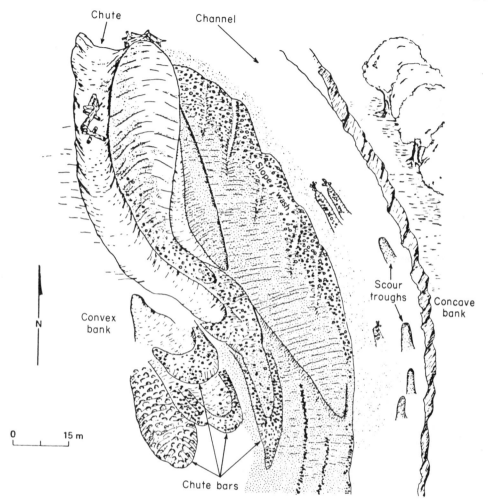

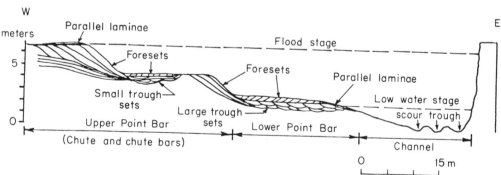

Figure 5.7   Plan view and cross section of a coarse-grained point bar with a stepped profile showing distribution of textures and sediment bodies (after McGowen and Garner, 1970).

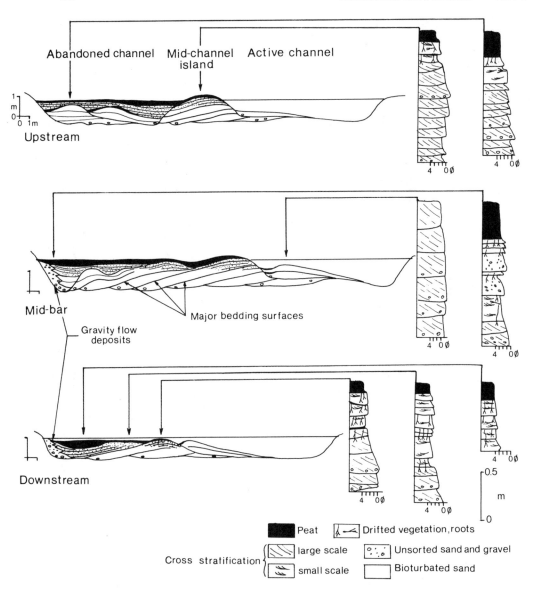

Figure 5.8 Cross-sectional profiles and types of depositional sequences found in a low-sinuosity stream (after Bridge and others, 1986).

mulate. Crossbedded sands normally are only a minor component of sequences from downstream ends of abandoned channels, where thick accumulations of peat are deposited instead.

The third major component of the alluvial plain is the valley floor or flood-plain, where overbank deposition occurs on levees, splays, and floodbasins (Fig.

5.6). Levees are ridges separating the channel from the adjacent floodbasin. They form where the coarse fraction of the suspended load is deposited when flow decelerates as it leaves the confines of the channel. Typical deposits consist of silt deposited during peak flow in climbing ripples and plane beds that are draped by mud during slack water phases.

Splays form when floodwaters leave the channel through a gap (crevasse) in the levee. When this occurs, coarse bedload material is introduced onto the floodplain in a lobate mass that is wedge-shaped in longitudinal profile and convex-upward in transverse profile (Fig. 5.6). Splay sediment may be deposited in massive or graded beds, plane beds, or beds of climbing ripples in sequences indicative of deposition from rapidly waning flow. Typically, such sediments are interbedded with muds that make up the bulk of the floodplain sediment (Farrell, 1987).

The finest-grained sediments of the fluvial component are deposited in the floodbasin away from the alluvial ridge. Most sediments are deposited by settling from suspension, and only during the most severe flooding is traction transport of coarse sediment important. The deposits tend to consist of groups of fine laminae of more silty or clayey composition, but primary structures are commonly obscured by root mottling. Floodplain sediments are variably organic. Peat can accumulate where substrates are elevated sufficiently to support abundant plant growth, but not so high as to promote oxidation. Low areas on the floodplain tend to remain perenially submerged and to collect relatively little organic material. Such areas may occur in minor depressions that collect floodwaters, or they may occur in more established depressions that form in response to changes in channel position.

The basic sedimentation unit on the floodplain is centimeter- to decimeter-thick bedsets produced by discrete overbank flooding events (Bridge, 1984). The units usually fine-upward from an erosive base, reflecting erosion of the floodplain prior to deposition during waning flow. Absolute grain size and internal structures depend on local flow conditions and sediment availability. Sediments of the alluvial ridge may be comparable to channel bar deposits but progressively decrease in size away from the ridge. Units may be sheetlike on levees and in floodbasins, but display channel-fill geometry in crevasses, splays, and drainage channels.

Sedimentation units are stacked into sequences in response to channel belt evolution. Coarsening-upward sequences develop as the alluvial ridge matures and levees and splays prograde into floodbasins (Bridge, 1984; Farrell, 1987). Fining-upward sequences, on the other hand, may develop when an individual crevasse channel is abandoned in favor of an adjacent one.

Alluvial channels are inherently unstable as they constantly readjust their patterns and profiles to changes in discharge and sediment load. A river may steepen its gradient by shortening its course through chute cutoffs or meander cutoffs. Chute cutoffs occur when a swale or point-bar chute gradually assumes an increasing proportion of flood flow until the old channel is abandoned and gradually filled in. Meander cutoffs occur when the river establishes flow across the neck of a meander either catastrophically through a breach in the levee during flood or by migration of the cutbanks of tortuous meanders toward each other (Fig. 5.9). In both

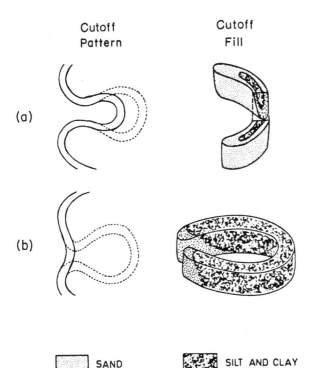

Cutoff   Cutoff
Pattern   Fill

(a)

(b)

SAND   SILT AND CLAY

Figure 5.9 Channel shape and pattern of infilling of (a) a chute cutoff, and (b) neck cutoff (after Allen, 1965a; Reineck and Singh, 1980).

cases, fine-grained sediment fills the abandoned channel, forming a cohesive mass that is resistant to erosion and an impediment to further channel migration.

Braided channels.   Braided channels consist of complexes of shifting bars and interbar or braid channels (Fig. 5.10). They are characterized by large width-to-depth ratios, relatively straight channels, and a large bedload-to-suspended load ratio. They occur commonly in regions with steep regional slopes in climatic regimes that promote flood discharges that peak and subside quickly. Because banks are usually composed of easily eroded materials, increased discharge is accommodated by channel widening rather than channel deepening.

None of these characteristics is exclusive to braided channels, and some braided channels may display their opposite. Although a few braided channels may be the result of the dominating effects of one of these characteristics, most often they appear to be complexly interrelated. Steep slopes, for instance, will contribute coarse material to a stream, at the same time encouraging flashy discharge. The fact that sparse vegetation promotes flashy discharge, heavy sediment loads, and easily erodable banks explains why braiding is favored in arid or cold climates.

Although these characteristics may be immediate causes of braiding, the ultimate cause lies in the inability of a stream to move its imposed load. Bars form during flood flow and become emergent as the flood wanes and the flow is split

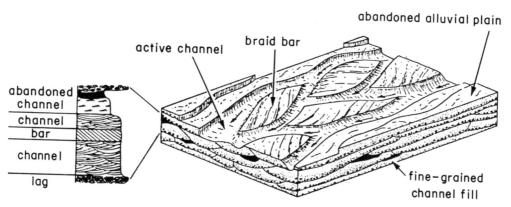

Figure 5.10 Morphology, internal organization of sediment types, and one type of vertical sequence in a braided stream (after Allen, 1970; Cant and Walker, 1976).

into a characteristic series of braid channels. Unit bars may emerge in a relatively unmodified form, but most often they are modified by localized late-stage flow or succeeding floods that form multistory bars with complex depositional histories (N. D. Smith, 1972).

Numerous bar types have been recognized in the modern environment, but most are undoubtedly of limited occurrence, having formed under unique combinations of hydraulic conditions. Most bars belong to one of four main types: longitudinal, transverse, lateral, and diagonal (Smith, 1978), and of these only transverse and longitudinal bars can be confidently recognized in the rock record (Miall, 1977).

Longitudinal bars are diamond-shaped in plan view and elongate parallel to flow (Fig. 5.11). The upper surface is convex-upward in transverse profile and inclined downstream in longitudinal profile. They may form around a nucleus of the coarsest fraction of the bedload during waning flow (Leopold and Wolman, 1957), or they may be stable bedforms during flood flow that cease movement and become emergent during waning flow (Hein and Walker, 1977).

Longitudinal bars are characteristic of rivers with a gravelly bedload. They consist generally of crude horizontal beds differentiated on the basis of varying grain size or matrix. The bars tend to be coarsest at the upstream end, and at the downstream end the bar margin may consist of a low-angle slipface. Lateral margins may slope gradually into coarse channel bottoms, or they may show steeper erosional margins against which wedges of relatively fine-grained sediments may accrete with high-angle foresets during waning flow (Fig. 5.11).

Transverse bars are broad sand bodies with downstream slipfaces oriented more or less perpendicular to flow. They are characteristic of sandy braided rivers where they migrate in echelon trains during flood. Bars migrate as sand in transported up stoss sides during flood and avalanches down lee sides, forming high-angle planar crossbeds that constitute the bulk of their primary internal structure.

Transverse bars are commonly modified extensively during falling stage (N.

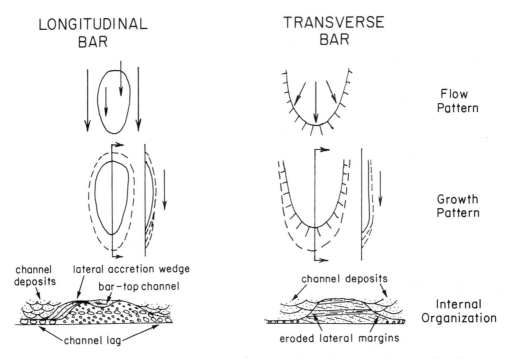

Figure 5.11 Plan views and cross sections of longitudinal and transverse bars, showing orientation to flow, manner of growth, and typical internal organization of textures and structures (flow pattern and growth pattern after Smith, 1974, by permission of the *Journal of Geology,* University of Chicago press).

D. Smith, 1972). Portions of bars become exposed as water depth decreases and flow is confined to small channels that dissect the bar surface. Where flow from these channels enters the main channel, accretionary wedges are built out from bar margins, and these, in turn, may themselves be channeled and reworked as flow continues to wane. Bar-top channels are filled with smaller-scale transverse bedforms, and the remainder of the bar surface is commonly veneered with ripples or plane beds. Bar margins are eroded during falling stage, producing sharp lateral contacts between bar and channel deposits. Downstream slipfaces may be eroded during waning flow and remobilized during a subsequent flood, producing reactivation surfaces, and late-stage flow may even be directed parallel to slipfaces, superimposing secondary bedforms on foresets.

Sediments in most streams fine downstream, and because braid patterns are different for gravelly and sandy sediments, it is not surprising that a change occurs in bar morphology in a downstream direction (N. D. Smith, 1970). Longitudinal bars dominate the relatively proximal gravelly channels, whereas transverse bars become more prevalent with increasing distance from the source. Lunate dunes migrate in interbar channels during low-flow conditions and, therefore, show relatively little downstream change in abundance.

In relatively deep sandy streams, sand flats may be an important depositional component of braided channels (Cant and Walker, 1978). They are large, complex sand bodies that are constructed of superimposed bedforms, each of which forms in response to changing hydraulic conditions during aggradation of the bar. Deposition is initiated as a unit bar forms during flood, and as flow wanes the current is redirected around the bar, causing barchanlike projections to grow downstream on either side of the nucleus. The bar also accretes upstream as smaller scale bedforms are driven up out of the channel onto its surface. As bars coalesce, a large wave or sandflat is formed that may be stable over a period of several years. Internal structures consist mainly of planar cross sets that decrease in scale upward, where they commonly become interbedded with plane beds. Sand in trough cross sets accumulates in adjacent channels, and as channels aggrade and are abandoned, they and the associated sandflats become the sites of overbank deposition of fine-grained sediments.

Anastamosing channels.   The terms anastamosing and braided were long used interchangeably in describing streams with multiple channels. Schumm (1968), however, pointed to certain qualitative differences among such streams that allowed their classification into two types. Whereas the channel system (both braid channel and main channel) in braided streams is inherently unstable, the channel system of an anastamosing stream is much more stable, showing at best only slow rates of lateral migration (D. G. Smith and N. D. Smith, 1980) (Fig. 5.12).

Channels of anastamosed streams tend to be narrow and deep. They follow tortuous paths as they bifurcate and rejoin around islands that are not necessarily streamlined parallel to down-valley flow. The islands may be considered part of the floodplain; they often are densely vegetated and are the site of overbank sedimentation. Three wet environments occur on the floodplain, including peat bog, backswamp, and floodpond, which are the depositional sites for organic matter, organic muds, and laminated muds. Adjacent to the channel, somewhat coarser sediments are deposited on levees and crevasse splays, and the coarsest sediments are deposited in the channel (D. G. Smith and N. D. Smith, 1980).

The type of sediment and lateral distribution of these facies are very similar to those of meandering streams. But the unique characteristic of anastamosed channels is their relatively long stability, which precludes the formation of lateral accretionary deposits.

Intrinsic and extrinsic controls on alluvial channel sequences.   Meandering channels migrate laterally across valley floors as sediment is added to the point bar on the convex bank and eroded from the concave bank. Erosion also occurs on the downstream bank of the crossover point, resulting in a down-valley migration of the meander pattern as well. The basic unit deposited by this process consists of relatively coarse-grained point-bar and/or channel sediments at the base overlain by fine-grained floodplain deposits. If the channel is aggrading at the same time, the

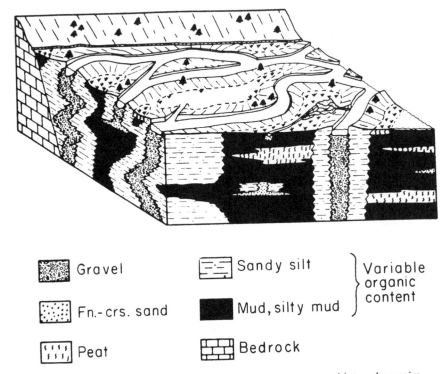

$$\boxed{\text{Gravel}}$$

Gravel

Fn.-crs. sand

Peat

Sandy silt

Mud, silty mud

Bedrock

} Variable organic content

Figure 5.12 Diagrammatic representation of the geometry and internal organization of sediments of an anastamosing river (after Smith and Smith, 1980).

resulting deposit is a ribbon of sand wider than the channel and thicker than it is deep.

The ultimate shape of the sand body with respect to the overbank sediments is dependent on the rate of aggradation of the alluvial plain. If the rate of aggradation is slow, a stream's residence time at one area of the floodplain may be sufficiently long for it to alter its shape numerous times by alternately forming meander loops and cutting them off as it tries to maintain an equilibrium gradient in the face of changing demands on its ability to transport sediment and water. Cutoff plugs may form more or less continuously on either side of the meander belt, limiting the extent to which the channel can migrate.

Under these conditions, the meander belt aggrades slowly above the level of the surrounding floodplain until a threshold of instability is reached and a radical shift of the channel position takes place through avulsion. The period between episodes of channel avulsion is long, and the channel can sweep back and forth across the channel belt several times and, in the process, erode parts of previously deposited point-bar sediments. The resulting channel belt deposit is a relatively thick sand body consisting of stacked sequences of channel and lower point-bar sediment bounded laterally by cutoff plugs. Multistory channel deposits formed in this manner are relatively common in the rock record (Campbell, 1976; Puigdefabregas and van Vliet, 1978; Nijman and Puigdefabregas, 1978).

Channel belts may also become superimposed, leading to partial erosion of previously deposited sand bodies (Bridge and Leeder, 1979). But because buried sand bodies tend to become positive features because of the relatively greater compactibility of enclosing floodplain deposits, channel belt deposits may avoid forming stacked sequences (Allen, 1978).

Facies models that attempt to generalize the extreme variability that exists in braided streams in different depositional settings may be constructed (Miall, 1977). But these models do not address the real problem of isolating and understanding the processes resulting from the episodic nature of the flow, which is a dominating characteristic of braided streams. The shortest flow cycles that can be identified in any braided stream sediments are diurnal meltwater cycles in proglacial settings. Short-term cycles in nonglacial settings form in response to storms or to channel abandonment and reoccupation. Sediment response includes minor bar modification, reactivation surfaces on foresets, and vertical and lateral juxtaposition of sediments formed under greatly different hydraulic conditions.

Longer-period cycles may occur as a response to yearly meltwater cycles or periods of maximum storm frequency and intensity. Sediment response to longer-period cycles consists of major changes in positions of bars and channels, with an overprint of sediment response to the shorter-period cycles. A typical cycle might consist, therefore, of a basal sequence of flood deposits slightly modified during periods of relative low flow, overlain by sediments deposited during an extended period of low flow interrupted intermittently by minor flood episodes.

The basic units of braided channels are fining-upward sequences deposited in response to waning flow of flood episodes or channel abandonment. Braid channels are inherently unstable, however, severely reducing the potential for preservation of complete sequences. Most often, erosional remnants of bar sediments truncated by channel deposits are seen in braided channel sequences.

Larger-scale depositional units are the main channel deposits, consisting of bar and braid channel sediments. The relative absence of significant thicknesses of cohesive overbank muds enhances the ability of braided channels to erode their banks and migrate laterally. Previously formed channel deposits, therefore, also possess relatively little chance of surviving as intact sequences. The resulting architecture of a braided alluvial plain will consist of partially preserved bar and braid channel sediments in valley-wide sheets, consisting of discrete channel systems separated laterally and vertically by erosion surfaces and interspersed irregularly with remnants of overbank deposits (Fig. 5.10).

The few anastamosing streams that have been studied showed patterns of nearly vertical aggradation of channel and floodplain facies (Smith and Putnam, 1980). This is in part due to rapid rates of aggradation characteristic of the streams, and in part to the stability of channel position in the floodplain. Under these conditions, the various facies associated with anastamosing channels have an excellent potential for preservation.

Controls on alluvial plain sequences. Alluvium is deposited in valleys bounded by walls that may restrict lateral movement of channels or on laterally extensive alluvial plains where channels migrate freely. Interior alluvial plains occur as broad aprons that fringe uplands and depositionally smooth preexisting topography with thick sequences of alluvial sediments. They may also occur as coastal plains. These are similar to interior plains except that changes in sea level are a primary factor in their evolution, and they may prograde seaward over a low-relief platform of marine and transitional marine sediments.

The way in which sediments fill valleys depends on the nature of the source materials, the channel pattern of the stream, the caliber of the load, and the pattern of floodplain aggradation. Floodplains may aggrade in three ways (Schumm, 1977). If the gradient is reduced through a change in base level or by channel extension at the mouth, the valley will be progressively backfilled. The coarsest material will be deposited in the upstream channel, which bypasses finer material to be deposited farther downstream. As the wave of aggradation passes upstream, finer material is thus deposited over coarser.

Downfilling will occur if an increase in sediment yield is produced through climatic change or source area uplift. Coarsest sediment is deposited in upstream areas, and fine sediments are bypassed downstream, but a sequence produced by downfilling will coarsen upward as the wave of aggradation passes downstream.

Lowering of base level will produce an initial phase of downcutting. Rejuvenation of tributaries, however, will cause reaggradation of the valley in a pattern that will be entirely dependent on the nature of the sediment in the tributaries.

These basic valley-filling patterns are represented in the river systems of the Catskill magnafacies (Devonian) in New York State (Gordon and Bridge, 1987). The alluvial deposits show systematic spatial variations with increasing distance from the paleoshoreline that reflect an increase in valley slope and an increase in the ratio of channel-belt width to floodplain width.

These spatial changes are also represented in vertical sequence. Regressive sequences reflect a downvalley translation of facies (downfilling) caused by increased sediment yield to rivers during periods of uplift. Transgressive sequences, on the other hand, represent episodes of backfilling and were deposited during periods when subsidence exceeded sediment yield.

The nature of the valley fill is also determined to a large degree by the caliber of the load, which is, in turn, dependent on the source material and stream gradient. Bedload streams carry greater than 11% bedload material and deposit a sequence consisting predominantly of coarse sediment interbedded with thin, discontinuous deposits of finer-grained alluvium. In the active channel, the bedload is deposited in horizontal or convex-upward layers that coalesce into sheet sands as the channel sweeps across the valley floor (Fig. 5.13).

Mixed-load streams transport between 3% and 11% bedload material. In upstream reaches, coarse material is deposited in the center of the valley and finer material is bypassed downstream where lateral deposition predominates. As the valley aggrades, lateral processes of deposition move upstream and enclose the coarse

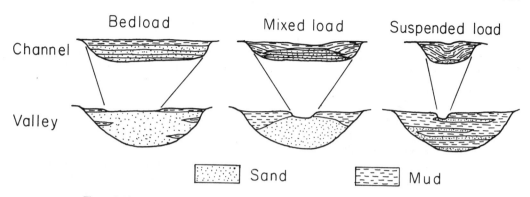

Figure 5.13   Cross sections illustrating the way in which bed load, mixed load, and suspended-load rivers may fill valleys (after Schumm, 1968).

central valley fill in fine alluvium (Fig. 5.13), and fine alluvium eventually caps the sequence.

Lateral depositional processes predominate in suspended-load streams that carry less than 3% bedload. Lateral channel migration occurs preferentially in the central part of the valley, depositing channel sands that are overlain by laterally accreting point-bar deposits and vertically accreting overbank deposits. Because channel migration tends to occur in the center of the valley, overbank fines dominate the sequence along the valley flanks (Fig. 5.13).

The basic valley fill sequence fines upward. Coarse sediments are ordinarily delivered first in response to uplift; but as slopes and gradients are reduced and storage of coarse material in upland valleys becomes important, an increasing proportion of fine-grained sediments is delivered to lowland valleys. The basic response can be complicated, however, by a number of other factors. Isostatic readjustment in the source area, for instance, can cause a complex response that has been previously described for alluvial fan sequences. The effects on the alluvial plain would be similar, although somewhat dampened by distance.

Fining-upward sequences can result from intrinsic factors in the system as well. In many cases the initial phases of valley filling take place when flow is confined within a bedrock valley. Flow expansion cannot occur under such conditions, and coarse sediments accumulate as channel deposits. Fine alluvium bypasses the valley because broad floodplains where fine sediments may be deposited do not occur in such situations.

As the valley aggrades, however, valley floors broaden, lateral migration of channels becomes possible, and fines can be deposited on floodplains during overbank flooding. The change in depositional style in this case would not be caused by extrinsic influences such as tectonism or climate, but by a change in channel pattern made possible by the aggradational activities of the stream itself.

Bridge and Leeder (1979) have simulated alluviation in aggrading valleys. In their model the proportion of channel to floodplain sediments, as shown in cross section, is dependent on channel belt size, floodplain width, mean avulsion rate, rate of channel belt aggradation, and tectonism.

For braided streams, the channel belt size is the stream width at bankfull stage, and for meandering channels, it is the maximum amplitude of the river bends. Long-term trends in rates of valley floor aggradation and channel avulsion are controlled by the amount of water and sediment provided by source areas and the ability of the river system to transport it. Tectonically induced changes in gradient, for instance, will modify the transport capability of the stream. They may also alter the sediment yield, as will changes in climate.

The frequency of avulsive events is recorded in the overbank deposits that occur between successive channel belt sequences in vertical section (Bridge, 1984). In an aggrading valley, a channel belt will return, at a higher elevation, to its former position after a finite number of avulsions occur. Experiments suggest that this number is approximately equal to the ratio of the floodplain width to the channel belt width multiplied by a factor ranging from 0.6 to 1.0.

Avulsive events may be recorded in floodplain sequences by an increase in number, thickness, and grain size of sedimentation units that occurs in response to unusually large floods that can accompany the process of channel diversion (Bridge, 1984). Evidence of avulsion may also include a persistent change in thickness and grain size of sedimentation units that occurs when a channel belt moves. Such a change might appear suddenly if the avulsion is rapid, but if it occurs through a slow diversion of flow, the change may be gradual and difficult to differentiate from changes in sedimentation patterns of a fixed channel belt.

Changes in any one of the variables or, more likely, changes of any of them in combination will alter not only the dimension and shape of the channel, but also the pattern of alluviation. But even in the absence of significant changes in extrinsic variables, changes in the pattern of alluviation could take place, even though gross changes in river morphology did not occur.

Bridge and Leeder (1979) found that both the density of channel deposits and the degree of their interconnectedness increase as the ratio of channel belt width to floodplain width increases. Where rivers are constricted in narrow bedrock valleys, therefore, it might be expected that channel deposits might dominate the alluvial sequence. They found also that density and interconnectedness of channel deposits decreased as the rate of aggradation and channel avulsion decreased.

Most often, when the effects of tectonism on fluvial components are considered, they are presumed to act perpendicularly to flow with uplift in the source area or change in base-level elevation. Miall (1981), however, showed that a significant group of streams flow parallel to tectonic strike, and this suggests that sideways tilting of the valley may be another factor controlling alluvial architecture.

Bridge and Leeder (1979) factored such tilting into their model and showed that as the degree of tilting increases, either by decreased rate of aggradation or shortened period between tectonic events, there is an increasing tendency for channel belts to cluster on the downwarped side of the valley (Fig. 5.14). Examples of such clustering may be seen in the channel sand distribution in the Westwater Can-

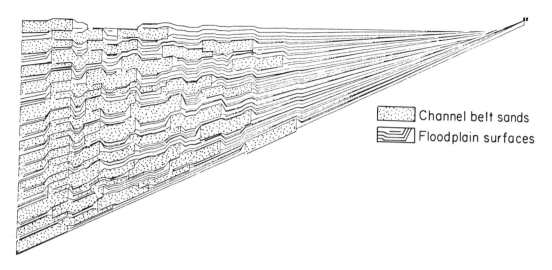

Figure 5.14 Simulated cross section showing distribution of channel deposits in a tectonically tilted valley with a short-period cycle of tectonism (after Bridge and Leeder, 1979).

yon Member (Campbell, 1976) and in the Tertiary fluvial deposits of Pakistan (Behrensmeyer and Tauxe, 1982).

Coastal alluvial plains are qualitatively different from interior alluvial valleys and plains. Streams almost always flow at right angles to structural and depositional strike, and stream gradients tend to be low. Streams crossing the plain flow over a constructional platform built of marine and transitional marine sediments, and they establish zones of influence within which they migrate.

The distribution of sand bodies within an alluvial suite on a prograding coastal plain may be affected mostly by the rate at which subsidence is taking place (Allen, 1978). Thickness of channel sand bodies is dependent on channel depth and thickness of sand under the channel, and the width of the sand body is a function of the meander belt width at a given time.

The arrangement of sand bodies in vertical section can be described by the number of sand bodies, the mean number of sand bodies in immediate contact, the mean proportion of sand body perimeters in contact with surrounding sand bodies, and the number of sand bodies forming clusters in overbank muds (Allen, 1978).

Sand bodies are few and the degree of connectedness is small where rates of subsidence are large (Fig. 5.15). Conversely, as rates of subsidence slow, the number of sand bodies, as well as the number of bodies in contact, increases. In addition, the average fraction of the sand body perimeter in contact with other sand bodies increases, giving rise to multistory sands of relatively great lateral extent (Fig. 5.15).

As coastal plains prograde, the hinge line migrates seaward so that at any given point on the plain the rate of subsidence will slow through time. Changing rates of subsidence will lead to gross coarsening-upward sequences that in the past have been

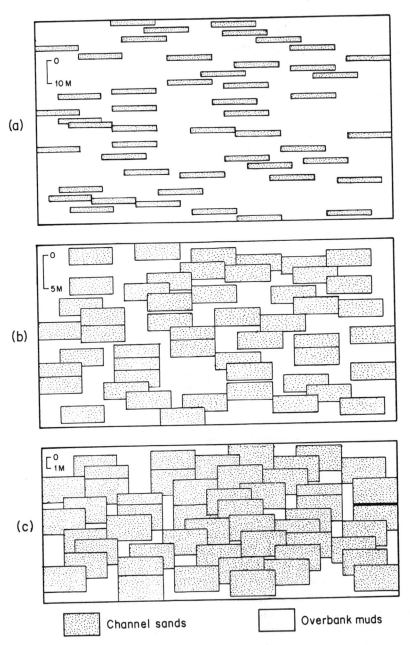

Channel sands   Overbank muds

Figure 5.15 Simulated cross sections showing distribution of channel deposits within overbank facies as the rate of subsidence slows from (a) to (c) (after Allen, 1978).

attributed to a change in channel pattern in response to tectonism or a change in climate. But upward coarsening can also result from a response of the system to a reduction in the rate of subsidence (Allen, 1978).

Analysis of the geometry of the sandstone bodies in the Upper Triassic Chinle Formation of Colorado (USA) by Blakey and Gubitosa (1984) showed that the major control on sand body geometry was exerted by varying rates of subsidence and that secondary control was exerted by the quantity and caliber of the sediment yield to the rivers. When the ratio of the rate of subsidence to episodes of avulsion was small and sediment influx was large, braided stream deposits in a sheet geometry were the dominant architecture (Fig. 5.16). When the ratio of subsidence to episodes of avulsion was large, however, and when the rate of sediment influx was small, single-story sand bodies in sand ribbons encased in thick mudstones were deposited (Fig. 5.16).

Alluvial plain sequences in the rock record.   A detailed analysis of the lateral facies changes and variations in vertical profiles has been made of a relatively large meander belt sandstone in the lower Cretaceous Sudmoor Point Sandstone Member of southern England (Stewart, 1981). The sand body is approximately 6 m thick and its exposed portion is 1.8 km wide. It rests on an erosional base and can be separated into six subunits that are also bounded by erosional surfaces (Fig. 5.17).

Interformational conglomerates at the base of the sequence probably formed in riffles at crossover points between meanders or as scour pool bars. They are overlain by crossbedded sandstones that probably formed on the lower part of point bars, and the flat-bedded or ripple-bedded sandstones that occur near the top of the sequence formed in shallow water on the upper part of point bars. Mudstones are intercalated throughout the sequence, suggesting that discharge was periodically very low, and the sand body is enclosed within variegated floodplain mudstones.

The six individual sand bodies of the meander belt represent individual point-bar deposits consisting of lateral accretionary units. Vertical sequences within these bodies vary as a function of their position in the point bar in a manner similar to that shown by Jackson (1975) on the Wabash River.

Most sequences consist of interformational conglomerates overlain by cross-bedded sandstones and mudstones or plane- and ripple-bedded sandstones (Fig. 5.17) and probably represent fully developed point bars. One sequence, however, consists of uniform sands that may have accumulated in a straight reach between meanders, and a second sequence consists primarily of muds that probably accumulated in an abandoned channel.

A point bar remarkably similar to the coarse-grained point bars of McGowen and Garner (1970) is found in the Tertiary Castinet Formation of Spain (Nijman and Puigdefabregas, 1978). Cobble conglomerates at the base of the sequence overlie an erosional surface incised deeply into underlying sandstone (Fig. 5.18). They probably formed in scour pools and are overlain gradationally by crossbedded conglomeratic sandstones that may have formed near the toe of the point bar. These, in

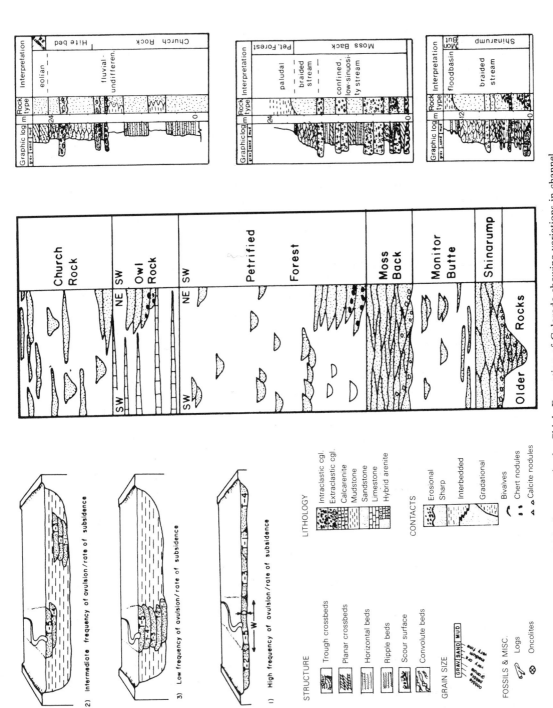

Figure 5.16 Sequences in the Chinle Formation of Colorado showing variations in channel shape and nature of channel deposits in response to variations in rate of channel avulsion and subsidence (after Blakey and Gubitosa, 1984).

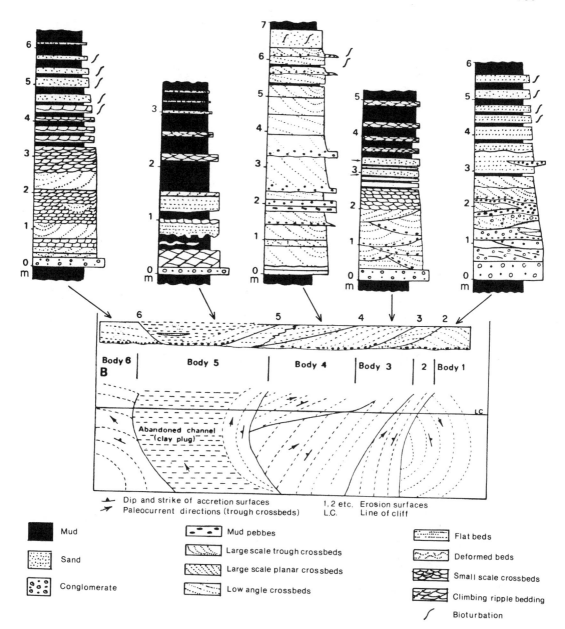

Figure 5.17 Reconstruction of a portion of the Sudmoor Point Sandstone meander belt showing the stratigraphic sequences that occur in different portions of the meander. The cross section shows the lateral relationships of the sediment types to erosional surfaces, and the plan view shows their areal distribution (after Stewart, 1981).

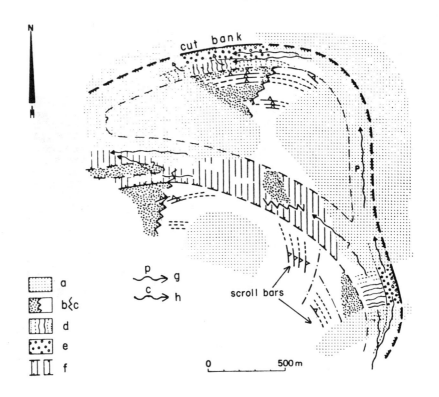

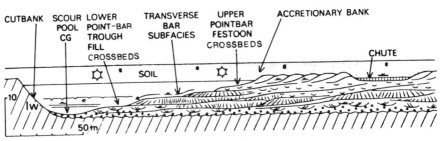

Figure 5.18 Reconstruction of the Castinet point bar in plan view (upper) and in diagrammatic cross section (lower), showing distribution of sediments and sedimentary structures. (a) covered; (b) transverse bar below point bar surface; (c) trough crossbeds; (d) transverse bar facies in point-bar slope surface or below cutoff fill; (e) scour pool conglomerate; (f) chute base and/or chute fill; (g) inferred pool current; and (h) inferred chute current at bank-fill discharge (after Nijman and Puigdefabregas, 1978).

turn, are overlain by planar crossbedded sandstones that probably accumulated as transverse bars. These are commonly associated with ripple-bedded sandstones and rill structures that suggest the occurrence of late-stage modification of bar surfaces during waning flow.

Festoon crossbedded sandstones formed by migrating dunes occur at a higher structural level on the point bar, where they are overlain by mottled sandstones in accretionary beds that dip at right angles to mean stream flow. They probably formed as scroll bars, and they are locally cut by channels floored by crossbedded sandstones and filled with mudstones that probably represent chutes. Overbank mudstones displaying several paleosol horizons cover the point-bar lithosome.

Meandering stream deposits from the Tertiary of the southern Pyrenees consist of floodplain mudstones and channel sandstones that occur as isolated bodies or as amalgamated sand bodies that may have formed as meander belts (Puigdefabregas and van Vliet, 1978) (Fig. 5.19). Both types of channel deposits display, at least in part, features consistent with the "classical" point-bar model. Differences between the two may represent varying rates of floodplain aggradation, with single sand bodies forming under conditions of rapid aggradation and frequent channel avulsions and amalgamated bodies forming in meander belts of more stable position on the floodbasin.

Pleistocene valley train deposits of the Upper Midwest, USA, characteristically formed in gravelly braided stream environments. Deposits consist commonly of coarse gravels in crude horizontal beds deposited as longitudinal bars, planar crossbedded sand and gravel deposited as transverse bars, and trough crossbedded sands that accumulated in interbar channels (Fraser and others, 1983).

Ubiquitous within these deposits are features indicative of episodic flow (Fig. 5.20). Short-term cycles are evident where lateral accretionary deposits on bar margins, reactivation surfaces, mud drapes, and evidence of shallow reworking of bar surfaces occur. In a glacial environment, these might be the result of a diurnal melting cycle, but in a nonglacial setting they more probably result from storm events.

Evidence of longer-period cycles include interbedding of longitudinal and transverse bars and extensive channeling of bar deposits. Such sequences probably form in response to seasonal flow fluctuations. Thick fining-upward sequences may also result from seasonal cycles, but they could also be the result of channel abandonment or channel reoccupation where coarsening-upward cycles occur (Costello and Walker, 1972).

The Upper Member of the Mississippian Cannes de Roche Formation of eastern Canada may be an ancient counterpart of distal braided streams (Rust, 1978). Plane-bedded sandstones tend not to accumulate in such a setting, and trough crossbedded gravels are the most abundant sediment in the formation. Conglomerates occur in multiple trough sets that thin and fine upward (Fig. 5.21). They rest on erosional bases and are intercalated with only minor amounts of plane-bedded conglomerate and sandstone. Sequences of trough sets are capped with silty mudstones containing abundant plant fossils and occasional beds of trough crossbedded sandstones.

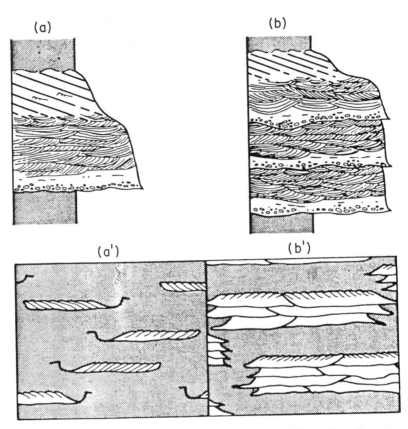

Figure 5.19 Types of channel sequences in the Tertiary of the southern Pyrenees. Vertical sequence in a single channel (a) in the Lower Montanana Group as it occurs within the alluvial succession (a'). Vertical sequence in amalgamated channels (b) in the Oligocene near Puerto de Monrepos as it occurs as a meander belt in an alluvial succession (b') (after Puigdefabregas and van Vliet, 1978).

Each thinning- and fining-upward sequence probably represents a cycle where gradual infilling and abandonment of an active channel complex in a distal braided stream setting occurred. Trough crossbedded sandstones and conglomerates accumulated in channels, whereas mudstones were deposited in areas of the floodplain abandoned after migration of the active channel to a new position.

Ancient counterparts of the sand wave deposits of the South Saskatchewan River can be found in the lower part of the Battery Point Sandstone of Canada (Cant, 1978). Channel deposits in the formation are represented by trough crossbedded pebbly sandstones. Large sets probably formed as flood stage bedforms in deep water, whereas smaller sets accumulated in shallower and topographically higher channels (Fig. 5.22).

Associated with both types of trough sets are large planar cross sets that may

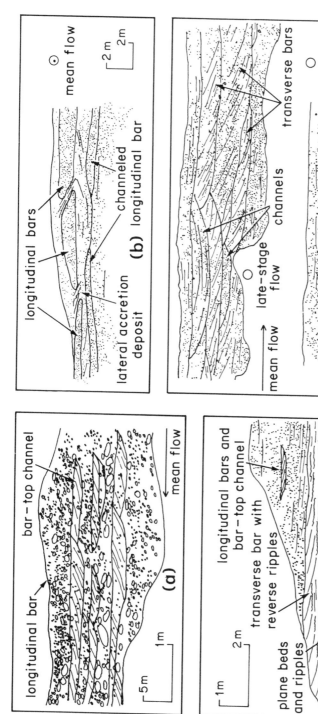

Figure 5.20 Outcrop drawings of Pleistocene valley train deposits of the Wabash River Valley, Indiana, showing relationship of bar and channel sediments deposited under conditions of episodic flow of varying periodicity (after Fraser and others, 1983).

111

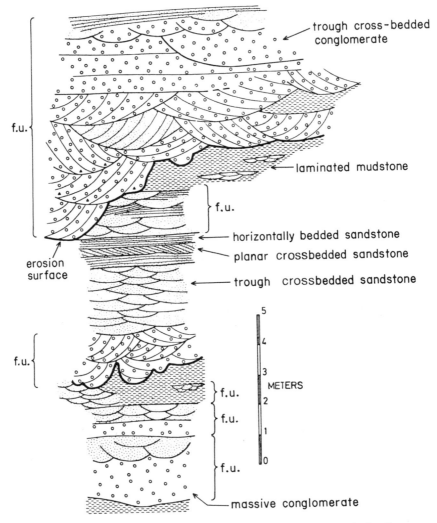

f.u.

trough cross-bedded
conglomerate

laminated mudstone

f.u.

horizontally bedded sandstone
planar crossbedded sandstone
trough crossbedded sandstone

erosion
surface

f.u.

f.u.

f.u.

5
4
3
METERS
2
1
0

f.u.

massive conglomerate

Figure 5.21 Vertical sequence in the Upper Member of the Mississippian Cannes
de Roche Formation, Canada, showing the internal arrangement of textures and
structures as they occur in fining-upward sequences in the succession (after Rust,
1978).

be analogous to the cross-channel bars of the South Saskatchewan River. The small
cosets that overlie the larger sets probably correspond to the sand waves that are
accreted onto the channel bar core. The upper part of the Battery Point sequence
consists of fine-grained ripple-bedded sandstone that commonly is associated with
mudstone. In the South Saskatchewan River, sand waves are commonly topped by
ripple-bedded sands, and the mudstones probably accumulated as accretion deposits
of the floodplain.

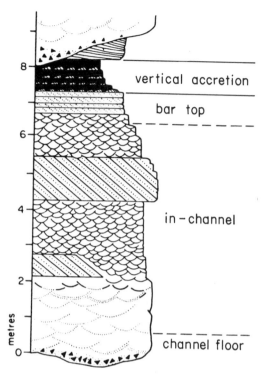

Figure 5.22 Typical sequence in the Battery Point Sandstone showing the relationships among channel, sand wave, and vertical accretionary deposits. Trough sets formed in channels whereas, planar cross sets probably accummulated as cross-channel bars (after Cant, 1978).

The Battery Point Sandstone is composed of at least 10 fining-upward and thinning-upward sequences that represent both channel aggradation, where the sequence consists almost entirely of trough sets, and sand wave aggradation where planar crossbedded sandstone are included in the sequence (Fig. 5.23).

An ancient example of an aggrading rock-bound valley is the Kissinger Sandstone (Shelton, 1973). It is a long, narrow sand body set within a channel locally eroded in shales and limestones to a depth of over 50 m. Erosion began with a drop in sea level, and deposition occurred in what must have been dominantly a bedload stream as sea level began to rise.

The lower 40 m of the formation consists of fining-upward sequences up to 10 m thick, consisting of conglomerate at the base that is overlain by conglomeratic sandstone and medium-grained sandstone (Fig. 5.24). Although internal structures suggest that these sequences accumulated in a braided stream setting, their thickness indicates that the stream must have been relatively deep. Lack of fine-grained overbank sediments may be attributed to the confinement of the channel in a narrow, rock-bound valley.

The upper 10 m of the formation is a single fining-upward sequence consisting of crossbedded conglomeratic sandstone that is overlain by sandstone in plane beds and ripple beds and capped by parallel-laminated mudstones. Filling of the valley, coupled with rising base level, may have produced an unconfined, low-gradient

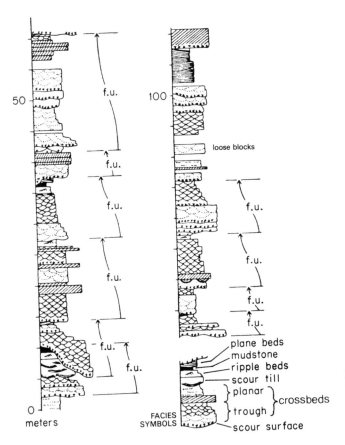

FACIES SYMBOLS

plane beds
mudstone
ripple beds
scour till
planar ⎫
        ⎬ crossbeds
trough ⎭
scour surface

Figure 5.23 Section in the Battery Point Formation showing the succession of fining-upward sequences defined by scour surfaces (after Cant and Walker, 1976).

stream capable of lateral migration and overbank flooding that deposited the fining-upward sequence at the top of the formation.

Thus far, discussion has centered on descriptions of valley fill sequences and alluvial plain sequences as separate entities or as entities that succeed each other through time. Patterns of alluviation, however, may be complicated by the coexistence of these two types. Such a pattern has been described from the Siwalik Group of northern Pakistan by Behrensmeyer and Tauxe (1982).

Two sandstone types are readily apparent, a gray sheet sandstone and a buff sandstone that occurs in laterally restricted bodies (Fig. 5.25). They are bounded at the top by a laterally persistent facies boundary that nearly parallels a magnetically established isochron, suggesting that the two sandstone types are laterally equivalent.

The blue-gray sheet sandstones contain only minor amounts of interbedded muds and marl and were probably deposited in a wide braided channel system dominated by complex lateral channel migrations that resulted in cyclic episodes of cut

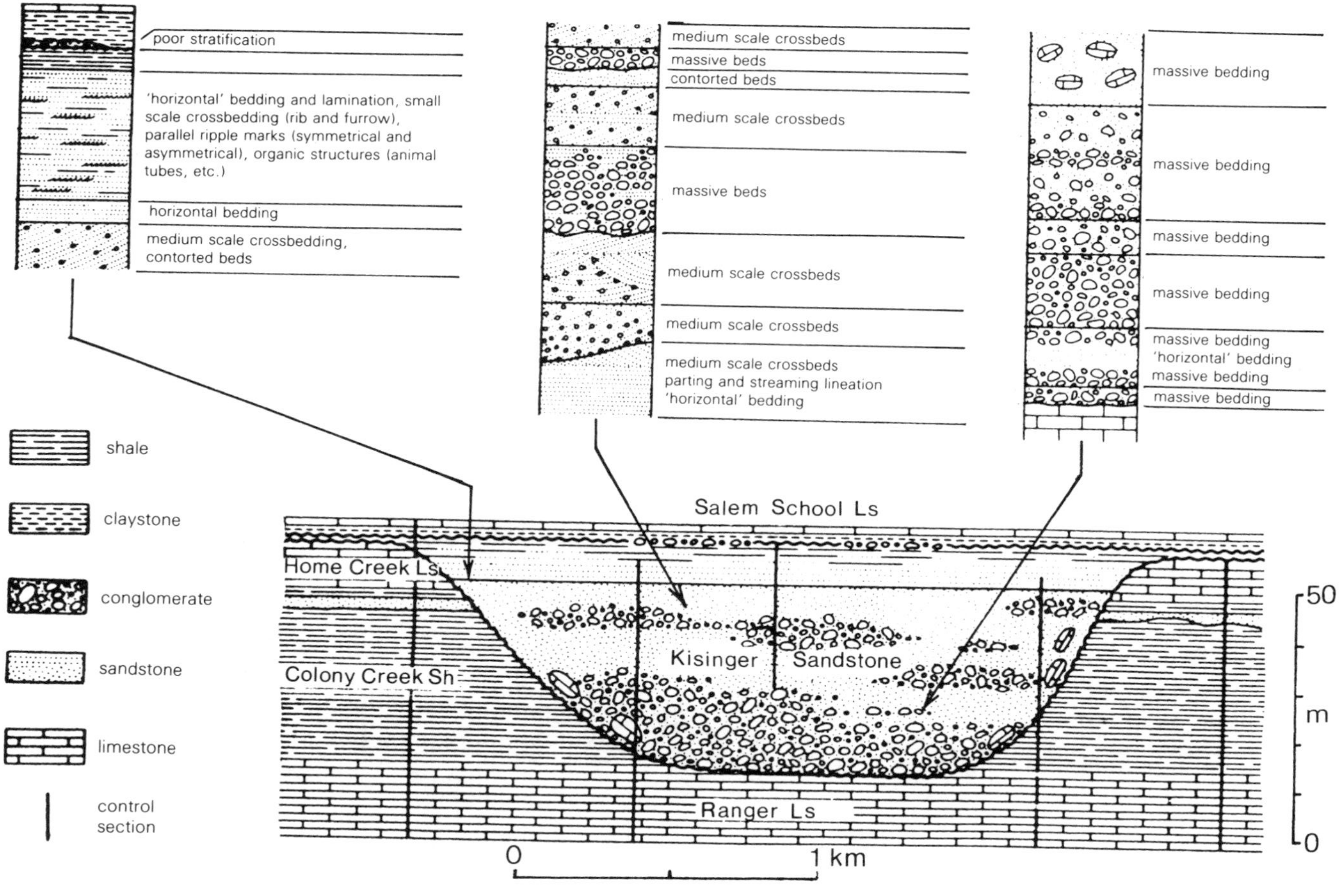
poor stratification
'horizontal' bedding and lamination, small scale crossbedding (rib and furrow), parallel ripple marks (symmetrical and asymmetrical), organic structures (animal tubes, etc.)
horizontal bedding
medium scale crossbedding, contorted beds
medium scale crossbeds
massive beds
contorted beds
medium scale crossbeds
massive beds
medium scale crossbeds
medium scale crossbeds
medium scale crossbeds
parting and streaming lineation
'horizontal' bedding
massive bedding
massive bedding
massive bedding
massive bedding
massive bedding
'horizontal' bedding
massive bedding
massive bedding
shale
claystone
conglomerate
sandstone
limestone
control section
Salem School Ls
Home Creek Ls
Kisinger Sandstone
Colony Creek Sh
Ranger Ls
0
1 km
50
m
0

Figure 5.24 Typical vertical sequences in the Kisinger Sandstone (Pennsylvania) as they occur within various portions of the alluviated valley (after Shelton, 1973).

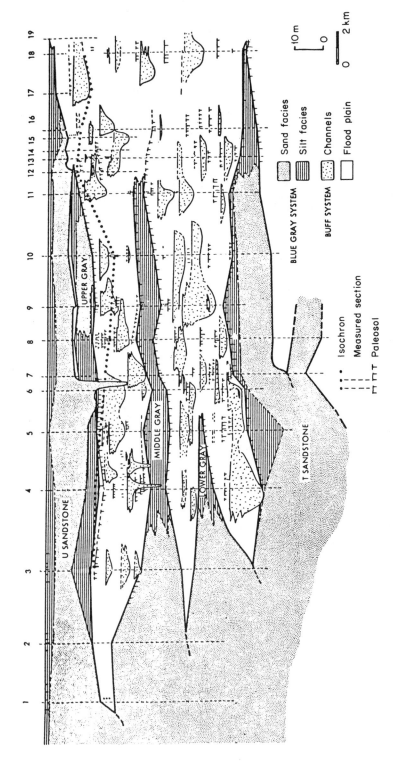

Figure 5.25 Cross section of Tertiary alluvial deposits of Pakistan showing the relationship of the blue-gray sheet sandstones and the buff channel sandstones (after Behrensmeyer and Tauxe, 1982).

and fill. The buff sandstones, on the other hand, occur in channels that make up only 25% of the total rocks of the facies. Small channels fill shallow scours and larger channels fill deep erosional channels. Flow was confined in the eroded channels until they were aggraded above the level of initial scour, and lateral flow divergence became possible. The channel fill sequences are neither obviously braided nor meandering in origin, and it may be that the dominant process of channel shifting was avulsion rather than lateral migration.

Behrensmeyer and Tauxe (1982) interpreted the gray system to be the major drainage of a high hinterland, whereas the buff system was deposited in a complex of small tributary channels that drained laterally from the mountain front into the trunk stream. Valley formation was apparently induced by tectonism or by climatic changes, and downcutting produced interfluves on which paleosols were formed. As valleys were alluviated, lateral spreading of stream facies occurred and interfluves were covered (Fig. 5.25).

When attempting to model coastal plain alluviation, Allen (1978) assumed steady and uniform subsidence of a plain occupied by rivers of similar load, discharge, and channel shape. These conditions, of course, are never attained in nature, and the shape and distribution of alluvial sand bodies in three dimensions are dependent on a set of complexly interrelated factors.

The Tertiary Gulf Coastal Plain of Texas, for instance, was crossed by rivers of greatly different character (Galloway, 1981). Some rivers originated extrabasinally, and they tended to be large because they tapped integrated drainage basins that extended inland beyond the fall line. Discharge was relatively steady in these rivers, and they entered the coastal plain at structural or topographic lows, where they tended to remain over long periods of time and where they formed laterally restricted, dip-oriented belts of channel-fill facies.

Over topographic or structural highs, stream plains were formed that were traversed by intrabasinal rather than extrabasinal streams. Flow tended to be flashy and flood-prone because of the small size of the catchment basins, and streams were small. They deposited sand bodies of limited lateral and vertical dimensions that were enclosed by fine-grained sediment.

Extrabasinal rivers entered the coastal plain and attempted to aggrade it uniformly. Widely spaced rivers fanned out in a distributary manner by nodal and downchannel avulsion at points where channel belts aggraded to thresholds of instability. Closely spaced systems, on the other hand, formed complexely interwoven channel trends.

Distributary channel axes consisted of both bedload streams (Gueydan and George West Axes) and mixed-load streams (Chita Axis) (Fig. 5.26). Bedload channels formed thick, stacked multilateral sand bodies in belts tens of miles wide. Mixed-load channels, on the other hand, bifurcated and intersected at oblique angles within the confines of a uniformly broad belt. Both multilateral and stacked channel sequences were formed, but in mixed-load axes, they are separated by mud-rich facies containing only thin, discontinuous sand units deposited in floodplain and lacustrine settings.

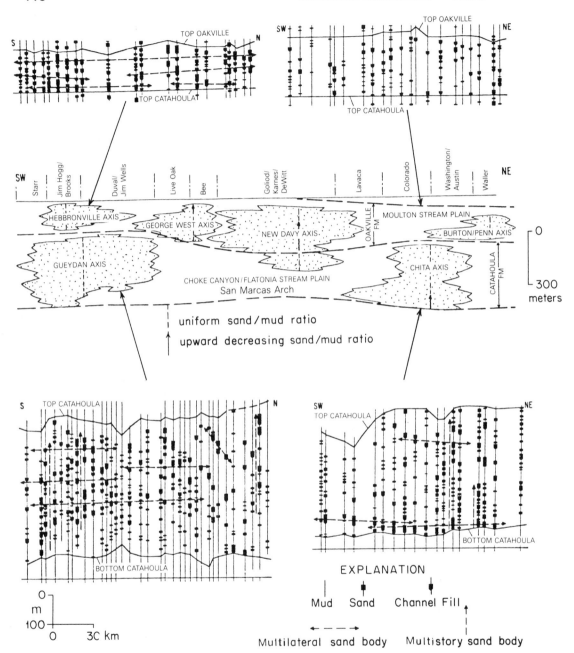

Figure 5.26 Cross section of the Tertiary alluvial deposits of the Texas Gulf Coastal Plain showing the internal arrangement of channel sands and floodplain muds and the location of the major river systems relative to structural elements of the plain (after Galloway, 1981).

Variations occurred within axes as well. The Gueydan, Hebronville, and George West axes show no systematic vertical variation in texture or facies composition. The Chita and Burton-Penn Axes, however, show a decreasing bedload:suspended load ratio that may reflect an increased subsidence rate or a change in channel load.

Lakes

Sedimentary processes and products.   The geomorphic setting of lakes in humid regions is similar to that of saline lakes and playas in arid regions. They occur as landlocked bodies and they persist because the amount of water entering the lake exceeds the amount lost. Sources of water for lakes in humid areas include surface drainage and subsurface seepage, as well as precipitation directly on the lake surface. Losses occur chiefly through outflow and evaporation.

Lake basins commonly occur in youthful terrains. Large, deep lakes occupy structural basins formed by downfaulting or crustal warping. Smaller lakes are normal components of terrains that were aggraded rapidly and where drainage networks are poorly developed, such as on valley bottoms, coastal plains, or glaciated areas. Abundant, relatively small lakes on alluvial plains may signify continuing regional tectonic movement in the area (Tanner, 1974).

Among the sedimentologically important criteria by which lakes are classified are climate in which the lake occurs, depth, and biological activity. Sediment distribution in a lake is dependent on the temperature profile in the lake, which is a function of the climatic regime in which the lake occurs, and the depth of the lake. Lakes in temperate regions differ considerably from those in tropical regions, and within temperate zones, nonglacial lakes differ from proglacial lakes in the path sediment takes from source to ultimate site of deposition.

Biological activity within lakes is also important in determining composition. Activity is minor in oligotrophic lakes, which are deficient in nutrients, and is somewhat greater in mesotrophic lakes, where a moderate nutrient supply is available. An abundant supply of nutrients is available in eutrophic lakes, and biological productivity may be great enough in surface waters to cause deoxygenation of bottom water where decaying organic material collects. In some eutrophic lakes, algal blooms may lower pH sufficiently to precipitate carbonate, which can collect in layers where the supply of terrigenous sediment is low.

In shallow temperate lakes, the water column is thoroughly mixed by the wind, but in lakes even a few feet deep a circulation pattern is imposed on the lake by a pronounced temperature stratification that is seasonally established (Fig. 5.27). In summer, the water column is stably stratified with warm waters of the epilimnion separated from cooler water of the hypolimnion by the metalimnion, where a steep temperature gradient (thermocline) occurs. Water in the epilimnion is mixed by wind and it remains well-oxygenated. Water below the metalimnion is effectively separated from the atmosphere, however, and available oxygen can be depleted by fish and decomposing organic debris. Anaerobic bacteria may produce H_2S in bottom sediments where pore water may become anoxic.

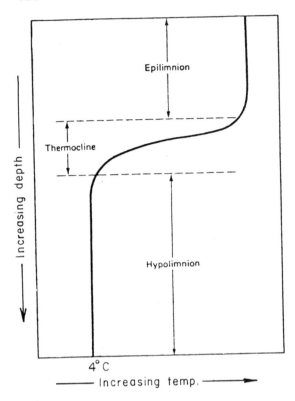

Figure 5.27 Temperature-stratified water column of a typical lake. Position of the thermocline varies according to climate, lake size, degree of turbulence in the epilimnion, and time of year (after R. A. Davis, 1983).

During the fall, surface waters cool, become dense, and sink to the bottom, mixing the entire water column. When this occurs, nutrient-rich bottom waters rising to the surface may cause a bloom of phytoplankton; but when bottom waters contain H_2S, mass mortality of lake inhabitants may occur.

Lake water is separated from the atmosphere in the winter if ice forms on the surface; but even if ice does not form, stable stratification may be established if surface waters are cooled below the temperature of maximum water density at 4°C. During the spring, as surface ice melts or water warms to 4°C, lake waters turn over again before summer stratification is reestablished.

Temperature stratification within mid-latitude lakes controls the way in which sediment is distributed in a lake (Fig. 5.28). Low-density inflows that are warm or that transport little sediment may move over the surface of the lake, carrying a plume of sediment basinward. Incoming waters that are somewhat cooler or that transport a greater suspended load may sink to the thermocline before flowing out into the lake as interflows, and very cold or turbid waters may move downslope as underflows or turbidity currents.

Where incoming water flows into the lake, most coarse material drops out, quickly forming a delta, although some may be carried along the lake bottom by density currents. Coarse sediment transported by turbidity currents may settle out

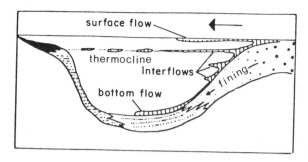

Figure 5.28 Sediment transport paths from point of influx into a temperature-stratified lake (after Leeder, 1982).

quickly, whereas muds carried in suspension in interflows may settle out slowly; and fines carried by overflows may remain in suspension for very long periods because of wind-induced turbulence in the epilimnion.

The segregation caused by these processes produces a concentric distribution of sediment types in lake basins (Picard and High, 1973). Coarse sediments are most abundant at the point of influx and along the lake margin, where they are distributed by wave activity and longshore currents (Fig. 5.29). The marginal band of coarse sediments is normally narrow because of the low-energy condition encountered in most lakes. Coarse-grained coastal sediments are more extensive in large

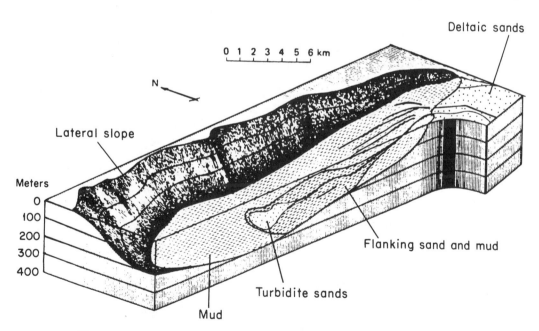

Figure 5.29   Model of sediment distribution in Lake Geneva. Coarsest sediments occur at the lake margin and as turbidite deposits on the lake plain. Finest sediments occur on lateral slopes, and sediments of intermediate texture occur on the lake plain flanking the turbidite fan (after Houbolt and Jonker, 1968; Reineck and Singh, 1980).

lakes, however, where they may approximate the size of some marine coastal deposits (see G. K. Gilbert, 1885, for instance).

Turbidity currents flowing along deep axial parts of lakes may deposit coarse sediment along the flow path. The remainder of the surface of the deep basin, however, is commonly covered by silty sediments deposited by distal density flows, overflows from turbidity currents flowing in axial channels, and from suspension out of the epilimnion and metalimnion.

Finest-grained sediments occur on marginal slopes because the only sediment available is that carried in suspension. On relatively steep marginal slopes, these sediments are subject to mass movement, especially in tectonically active areas.

In addition to terrigenous clastic debris, carbonate sediments may accumulate in some lakes. Photosynthesis during seasonal algal blooms can reduce pH sufficiently to precipitate $CaCO_3$ in basinal settings. In addition, algal carbonates and tufas may form in shallow water near the margin, where waters can be significantly warmer than in basin centers. Carbonate layers are commonly interbedded with layers rich in organic detritus that accumulate during periods of mass mortality and they may also be interbedded with terrigenous clastics deposited by density currents.

Proglacial lakes are distinguished from other temperate-region lakes by the dominance of underflows as the main transporting agent. Water from melting ice flowing into proglacial lakes is usually very near the maximum density of water, and even a small addition of sediment ensures that the incoming water will be denser than lake water.

During summer months, sands and gravels are deposited at the point of influx as "Gilbert-type" deltas with high-angle foresets. Finer-grained sediment is transported into the lake as a more or less continuous underflow, whereas the finest-grained sediments normally remain in suspension, especially in those cases where the density difference between lake and stream water is small.

During the winter, when the lake is ice-covered, the fines settle out of suspension, forming the fine-grained part of a varve couplet. Not every fine-grained layer is the result of winter deposition. In the summer, some density currents may be so dense that little mixing occurs between lake waters and the current. In these cases, fine-grained sediments are carried downslope with the current, and they settle out of suspension as the current slows. The resulting fine layers can be misidentified as winter layers of varves, causing errors in estimating the age of some proglacial lakes.

Annual temperature ranges in low-latitude lakes are small, and tropical lakes may be permanently stratified because the warm upper layer may never cool sufficiently to induce overturning. Oxygen exchange between upper and lower layers may be stopped permanently and, where sufficient nutrients are available, eutrophication can occur.

Bottom sediments in tropical lakes consist commonly of rhythmic alternations of mud layers and layers rich in organic debris. Mud layers are produced during flood season when influx of terrigenous sediments is large enough to overwhelm the organic production of the lake. Organic layers, on the other hand, are produced during the dry season when sediment influx is small and organic material is not diluted by terrigenous clastic debris.

Intrinsic and extrinsic controls on humid system lacustrine sequences.   The constructional features along lake margins tend to be relatively narrow, and even in very large lakes they tend not to prograde very far lakeward from basin margins. The only major exception to this occurs at the point where rivers enter lakes and build deltas. Lakes fill as the lake bottom builds upward through deposition by turbidity currents and as deltas prograde longitudinally into the lake (Visher, 1965) (Fig. 5.30). Lateral progradation of shore zones appears not to be a major factor in lake filling, and lacustrine sequences most often consist of basal turbidities overlain by deltaic foresets and topsets. Shore zone deposits, such as spits, barrier islands, and attached beaches, although common features of large lakes, occur in the sequence only in a band adjacent to the margin of the basin.

Superimposed on simple lake-filling sequences are cycles caused by climatic fluctuations and tectonic episodes. Short-term climatic cycles are related to storm episodes and result in varves composed of various combinations of clastic and non-clastic layers formed in response to flood-induced influxes of sand and mud, followed by periods of low terrigenous input.

Longer-period and more severe climatic fluctuations also occur in many lacustrine sequences. The Lockatong Formation (Triassic) of New Jersey, for instance, is composed of cycles several meters thick, consisting of "detrital" cycles and "chemical" cycles (Van Houten, 1964) (Fig. 5.31). Detrital cycles consist of black mudstone that grades upward into dolomitic mudstone that usually displays evidence of emergence. These cycles probably formed as small-scale regressive units

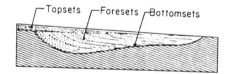

Figure 5.30 Diagrammatic representation of deltaic infilling of a lake basin (after Wagner, 1950; Reineck and Singh, 1980).

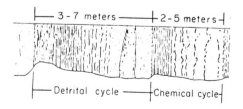

Figure 5.31 Typical stratigraphic sequence formed during a change from humid to more arid conditions in the Lockatong Formation, New Jersey. Chemical cycles typically are 2 to 5 m thick, whereas detrital cycles tend to be slightly thicker (after Van Houten, 1964).

that accumulated during humid periods when through-flowing drainage was established. Chemical cycles consist of an upward transition from dark mudstone into analcine-rich lenses, and they probably record arid episodes characterized by internal drainage into the lake.

In arid regions, salts precipitate during dry periods when yield of terrigenous clastics is reduced, and clastic layers are deposited during wet periods when valleys are flushed of sediments. In humid regions, however, the reverse may be true. Donovan (1975), for instance, proposed that limestone deposition in a Devonian lake of the Orcadian Basin of Scotland occurred during high lake level stands when sediments were trapped in valleys by rising base level. When lake level dropped during drier periods, valleys were eroded and greater amounts of clastic material were added to the basin, effectively masking carbonate deposition.

The reason for the variation lies in the different way source areas and rivers respond to changes in precipitation in arid and humid climates. In arid climates, a small increase in precipitation causes increased slope erosion, resulting in an increased delivery of sediments to rivers (Schumm, 1977). Valley profiles steepen to handle the load, and valleys are flushed of sediment stored during drier periods. Apparently, this response is able to overwhelm a river's tendency to aggrade itself when base level rises.

In humid climates, however, a small increase in rainfall might cause lake levels to rise and rivers to aggrade, because vegetated slopes would not respond as dramatically to the increase as would slopes in arid regions.

The effects of uplift in source areas or a drop in elevation of the basin would not differ significantly between arid and humid climates. The reader is referred to tectonic influences on saline lake sequences, which have already been discussed in detail.

Humid system lacustrine sequences in the rock record.   Intracontinental rifting during the Triassic formed the Hartford Basin of Connecticut in which the East Berlin Formation was deposited in a climatic regime that shifted from humid to semiarid (Hubert and others, 1976). During drier periods, the basin was occupied by floodplains on which small shallow lakes occurred. The sediments deposited in these lakes consist of oxidized ripple-bedded sandstones and coarse siltstones that are interbedded with floodplain mudstone, attesting to the short-lived nature of these lakes.

During humid periods, the basin was occupied by large perennial lakes in which black shale and gray mudstones were deposited in cycles separated by red floodplain mudstones. The cycles are 2 to 7 m thick and consist of a central layer of black shale bounded at the top and bottom by gray mudstone and siltstone (Fig. 5.32). The gray mudstone and siltstone contain features indicative of shallow water, whereas the black mudstone probably accumulated in relatively deep water. Planar crossbedded sandstone at the top of some cycles may have formed along lake margins, and the whole cycle records a period of lake expansion when rainfall was abundant, followed by contraction of the lake as it filled and/or the climate became drier

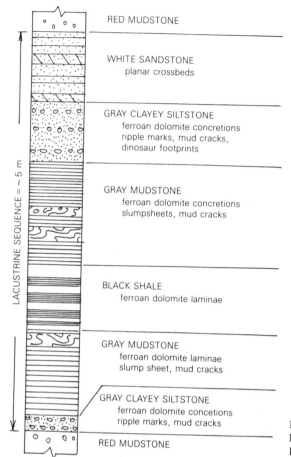

Figure 5.32 Typical sequence of lacustrine sediments of the East Berlin Formation, Connecticut (after Hubert and others, 1976).

(Hubert and others, 1976; DeMicco and Kordesch, 1986). Within these cycles are couplets of light gray dolomite and black shale or gray mudstone that probably reflect shorter-period rhythmic alternations of climatic conditions.

During part of the time that the Green River Formation was being deposited in Eocene Lake Gosiute, the climate was dry and deposition took place in large saline lakes or playas. During the early and late periods of lake history, however, the climate was humid. The Laney Member was deposited during the waning phase of Green River deposition and during that phase Lake Gosiute evolved from a saline, alkaline lake to a freshwater one (Surdam and Stanley, 1979).

The open lacustrine facies of the Laney Member consist of fossiliferous mudstones and kerogenous laminated mudstone (Fig. 5.33). Fossiliferous mudstones are intensely bioturbated and probably were deposited in an oxygenated, freshwater environment. But the laminated mudstones probably accumulated in an anoxic environ-

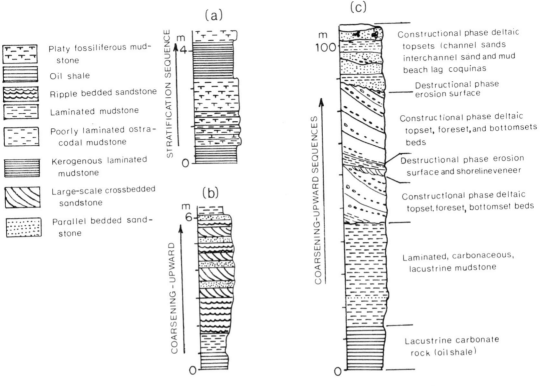

Figure 5.33 Stratigraphic sequences of the Laney Member of the Green River Formation representing sediments deposited in open lacustrine (a), marginal lacustrine (b), and deltaic (c) settings (after Surdam and Stanley, 1979).

ment that may have been somewhat saline and probably received less terrigenous clay.

Marginal lacustrine facies include both shoreline-deltaic and Gilbert-type deltaic sequences. The shoreline-deltaic sequence consists of an upward-coarsening series of kerogenous mudstones, wave-rippled fine sandstones, and fine- to medium-grained sandstone in large-scale crossbeds (Fig. 5.33). Shoreline-deltaic sequences are relatively thin, but where the lake was deeper, Gilbert-type foresets as much as 25 m thick were deposited. These delta sequences consist of a constructional phase, where foresets and topsets prograded over prodeltaic sediments, and a destructional phase, where delta topsets and foresets were reworked and overlain by a veneer of shoreline sandstones (Fig. 5.33) during transgression.

Broad mudflats and swamps bounded the lake along some portions of the shore. Most streams flowed into these flats rather than directly into the lake, and the muds carried by the streams settled out in ephemeral ponds on the flat or were trapped by vegetation. Clays were stopped from entering the lake directly near those portions of the shore, and kerogenous laminated carbonate was deposited in the open lacustrine setting. Repetitive alternations of laminated, kerogenous carbonate and shallow-water carbonate result from alternating high and low lake level stands. Even during the time when Lake Gosiute was evolving into a freshwater lake, climatic fluctuations changed water levels sufficiently for transgressive and regressive phases that were felt broadly over the whole lake.

The Humid System in the Rock Record

The late Cretaceous and early Tertiary rocks of the Uinta Basin, USA, record the changes in sedimentation that occurred during the evolution of a humid system. During the late Cretaceus, sedimentation took place on a coastal plain that was prograding into the Cretaceous epicontinental seaway. The alluvial and transitional marine rocks deposited at this time are formations of the Mesaverde Group, which intertongues complexly with the marine facies of the Mancos Shale (Fig. 5.34) due to intermittent transgressions that interrupted this regressive phase.

The Castlegate Sandstone is a representative unit of this regression (van de

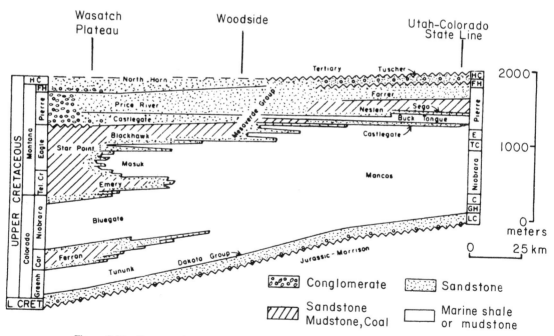

Figure 5.34 Cross section of the Upper Cretaceous rocks of the Uinta Basin showing the relationship of the alluvial fan sediments of the Price River Formation to the alluvial and deltaic facies of the Castlegate Sandstone and the marine Mancos Shale (after van de Graaf, 1972).

Graaf, 1972). To the west, the Castlegate grades into the alluvial fan facies of the Price River Formation, which consist of poorly sorted conglomerates with rare crossbedded sandstone and mudstone lenses and few bedding planes. These conglomerates grade eastward into the fluvial facies of the Castlegate, consisting of trough crossbedded sandstones and abundant conglomerate lenses that were probably deposited in a network of braided streams. The fluvial facies, in turn, grades eastward into distributary sandstones, coal, and marsh mudstones of the delta plain facies of the Castlegate.

This regressive phase was terminated when a large basin of interior drainage was formed on the coastal plain. Initial deposits consisted mainly of alluvial siliciclastic sediments, but along an axis of maximum subsidence, claystones and argillaceous sandstones were deposited in relatively small lakes (Ryder and others, 1976) (Fig. 5.35).

In the center of the basin, a core of open lacustrine carbonate and calcareous claystones with local interbeds of sandstone, siltstone, and carbonate packstone was deposited (Fig. 5.36). Organic debris in these rocks consists of disseminated particulate matter, kerogenous laminae, and algal coal beds. Away from the center of the lake, the open lacustrine facies decreases in carbonate content and organic richness, and increases in terrigenous clastic content.

Deltas and carbonate flats occurred along the margin of the lake. The carbonate flats consisted of fossiliferous and oolitic carbonates that graded lakeward into open lacustrine ostracodal mudstones. These rocks are deposited in cycles 2 to 5 m thick that record shoaling episodes of the lake. They consist of open lacustrine carbonate mudstones at the base grading upward into horizontally bedded and crossbedded carbonate grainstone that is capped by a thin bed of algal-coated ostracod grains.

Delta sequences are 20 to 40 m thick and may extend up to 45 km along the lake margin. They consist of channel-form sandstones that truncate adjacent claystones, siltstones, and fine-grained sandstones. Sand bodies are up to 15 m thick and consist of crossbedded sandstones that fine and decrease in set thickness upward. The sand bodies were probably deposited in distributary channels, and finer-grained sediment probably represents overbank deposition. The delta sequences form cycles with carbonate flat sediments that accumulated when distributary channels became inactive following transgression, channel abandonment, or a reduction in terrigenous sediment supply.

Interdeltaic settings occurred along depositional strike from delta complexes. Claystones and siltstones were deposited in quiet water settings, and channel sand bodies are smaller and fewer than in delta complexes. Influx of terrigenous sediment was relatively small, although it was sufficient to inhibit carbonate deposition.

Updip from marginal lacustrine facies are rocks of the alluvial component consisting of delta plain, high mudflat, and alluvial fan facies. Delta plain sequences also consist of channel-form sand bodies and overbank siltstones, sandstones, and claystones. But sequences are up to 100 m thick and channel sands are thicker and wider than those of the delta front. Fine-grained units of the delta plain accumulated

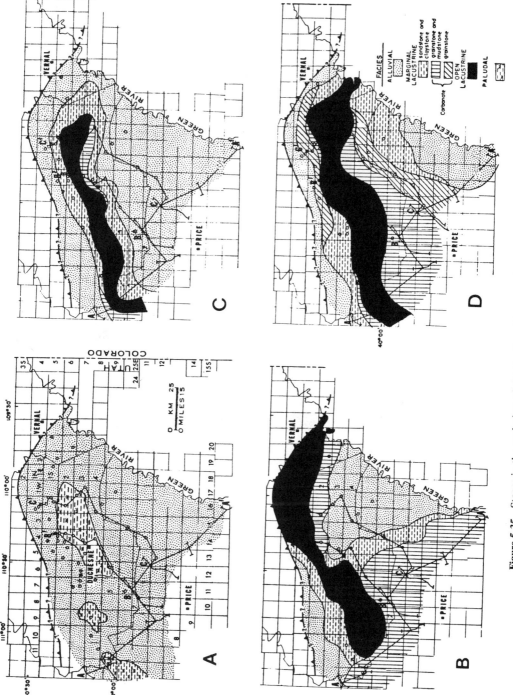

Figure 5.35 Stages in the evolution of the stratigraphic succession of the Uinta Basin during the Late Cretaceous (A), middle Paleocene (B), late Paleocene (C), and early Eocene (D) (after Ryder and others, 1976).

129

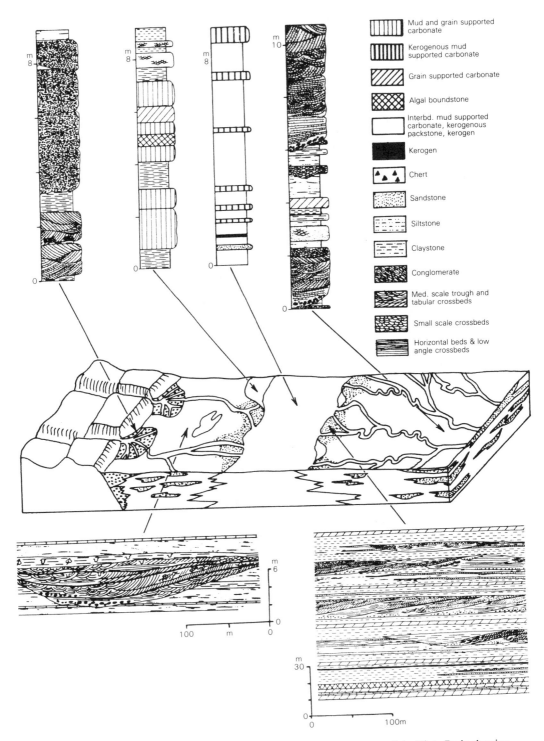

Figure 5.36 Relationships among the humid-system components of the Uinta Basin showing the stratigraphic sequences developed in various parts of the system (after Ryder and others, 1976).

130

as overbank deposits and crevasse splays, and they attained their greatest thickness during major transgressive episodes.

The high mudflat facies is laterally adjacent along strike to the delta plain facies. It contains fewer and smaller channel sand bodies than the latter facies, and its fine-grained sediments show more abundant evidence of subaerial exposure. The high mudflat was traversed by small meandering channels and occurred in a position between zones of influence of the major streams of the delta plain.

Alluvial fans occurred along the west flank of the basin adjacent to the thrust faults of the Sevier orogenic belt. Fan sediments consist mainly of conglomerates in anastamosing channel-form deposits that probably originated as braided streams. Minor sandstones, siltstones, and claystones also occur in the fan deposits, which are also intercalated with carbonates of the lake margin facies.

Lake Uinta was flanked on each side by marginal lacustrine and alluvial plain settings. On the steeper north flank of the basin, alluvial fan sediments were deposited, and alluvial plain and marginal lacustrine facies were confined to relatively narrow bands (Fig. 5.35). On the south side of the basin, however, alluvial and marginal lacustrine facies tracts were broader.

Regressive sedimentation almost filled the basin by late Paleocene, but during the early Eocene the lake expanded greatly in response to increased rainfall or a change in the rate of subsidence (Fig. 5.35). During this major transgression, relatively little terrigenous material was provided to the lake, possibly because the sediment alluviated floodplains in response to changes in base level. With equilibrium restored, however, regression resumed and, as the climate changed, the lake became hypersaline (Fig. 5.35). By the late Eocene, the lacustrine sediment of Lake Uinta were buried by coarse alluvial sediment derived from the Uinta Uplift.

Discussion: Sequence Evolution in Continental Environments

Like systems in other environments, humid and arid systems coexist in spatial juxtaposition. And, like other systems, they can succeed one another during the filling of a basin if climatic changes are pronounced. This is particularly true in the semiarid transition zone between humid and arid systems where even a relatively small change in the climatic regime can produce substantial changes in depositional style. Lakes, for example, may dry and their beds may then be occupied by ergs. Alluvial plains, especially, are liable to such changes because, once dried, they are excellent sources of sand suitable for eolian transport. Mainguet and Chemin (1983), for example, have shown that the largest modern ergs in North Africa derive most of their sediment from alluvial sources that were active during periods of semiaridity in the Pleistocene.

Such changes in depositional regime are also evident in the rock record. During initial stages of accumulation of the Big Bear depositional system in Canada, the depositional basin was occupied by a braid plain (Ross, 1983), but later a change in climate favored formation of an extensive sand sea. During later stages of evolution, however, eolian and fluvial deposition alternated in the basin, and finally the climate changed sufficiently for fluvial deposition to dominate. Mader (1981) also reports a change from a regime dominated by eolian processes in the Middle Buntsandstein (Triassic) of Germany to braided stream deposition in the Upper Buntsandstein.

Some climatic changes may be ascribed to nonrepeatable events, such as changes in the orographic factors that influence climate in an area, or to changes in latitudinal position of a depocenter caused by plate motions. Other changes, however, are cyclic and can be ascribed to variations in the amount of incoming solar radiation (insolation). A substantial body of data suggests that such factors as eccentricity of the earth's orbit, its axial tilt, and the precession of its equinoxes can result in cyclic changes of substantial amplitude in the amount of solar radiation reaching the earth. These cycles can range in periods from tens to hundreds of thousands of years and are reflected in the records of various climate-sensitive chemical, biotic, and sedimentological indexes (see Berger and others, 1984, for an extensive discussion of the mechanics and sedimentological impact of celestial forcing factors).

The rapidity with which climates can change imposes constraints on the way humid and arid systems form sequences. Systems in other environments succeed one another as their component parts migrate laterally and produce diachronous boundaries between deposits of successive systems. Humid and arid systems, on the other hand, can only migrate as climates change, because they are geographically defined by the climatic zones they occupy. Their components, however, are spatially restricted, because their rate of migration is much slower than the potential rate of climatic change. Dunes, for example, move at rates ranging from a few to perhaps 25 cm/year, and a dune field may require as much as 400,000 years to migrate from a playa of even moderate size (Rubin and Hunter, 1984).

Orbitally forced cycles of insolation with periods as short as 20,000 years, however, can induce substantial shifts of climatic zones. Systems succeed one another, under such conditions, by replacement rather than by lateral migration, and the system boundaries thus formed are nearly isochronous.

Because the components of humid and arid systems are spatially persistent, the deposits of lakes, alluvial fans, and alluvial plains are commonly alternately deposited under arid, semiarid, and humid conditions. Large Tertiary lake basins in Utah and Colorado, for instance, persisted even though conditions changed from dominantly subhumid to dominantly arid, and back to humid again. Lacustrine sedimentation prevailed even though the type of sediment deposited in the lake changed radically (Surdam and Stanley, 1979; Eugster and Hardie, 1975). A similar change in lacustrine sedimentation is documented by Hubert and others (1976) in the East Berlin Formation (Triassic) of Connecticut, and Van Houten (1964) and Olsen (1987) report the occurrence of a 40-million year record of climatically forced cyclic expansion and contraction of rift basin lakes in the Newark Basin of New Jersey.

Alluvial plains may persist through climate changes, even though the shapes of their rivers may be altered. During a dry period in New South Wales, Australia, for instance, the Murrumbidgee River maintained a nearly straight channel and carried a sediment load consisting primarily of sand. During more humid periods, however, the river flowed in a sinuous channel and carried a load dominantly of mud (Schumm, 1977).

A similar change in alluvial style can be seen in the Permo-Triassic New Red Sandstone of Scotland (Steel, 1974), where the basal sediments, probably representing braided stream deposition, give way upward to sediments deposited by meandering streams. Channel deposits near the top of the sequence are thicker, but finer grained, than those lower, and the floodplain deposits are also normally finer grained. In addition, the channel deposits are less laterally persistent in the upper part of the sequence, and the variance in paleocurrent directions increases upward. Such trends are consistent with a change upward from braided to meandering stream deposition, and the upward decrease in the occurrence of caliche beds signals a concomitant change to a wetter climate.

When systems shift laterally in other environments in response to changes in stress and succeed one another during basin evolution, the components shift as well. In continental environments, however, climate can change rapidly, and as humid and arid systems succeed one another, the response may be reflected only in a change in the depositional style of the same components.

Part III

COASTAL TRANSITION ZONE

Introduction

The coast is a transition zone between continental and marine depositional environments because it is there that the physical and chemical factors characteristic of each environment mix and are transformed. Waves and tides in the open marine environment are oscillatory movements of water. Waves possess a short period cyclicity measured in seconds, and tides possess a twice-daily cyclicity. The short-period, symmetrical oscillation of waves is transformed in shallow water, however, into one composed of a unidirectional flow that moves up the beach face and a unidirectional flow that returns seaward. When tides enter restricted basins, the normal periodicity may be transformed and the amplitude may be reduced or enhanced. In addition, flow velocities may become unevenly distributed about the cycle.

Flow velocities in rivers slow as their beds approach base level, and as river flow meets marine waters of greater density, the flow is lifted from its bed, effectively stopping bedload transport by the stream. In coastal areas of low wave energy, the buoyant flow may travel for long distances seaward before it gradually loses its identity through slow mixing with marine waters. Most often, however, wave energy is sufficient to cause rapid mixing relatively near the coast.

Mixing may occur inland from the coast when marine waters extend landward in estuaries. Marine waters can extend as an unmodified salt wedge far up some estuaries, producing a steep vertical salinity gradient. In other areas, tides and waves may cause thorough mixing of fresh and marine waters, resulting in sharp lateral salinity gradients. The rapidity with which mixing occurs causes a profound impact on sediment distribution within estuaries.

Sediments are also transformed in the coastal transition zone as the quantity and caliber of the load brought to the coast are altered and the timing of their delivery is changed. River beds are aggraded with coarse sediment as stream gradients flatten near the coast. Coarse bedload is also deposited rapidly at the point where stream flow is lifted off the river bed by the salt wedge. Even the suspended load is changed as clay flocculates in the mixing zone and settles out of the flow. Fine sediments are also trapped in lagoons and tidal flats, leaving only a small portion of the sediment to be delivered to the shelf. When rivers are in flood stage, however, channels are scoured, salt wedges are destroyed, and plumes of sediment can be sent far out onto the shelf.

Several systems can be defined within the coastal transition zone. Deltas are depositional systems that form irregular extensions of the coastline in front of rivers that provide most of the sediment to the delta. Nondeltaic depositional shorelines include barrier islands and lagoons, strandplains, and tidal flats. Shorelines of coasts dominated by such systems tend to be smooth and regular and the systems prograde seaward uniformly. Like deltas, they may also be supplied directly by rivers, but most of their sediment is provided by longshore drift, onshore transport, and erosion of preexisting coastal elements. Estuaries are ubiquitous elements of all these systems, and they form anywhere marine waters mix with fresh water in reentrants along the coast.

The shape of these transitional systems is a function of a complex interaction between the amount of sediment delivered to the system and the nature and intensity of the basinal processes acting on it. Galloway (1975) proposed a classification of deltas based on the response of the morphology to variations in the relative intensity of tide and wave action on incoming sediment. Hayes (1979) and R. A. Davis and Hayes (1984) also proposed a classification that relates tidal range to the morphology of nondeltaic coastlines. Tides on microtidal coastlines do not exceed 2 m in amplitude and are characterized by strandplain and barrier island systems. Tidal flats and their attendant components are poorly developed, and tidal inlets are widely spaced. Inlets are more closely spaced, and tidal deltas are better developed on mesotidal coastlines, where tides range in amplitude from 2 to 4 m. Tidal flats are well-developed, especially in back-barrier settings where areally they may become the dominant component along the coast. Tidal amplitudes are in excess of 4 m on macrotidal coasts. Strandplains and barriers are only irregularly developed and tidal flats dominate. River mouths act as sediment traps, where sand is remolded by intense currents into tidal shoals and ridges. These zones originally were defined on the basis of tidal amplitude. The degree to which any coast reflects this classification, however, is dependent on the interaction of wave energy as well as tidal amplitude, and no fixed limits can be set (R. A. Davis and Hayes, 1984).

Study of stratigraphic sequences deposited in the coastal transition zone is difficult but important. All systems in the zone are interactive either laterally along the coast or vertically through time. Deltas, strandplains, and barrier island/lagoon systems can exist in close proximity, and prograding deltas can be transformed into barriers with lagoons after abandonment. In addition, each system may use any other of the systems as components. Thus barrier islands may enclose lagoons with tidal flats and deltas, and large estuaries may enclose barrier/lagoon systems and tidal flats as well as deltas.

The coastal zone is a sensitive indicator of environmental conditions. Small changes in sediment yield from source areas can cause dramatic changes in the pattern of coastal sedimentation, and slight alterations in any of the hydraulic factors along the coast, including changing sea level, can also have a profound impact. Thus, an understanding of how coastal stratigraphic sequences accumulate can offer significant insight into the processes operating not only along the coast, but in source areas and marine basins as well.

The various systems of the coastal transition zone are considered in Part III. They are presented in order of their complexity and the degree to which they include elements of the other systems.

6

Tide-Dominated Coastlines

Sedimentary Processes and Products

Tides are caused by the rising and falling of water level under the gravitational attraction of the moon and sun. When the moon and sun are in direct alignment, higher-than-normal spring tides occur, and when they are in transverse alignment, lower-than-normal neap tides occur. The tide is in flood when water level is rising and ebb when the level is falling. The tide is slack at the end of each flood and ebb period when the water level is at a stillstand.

The amplitude of the tidal bulge as it moves is fixed by the gravitational constants of the earth, moon, and sun. The actual amplitude along a particular coast, however, is controlled mainly by the geometry of the basin and its alignment with respect to the incoming tidal bulge. Thus, tides along some coasts are negligible, whereas on others the amplitude may exceed 15 m.

Tide-dominated coastlines occur in areas with moderate to high tidal (meso- to macrotidal) range and relatively low wave energy (Fig. 6.1). Such conditions may occur along coasts where wave action is dampened by broad, shallow shorefaces, but most commonly they occur in protected areas such as estuaries, lagoons, and delta plains of tide-dominated deltas (Reineck and Singh, 1980).

Tide-dominated coasts may be divided into three coast-parallel zones based on frequency of tidal inundation. The supratidal zone, often occupied by a salt marsh, is only inundated during spring tide. The intertidal zone or tidal flat is submerged for at least a short period during every tidal cycle, and the subtidal zone is always submerged (Fig. 6.2).

The tidal flat may, in turn, be divided into three zones based on the duration

138

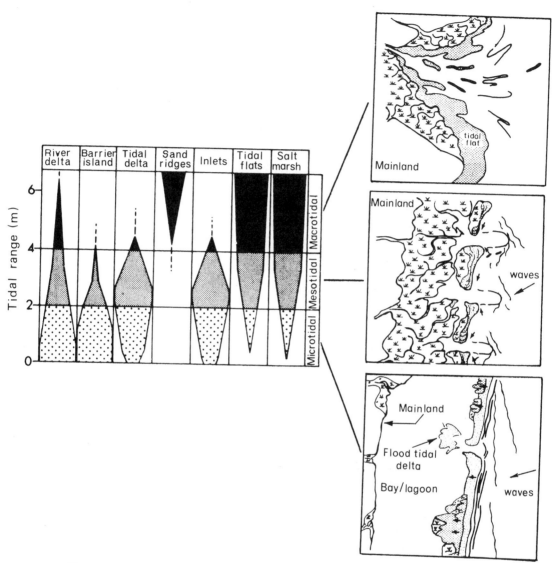

Figure 6.1 Variations in coastal morphology as a function of tidal amplitude and wave-energy flux (after Hayes, 1976).

of the flooding. The high tidal flat is submerged only during flood peak. The mid tidal flat is submerged during half of the period of the tidal cycle, and the low tidal flat is inundated for most of the cycle (Fig. 6.2).

Except along coasts with appreciable wave activity, tidal flats become finer-grained landward from the low water mark. Low tidal flats are submerged longest and are subject to wave action and traction transport by tidal currents for longer

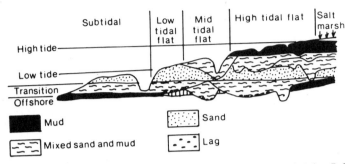

Figure 6.2 Sediment zones in a tidal flat as a function of tide levels (after Reineck and Singh, 1980).

periods of time. Also, the distribution of current velocities is asymmetrical during a tidal cycle, so velocities sufficiently low for muds to settle out of suspension occur for longer periods of time during flood than during ebb. Augmenting this process is the tendency of fine particles to move landward farther during flood than they move seaward during ebb because of settling lag and scour lag (Postma, 1967).

The supratidal zone lies above the mean high tide and is therefore only inundated during spring tide. In temperate regions, the zone is colonized by salt-tolerant grasses that are effective sediment baffles. Sediments consist mainly of laminated muds, but primary structures are commonly disrupted by roots and burrowing organisms. Macrophytes are scarce in arid climates, but the surface of the supratidal zone may be covered by blue-green algae. Although bioturbation is rare, primary structures are disrupted by the formation of mud cracks and the growth of salt crystals (Thompson, 1975).

Because high tidal flats are covered only for a short period at peak flood when current velocities are negligible, suspension deposition predominates and muds are concentrated. Normally, sediments consist of interbedded silt and clay, but on open coasts that experience appreciable wave energy, the more seaward parts of the high tidal flat may be covered by ripple-bedded sand and silt (Evans, 1975). Usually only remnants of primary sedimentary structures are preserved in the high tidal flat because of pervasive disruption by burrowing organisms.

The mid-tidal flat is submerged for about half the tidal cycle and therefore experiences both bedload and suspension transport and deposition. Alternation of sedimentary processes is reflected in the sediments, which consist of interbedded sand and mud in lenticular, flaser, and wavy beds (Fig. 6.3). Numerous features indicative of emergence occur in the mid-tidal flat. In muddy sediments, these consist of mud cracks and wrinkle marks, and in sandy sediments, flat-topped ripples, interference ripples, and rill marks occur. Primary sedimentary structures, however, occur mainly as remnants because of infaunal activity.

The orientation of bedforms on the tidal flat reflects the complex interaction of wave- and tide-induced currents, which can be highly varied in different parts of

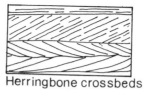

Herringbone crossbeds

Flaser beds (top) and
wavy beds (bottom)

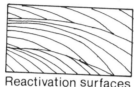

Reactivation surfaces

Figure 6.3 Primary sedimentary
structures commonly found on tidal
flats (after Reineck and Singh, 1980).

the tidal flat. In addition, currents generated just prior to emergence are controlled
largely by local topography, adding a further complicating factor.

Tidal creeks head in the high tidal flat, where they begin to collect the return
flow during ebb tide. Flow is concentrated in the creeks of the mid-tidal flat, where
channels are more fully developed. Current velocities in the channels are two to
three times greater than on adjacent mudflats, so sediments in the creeks tend to be
sandier and occur in larger-scale bedforms (Reineck, 1963).

Currents are somewhat simplified in tidal channels. Bipolar dip directions may
occur on foresets of megaripples, reflecting the occurrence of a balanced reversing
flow, but more commonly foresets dip predominantly in one direction because the
flow is either flood or ebb dominant. Reversing flow may only be suggested by the
occurrence of reactivation surfaces on sets of crossbeds (Fig. 6.3).

Depending on channel spacing and sediment type, tidal channels can migrate
in excess of over 100 m/year, and it has been demonstrated that over half the area
of a tidal flat can be reworked in less than 70 years (Bridges and Leeder, 1976;
Reineck, 1958), replacing sediments deposited on the flat with those deposited in
channels. Lags of mud clasts, shell debris, and wood collect at the base of the channels as they migrate. The lags are overlain by sandy channel deposits and lateral
accretionary deposits consisting of interbedded sand and mud (Fig. 6.4). The degree
of bioturbation is small near the base of the channel, but biogenic reworking increases up the margins of the channel where the substrate is subjected to bed shear
for shorter periods of time.

The low tidal flat is subject to bed shear throughout much of the tidal cycle.
Bedload transport is dominant and sediments consist mainly of sand in large-scale

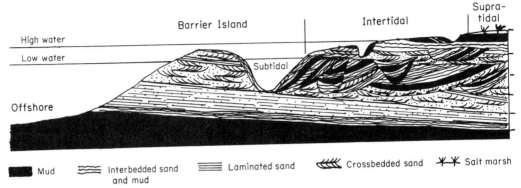

Figure 6.4 Sequence of sediments left by a laterally migrating tidal creek (after Reineck and Singh, 1980).

bedforms. Orientation of slipfaces may be bipolar, but normally either ebb or flood currents dominate, and slipfaces show a unimodal orientation with reactivation surfaces separating sets.

In the low tidal flat, tidal creeks coalesce into tidal channels. These channels can erode below mean low water, giving rise to channel bottoms that are not exposed at low tide. Tidal channels, like tidal creeks, migrate freely and deposit a characteristic sequence of sediments. Sand waves and dunes migrate over lags at the base of the channel depositing large-scale cross sets. As channels aggrade, finer-grained sand in smaller-scale bedsets is deposited, and the sequence is capped by tidal flat sediments deposited on the aggraded channel. Bioturbation is minimal in these deposits, except near the top of the sequence, because significant bed shear in the channel occurs through much of the tidal cycle.

Tidal flats grade seaward into muddy offshore zones along some coasts. In the Gulf of California, for example, large amounts of mud are provided to the tidal flats by the Colorado River. Wave energy is insufficient to concentrate sand to any great degree, and the tidal flats grade quickly offshore into muddy offshore sediments. Bioturbation is rare, apparently owing to the rapid rate of sedimentation (Thompson, 1968).

Along sand-dominated coasts, however, a sandy subtidal zone consisting of a shifting complex of sand bodies and channels may extend seaward from intertidal zones and mouths of estuaries. In the German North Sea, for instance, sand bars and channels occur in front of the Jade River, and sand tongues and channels extend from the tidal flats between the Weser and Elbe Rivers (Reineck, 1963) (Fig. 6.5).

The sand bodies of the Outer Jade are elongate parallel to current direction and consist of sand in plane beds and small- and large-scale crossbeds. Gravels occur at the base of the channels between the sand bodies, and medium to coarse sand occurs in the giant ripples that migrate in the channels. Crests of the giant ripples are oriented perpendicular to current direction, and the ripples themselves consist of amalgamated sets of smaller-scale crossbeds. Although the giant sets respond

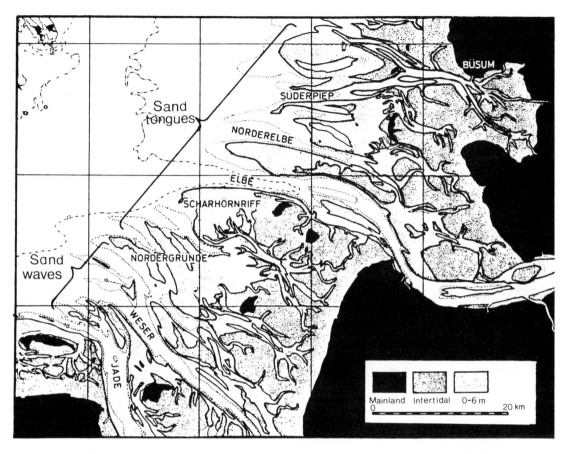

Figure 6.5 Sandwaves, channels, and shoals along the German North Sea coast (after Reineck and Singh, 1980).

only to the dominant flow, the smaller-scale bedforms superimposed on them commonly respond to both and produce bipolar dips on foresets (Fig. 6.6).

Sand tongues are extensions of tidal flats into the subtidal zone (Dorjes and others, 1970). They are as much as 18 km long and consist of sand in plane beds and small- and large-scale crossbeds. Bioturbation is rare above wave base, but is common below it and increases in abundance into deeper water (Fig. 6.6). Tongues are separated by channels that head in the intertidal zone. The bottoms of these channels are below wave base, and even though tidal current velocities are somewhat greater in the channels than in the shoals, the sediments are finer-grained because they are protected from wave activity. No clear-cut differentiation exists between the bedding styles of shoals and channels, although some systematic variation occurs according to position relative to wave base, with large-scale crossbedding more common in areas protected from wave action (Fig. 6.6) (Dorjes and others, 1970).

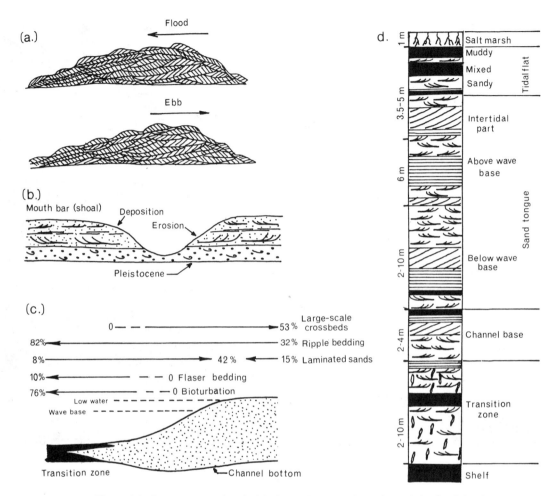

Figure 6.6 Structures associated with the sand waves, channels, and shoals of the German North Sea coast: (a) reversing megaripples superimposed on a sandwave dominated by ebb tidal flow (after Reineck, 1963); (b) sediments associated with a tidal channel of the outer Jade River (after Reineck and Singh, 1980); (c) distribution of primary structures in a sand tongue of the Nordergrunde (after Dorjes and others, 1970, in Reineck and Singh, 1980); (d) hypothetical vertical sequence developed by coastal progradation in the Nordergrunde area (after Dorjes and others, 1970; Reineck and Singh, 1980).

Intrinsic and Extrinsic Controls on Evolution of Tidal Flat Sequences

Historic evidence and modern surveys indicate that tidal flats are prograding seaward in areas with large sediment supply (Evans, 1975) or in protected areas that act as sediment traps. In most cases, a fining-upward sequence is produced, consist-

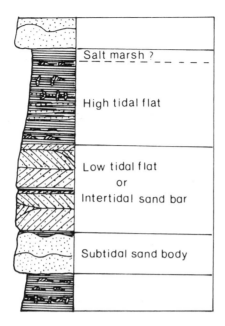

Salt marsh ?

High tidal flat

Low tidal flat
or
Intertidal sand bar

Subtidal sand body

Figure 6.7 Typical sequence produced by a prograding tidal flat (after Klein, 1970).

ing of subtidal sands at the base overlain by bioturbated muds and sands of the tidal flat and capped by supratidal muds (Fig. 6.7) (Klein, 1970; Reineck, 1972). On tidal flats where tidal creeks are closely spaced and the substrate is somewhat sandy, this sequence may be replaced by deposits of migrating tidal creeks and channels consisting of lag gravels overlain by crossbedded channel sands and lateral accretion deposits of interlaminated sand and mud (Fig. 6.4).

A hypothetical retrogradational sequence might consist of the reverse of a progradational sequence, with coarsening-upward tidal flat sediments resting on marsh muds and peat deposited over older sediments or rock during the early stages of transgression (Reineck and Singh, 1980).

In reality, however, a coarsening-upward retrogradational sequence is rarely achieved. At Mont-Saint Michel, for instance, fining-upward regressive tidal-flat sequences overlie thin marine deposits consisting of shells and sand deposited over fluvial sands and gravels during the Flandrian transgression (Larsonneur, 1975) (Fig. 6.8).

Tidal flats behind barrier islands also often consist primarily of regressive deposits. Along the Delmarva Peninsula on the east coast of the United States, the basal deposits of the regressive sequence consist of lagoonal muds on the landward side of the bay and sandy muds on the barrier side (Fig. 6.9). The regressive deposits overlie thin deposits of peat or bioturbated mud that were deposited over bedrock or older sediments early in the transgressive period (Harrison, 1975).

Regressive tidal sequences may be interrupted in their development by retrogradational episodes caused by sea-level fluctuations or reduction in sediment yield. The typical regressive sequence of tidal flat sediments in the Gulf of California was

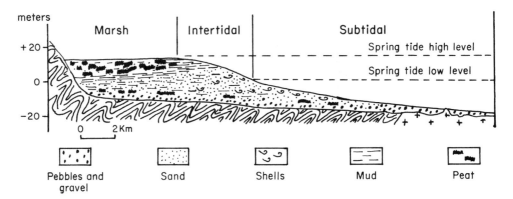

Figure 6.8   Tidal-flat sequence produced by progradation of the coast in the Mont St. Michel area (after Larsonneur, 1975).

deposited over coastal alluvial fan sediments during the latter stages of the post-Pleistocene sea-level rise. They consist of laminated subtidal silts and clays overlain by bioturbated muds of the intertidal zone and capped by silty supratidal muds (Thompson, 1975).

During periods of reduced sediment yield from the Colorado River, however, waves are given sufficient time to rework the mudflat sediments and winnow fine-grained sediment. During these periods, sand is concentrated in low berms and spread as thin veneers over the lower intertidal zone (Fig. 6.10). Periods of large and small sediment yield alternately resulted in episodes of progradation and retrogradation that are marked by zones of beach ridges at the surface and interbedded sands and muds in the subsurface.

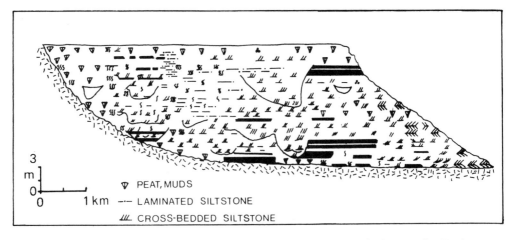

Figure 6.9   Tidal flat sequence produced by infilling of a back-barrier lagoon (after Harrison, 1975).

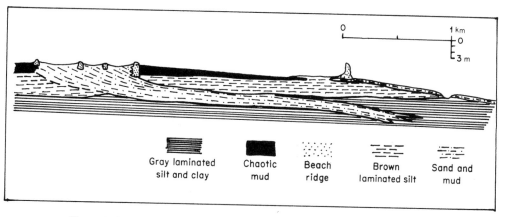

Gray laminated   Chaotic   Beach   Brown   Sand and
silt and clay   mud   ridge   laminated silt   mud

Figure 6.10 Stratigraphic sequence formed by alternating periods of tidal-flat progradation and beach ridge (chenier) formation during shoreline recession (after Thompson, 1975).

Tidal Flat Deposits in the Rock Record

The Dakota Group of Colorado is one of numerous regressive sediment bodies of the Cretaceous sequence in the Rocky Mountains. Near Denver, it consists primarily of deltaic sediments, but among the associated facies are nearshore marine and tidal flat deposits (MacKenzie, 1972, 1975) (Fig. 6.11).

The tidal flat sediments occur in an overall fining-upward sequence consisting of fine-grained crossbedded sandstones in the lower part and very fine grained sandstones in thin tabular cross sets at the top. Less than 10% of the section consists of mudstones, and these are restricted to the top of the section and to mud-filled channels that occur in the upper part of the section (Fig. 6.11).

The widespread occurrence of wave-formed ripples suggests relatively shallow water depths, and periodic emergence of the substrate is indicated by the presence of dinosaur tracks, mud cracks, and rooted horizons. The presence of interference ripples, alternating sets of small- and large-scale cross sets, and the occurrence of fine carbonaceous material on foresets are evidence of episodic flow, but reactivation surfaces are not recognized and herringbone crossbeds are rare. Mud-filled and sand-filled channels occur in the upper part of the sequence, and there are abundant trace fossils within the unit, including Planolites, Skolithos, and Trichichnus (Weimer and others, 1982).

These features taken together suggest a marginal marine setting where water level underwent short-period fluctuations and current directions were variable. The unit may have been deposited on a tidal flat that formed during the initial stages of transgression over an erosional surface.

Campbell and Oaks (1973) described possible tidal flat and channel facies in the lower Cretaceous Fall River Formation of Wyoming that closely resemble the unprotected tidal flats of the German North Sea. Tidal flat deposits consist of muds

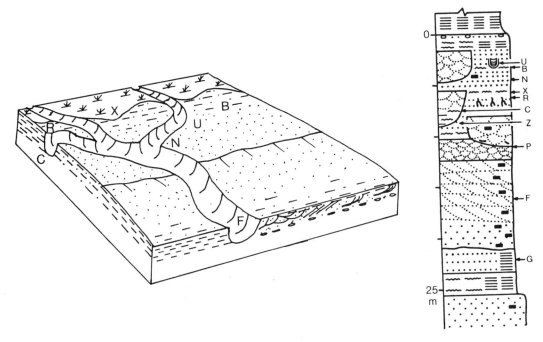

Figure 6.11 Tidal-flat deposits of the Colorado Group near Denver showing the vertical sequence of sediments and their proposed depositional setting. Letters on the vertical column are keyed to areas on the tidal flat where they may have originated (after MacKenzie, 1972).

and sand in thin wavy beds. Sands occur in flat-topped ripples, flaser beds, or ripples completely isolated in mud. Mudstones also occur in wavy parallel laminae, and both rock types are bioturbated. Tracks and trails are evident on bedding plane surfaces, and ripples have diverse orientations (Fig. 6.12).

Tidal flat sediments are overlain in places by marsh deposits consisting of carbonaceous mudstones. Faunal bioturbation is not common, but the beds are thoroughly disturbed by roots that penetrate into the underlying tidal flat sediments.

The tidal flat facies is cut by channels of two different scales. Small tidal channels are up to 3.3 m thick and 100 m wide. They occur as flat-topped lenses that are commonly filled with crossbedded sandstones. In some cases, however, they are filled with alternating beds of sandstone and mudstone similar to the deposits of the tidal flats. Lag gravels may occur at the base of the channels, and in some cases marsh deposits overlie the channel fills (Fig. 6.12).

Large-scale channels are up to 30 m thick and 1.5 km long. Fluvial sandstones occur in the most landward exposures of these channels, but channel deposits grade downdip into sandstones in large-scale planar crossbeds and then into mudstone interbedded with wave-rippled sandstone. Structures associated with large-scale crossbeds include overturned crossbeds and rare vertical burrows. Those associated with wave-rippled sandstones include flaser beds, current ripples, and abundant bur-

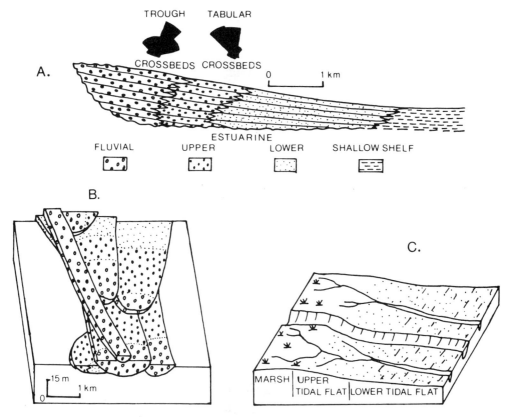

Figure 6.12   Tidal-flat and channel sediments of the Fall River Group in Wyoming showing (A) distribution of sediments, (B) spatial relationships within a complex of tidal channel sandstones, and (C) proposed depositional setting (after Campbell and Oaks, 1973).

rows. These channels form seaward-imbricated complexes in which each younger body fills scours cut into older ones in a manner that might occur where shifting channels occur seaward of tidal flats (Fig. 6.12).

The Bluejacket Sandstone of eastern Oklahoma was deposited in a deltaic complex that included marine, upper and lower deltaic plain, and interdeltaic components (Visher, 1968) (Fig. 6.13). During deposition of the sandstone, prograding delta lobes produced a sequence of prodelta sediments overlain by deltaic, marginal marine, and shoreline deposits (Visher, 1975). On the southern margin of the main delta complex, marginal marine units were deposited in an intermediate position between distributary and interdistributary units and dominantly prodelta units, and within these marginal marine strata are some tidal-flat deposits.

The vertical sequence within the tidal flat sediments consists of sandstone at the base and interbedded sandstone and mudstone at the top (Fig. 6.13). The sand unit at the base consists of large-scale cross sets separated by scour surfaces. Thin

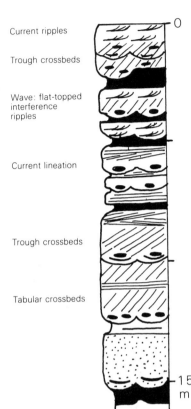

Current ripples

Trough crossbeds

Wave: flat-topped
interference
ripples

Current lineation

Trough crossbeds

Tabular crossbeds

0

15
m

Figure 6.13 Tidal-flat deposits
associated with deltaic sediments of the
Bluejacket Sandstone of eastern
Oklahoma (after Visher, 1975).

shales within the unit consist of interlaminated mud, fine sand, and silt. Scale of
bedding units decreases upward, and the sands at the top of the lower unit occur in
beds of interference ripples and flat-topped ripples with tracks and trails on bedding
surfaces. Mud cracks occur on interbedded shales in places. This basal unit was
probably deposited in a tidal channel that was filled with sand and overlain by tidal
flat and marsh sediments (Fig. 6.13).

The upper unit of the sequence consists of sands in large-scale cross sets at the
base overlain by laminated and ripple-bedded sand. It may have been deposited in
a small channel cut into the surface of the marsh deposits that cap the underlying
unit.

7

Muddy Coastlines

Sedimentary Processes and Products

Muddy shorelines develop along coasts where an abundance of fine-grained sediment is available and wave activity is slight. Because a delta would form otherwise under such conditions, their primary source of sediment cannot be directly from a river, but they do typically occur immediately downdrift from deltas prograding onto wide, low-gradient shelves. Muddy shorelines occur in the Firth of Thames in New Zealand, the Gulf of California near the Colorado River delta, and Broad Sound in Australia, which are all relatively enclosed bodies of water (Otvos and Price, 1979). Two of the best-known examples, however, occur on open coasts. These are the Louisiana–Texas coast west of the Mississippi delta and the Surinam Coast west of the Amazon delta.

Suspended sediment concentrations along muddy shorelines may exceed 5000 mg/l (Augustinus, 1980). At such concentrations, the sediment may actually form a gel that acts to dampen wave motion and further reduce the effects of wave activity on the shoreline. Some mud may settle out of suspension as individual particles, but at very large concentrations, when gels form, the sediment may settle out as a mass forming a supersaturated agglomerate that immediately begins to expel pore water (Wells and Coleman, 1981).

Some mud is not carried in suspension along the shore at all, but instead is transported in vast muddy shoals attached to the shoreline (Augustinus, 1980). They resemble sand waves in morphology, and they even migrate in a similar fashion, with simultaneous erosion of the stoss side and deposition on the steep lee side.

At the shoreline, silt is the dominant constituent, with lesser amounts of clay and minor amounts of sand and organic material. The sediments are massive to finely laminated, and where marshes are established above mean high water, the sediments are commonly root mottled. Further offshore, these sediments grade into shelf muds, which consist of silty clay with only traces of sand.

The sand that is present along muddy shorelines occurs in thin laminae or in packed burrows in the nearshore zone, but most is concentrated in low, very long ridges called cheniers. Normally only a few hundred meters wide, but up to 50 km long, they may form only a small part of a muddy coastal plain. Their presence, however, attests to the cyclic nature of deposition along such coasts.

Mud is deposited at the shoreline on tidal flats and marshes during periods of active progradation, but during periods of reduced sediment influx, waves rework the nearshore sediments and concentrate any available sand-sized material into low ridges (Hoyt, 1967) (Fig. 7.1). Some cheniers may also be built by a normal process of alongshore transport of sand at the high-tide level (Augustinus, 1980), and such ridges would be the counterparts of berms on sandy beaches (Fig. 7.2).

Ridges generally are only a few meters high. They consist of sand and shell debris and intercalated muds in laminae that may dip both landward and seaward (Fig. 7.3). Seaward margins are smooth and arcuate, but landward margins are irregular where washover fans have prograded into back-ridge marshes. Ridges curve sharply and fan out at river mouths, much like the downdrift ends of spits (Fig. 7.3).

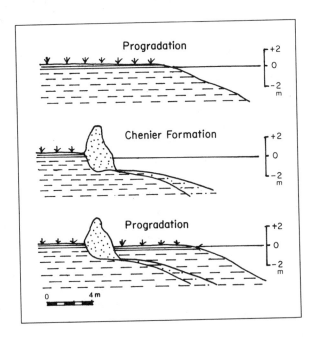

Figure 7.1 Sequence of events occurring during formation of cheniers (after Hoyt, 1967).

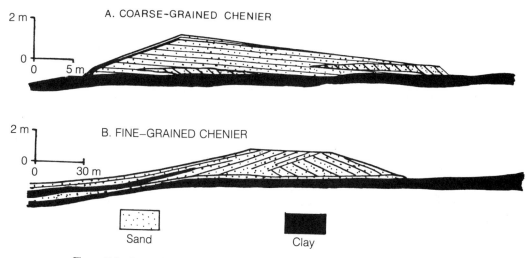

Figure 7.2 Internal structures of cheniers along the Surinam coast showing those built by reworking of tidal-flat muds (A), and those built by longshore transport of sand (B) (after Augustinus, 1980).

Intrinsic and Extrinsic Controls on Muddy Shoreline Sequences

A simple sequence deposited by a prograding muddy shoreline might consist of off-shore silty clays at the base overlain by tidal flat and nearshore clayey silts that are, in turn, overlain by organic muds deposited in marshes. A retrogradational sequence, on the other hand, would consist of a basal unit of marsh and swamp deposits that is overlain by offshore muds.

The characteristic presence of cheniers along muddy shorelines, however, indicates that simple sequence probably are not common. Instead, minor progradations and retrogradations caused by variations in sediment yield, superimposed on major eustatic sea-level changes, combine to produce complex internal architecture.

The chenier plain of the Texas–Louisiana coast, for example, consists of basal sediments deposited during a mid-Holocene transgressive phase and an upper sequence of sediments deposited during a regressive phase initiated about 2800 ybp (Gould and McFarlan, 1959) (Fig. 7.3).

Basal sediments of the chenier plain were laid down during the mid-Holocene transgression. They consist of organic clay and peat deposited in estuaries, shallow bays, and marshes that developed on the irregular surface of the transgressed landscape. Sediments predominantly are muddy, but sand layers a few centimeters thick occur within these deposits, and a bed of sand 30 cm thick occurs at the strandline, marking the furthest extent of the transgression. Gulf bottom silts and clays were deposited within about 1 km of the shoreline before the regressive phase began.

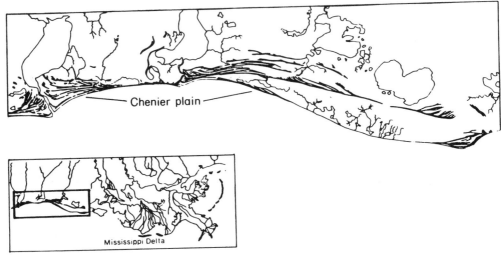

Chenier plain

Mississippi Delta

Location map

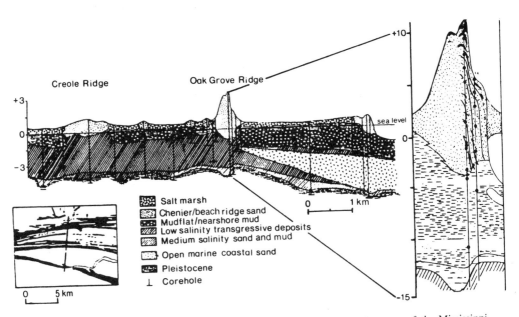

Creole Ridge   Oak Grove Ridge

sea level

- Salt marsh
- Chenier/beach ridge sand
- Mudflat/nearshore mud
- Low salinity transgressive deposits
- Medium salinity sand and mud
- Open marine coastal sand
- Pleistocene
- ⊥ Corehole

Figure 7.3 Cheniers of southeast Louisiana showing their location west of the Mississippi delta, the stratigraphic sequence developed during progradation of the chenier plain, and details of the sediment distribution within a chenier (after Byrne and others, 1959, in Otvos and Price, 1979).

The first "cheniers" were actually spits and barriers that formed relatively far from the shore. They enclosed a shallow, brackish-water embayment that was filled initially by clayey silts and later by organic marsh muds. These sediments were deposited over gulf bottom muds deposited during the culminating phase of transgression.

After the first cheniers were established, the shoreline began to prograde rapidly into the Gulf of Mexico. Tidal-flat and shallow-water sediments were deposited near the shoreline in front of each previously formed chenier, and bottom muds were deposited in deeper water (Fig. 7.3). At times of reduced sediment influx, however, wave attack was able to slow or halt shoreline advance and to build beach ridges or cheniers. During chenier construction, a distinctive suite of foreshore sands and muds was deposited in association with each ridge, and as each ridge grew, a salt marsh formed in the enclosed area shoreward of the chenier. As the shoreline prograded seaward, however, these gradually evolved into freshwater marshes.

Muddy Coastline Sequences in the Rock Record

Few rock units have been interpreted to be the deposits of muddy shorelines, probably because of lack of recognition rather than actual rarity. Few papers have been written about modern occurrences, and their poor potential as oil reservoirs has greatly limited the search for them. In addition, except for their lack of tidal channels, they are not greatly dissimilar from tidal-flat deposits, and only the presence of cheniers may be thought to be diagnostic.

Byrne and others (1959) advanced several rock units as possible muddy shorelines with cheniers. The "cheniers" are linear with smooth seaward margins and irregular landward margins, display biconvex cross sections, and possess stratigraphic relationships typical of cheniers. They are somewhat thick for cheniers, however, and only their biconvex cross-sectional shape distinguishes them from barrier islands. It may be that they are counterparts to the first cheniers of the Louisiana–Texas chenier plain that were built during the initial stages of shoreline progradation.

The Irish Valley Member of the Catskill Delta of central Pennsylvania is one possible example of a prograding muddy coastline without cheniers (Walker and Harms, 1971). The member consists of about 25 repeating sequences that represent periods of shoreline progradations interrupted by rapid transgression (Fig. 7.4).

Each sequence rests on a sharp basal contact that is burrowed and irregular. The transgressional phase of each sequence is represented by a thin, thoroughly bioturbated sand that is probably the erosional remnant of a retrogradational beach. Initial deposits of the progradational phase consist of fossiliferous green shales that grade upward into green siltstones and mudstones with very thin ripple-crossbedded sand layers. The upper part of the sequence was deposited during an emergent phase when red siltstones and mudstones with ripple-bedded sandstones were deposited.

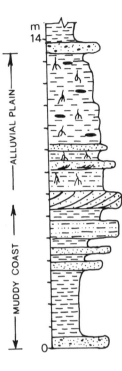

Figure 7.4 Typical sequence in the Irish Valley Member of the Catskill Formation showing sediments deposited by an aggrading muddy coastline overlain by sediments of an aggrading alluvial plain (after Walker and Harms, 1971).

They contain root traces, desiccation cracks, and calcareous nodules indicative of subaerial exposure in a relatively arid climate.

Major sand bodies are absent within the member, suggesting either that sediment influx was relatively uniform or that wave action was too weak to rework the shoreline into cheniers during periods of lesser sediment influx. In addition, the absence of channel sands indicates that tidal range was low on this coast.

8

Wave-Dominated Coastlines

Sedimentary Processes and Products

Wave-dominated coastlines are those where wave action is strong enough to cause significant reworking of incoming sediment but is insufficient to remove it completely, and where the effects of tides are masked by wave energy (Fig. 6.1). This classification eliminates rocky coasts that accumulate little or no sediment and tide-dominated coasts where wave energy may be considerable but where its effects are dissipated on the intertidal terrace during the tidal cycle. Wave-dominated strandlines by this usage include coasts with barrier islands, strandplains, and wave-dominated deltas, which are discussed elsewhere (Fig. 8.1). In all cases, wave energy dominates the coastline at least part of the time (Heward, 1981).

The nearshore can be divided into three shore-parallel zones with respect to depth and distance from shore (Swift and Niederoda, 1985). The innermost of these is the surf zone, where sedimentary processes are dominated by breaking waves.

Waves generated in deep water are movements of water manifested by an oscillatory rise and fall in the surface of the water. Water particles in a deep-water wave follow orbital paths whose diameter is dependent on the wavelength. The diameter of the orbits decreases exponentially with depth, so at a depth of about one-half wavelength only about 5% of the wave's surface energy is present.

As deep-water waves enter shallower water, frictional drag on the wave transforms circular orbits to elliptical ones. Orbits are no longer closed, and momentum is transferred to the substrate. Wave orbital velocities become asymmetric, so greater bed shear is produced on the forward wave stroke than on the reverse. As frictional drag increases, the onshore movement of wave base is slowed relative to

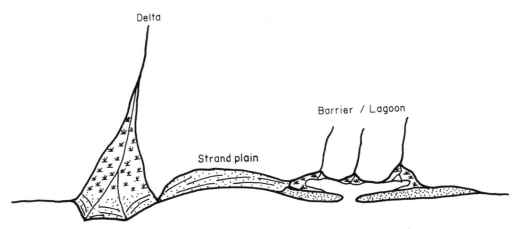

Figure 8.1 Types of depositional landforms on wave-dominated coasts.

the wave velocity at the surface, and the wave builds and becomes steepened eventually to the point where it becomes unstable and breaks. The wave at this point is transformed into a shoreward surge of water. Orbital velocity asymmetry increases into shallower water, so asymmetry is particularly pronounced in the surf zone. Bed shear is almost always sufficient to transport even coarse sand grades and the dominant transport direction is onshore. Most waves approach the shore obliquely, however, and they generate longshore currents that can transport sediment parallel to shore (Fig. 8.2). Return flows are concentrated locally into rip currents capable of transporting sediment away from shore, especially in downwelling situations. The offshore extent of the surf zone depends on the wave climate, but it normally does not extend into depths greater than 1 or 2 m.

   Seaward of the surf zone, the upper shoreface also is dominated by onshore transport of sediment under the influence of orbital velocity asymmetry, but sediment transport is one to two orders of magnitude weaker than that in the surf zone (Niederoda and others, 1984). Some sediment may be transported into the upper shoreface from offshore by shoaling waves, but for the most part, sediment is supplied by fallout from rip currents generated in the surf zone (Cook and Gorsline, 1972). The upper shoreface normally extends seaward to depths of about 5 m, and with the surf zone forms a closed system that continuously circulates sand (Niederoda and others, 1984).

   Wave orbital velocities are nearly symmetrical in the lower shoreface, but interaction of waves with steady or slowly varying wind-forced currents are sufficient to cause a landward creep of sediments (Swift and others, 1985). Coastwise-moving currents tend to reinforce the shoreward vector of the forward stroke of wave-orbital motion, but they partially cancel the return stroke, resulting in a net shoreward vector of bed shear and sediment transport. The lower shoreface may be thought of as a transition zone existing between an inshore zone dominated by fric-

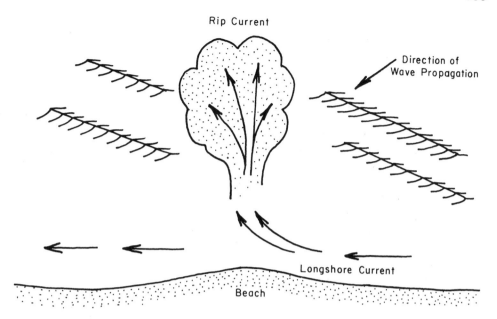

Figure 8.2   Circulation along a coast where waves approach the shore obliquely (after Ingle, 1966).

tional interaction of surface and bottom boundary layers (surf zone and upper shoreface) and one where a geostrophic core flow can occur (inner shelf).

At the shoreward limit of the nearshore zones on wave-dominated coasts is the swash zone. In the swash zone a shallow, fast flow runs up the beach, where the final stages of onshore momentum transfer occur. After a brief stillstand at the limit of wave run-up, flow moves down the beach and returns to the shoreface as a considerably weakened bottom flow under the onshore directed surge (Fig. 8.3).

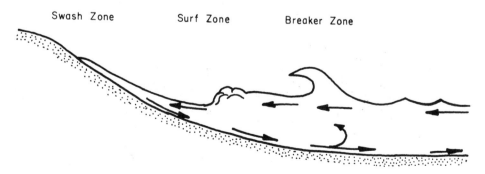

Figure 8.3   Shoreward change in the shape of a wave as it enters progressively shallower water (after Ingle, 1966).

Sediments of the upper shoreface and surf zone consist almost entirely of sand in a variety of bedding structures that reflect the interplay of storm and fair-weather waves, tidal currents, and biogenic processes.

Swash processes deposit seaward-dipping parallel laminae consisting of inversely graded couplets and heavy mineral layers (Clifton, 1969). Although appearing to be parallel over long distances, close inspection of these laminae reveals them usually to contain low-angle discordances, both perpendicular and parallel to the shoreline, formed in response to fluctuations in wave intensity and angle of attack.

Along many coasts, the upper shoreface topography is characterized by one or more coast-parallel bars. Bars consist of alternating layers of ripple crossbeds and plane beds formed in response to fluctuating bed shear during fair-weather periods (Davidson-Arnott and Greenwood, 1976). Plane beds dip seaward at low angles on the seaward slope and landward on the landward slope, whereas ripples are always directed onshore (Fig. 8.4).

Troughs between bars concentrate longshore currents. During periods of high wave energy, longshore currents may be sufficiently strong to transport sand in small- and medium-scale bedforms, but during periods of low wave energy, wave-induced currents directed onshore may overwhelm longshore drift and drive ripples landward across the trough.

Deeper water bars may be nearly continuous, but bars are cut by rip channels in shallower water. These channels concentrate return flow off beaches, and under certain conditions the currents in these channels may be able to transport sand seaward in large-scale bedforms (Ingle, 1966).

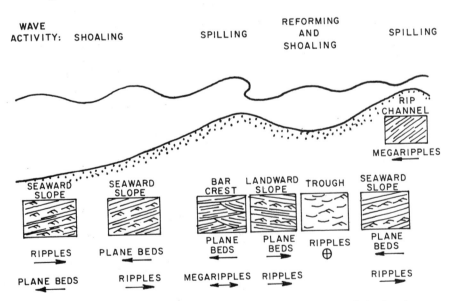

Figure 8.4 Distribution of stratification types along a coast with well-developed longshore bars (after Davidson-Arnott and Greenwood, 1976).

Longshore bars do not occur in the upper shoreface of some coasts. On low-gradient coasts experiencing low wave energy, ripple-laminated and plane-bedded sands extend to the foreshore. Even though sediment transport may be nearly continuous, the zone is populated by filter feeders that leave vertical burrows as traces (Howard, 1972). The degree of bioturbation tends to increase in an offshore direction (Howard and Reineck, 1972) (Fig. 8.5).

Bioturbation is less significant on high-gradient coasts experiencing high wave energy. On nonbarred coasts, sediment transport is dominantly onshore in response to shoreward-directed wave surge. Asymmetric ripples give way landward to lunate megaripples with shoreward dipping foresets (Clifton and others, 1971) (Fig. 8.6).

Some wave-dominated coasts are also affected by appreciable tides, resulting in the development of a foreshore zone. The foreshore occurs between mean low and mean high tide (Fig. 8.7). The surface of the foreshore is alternately exposed, then subject to swash processes, and then to wave surge during a tidal cycle. On mesotidal coasts, the effects of waves are minimized in the foreshore because the length of time the surface is subjected to wave action is shortened.

One feature characteristic of foreshores, especially those along low-gradient coasts, is the ridge and runnel system. During storm periods, foreshores are leveled, but during fair-weather periods, ridges form and begin to migrate shoreward. At low tide the ridges are exposed, and as tide rises the seaward side of each ridge

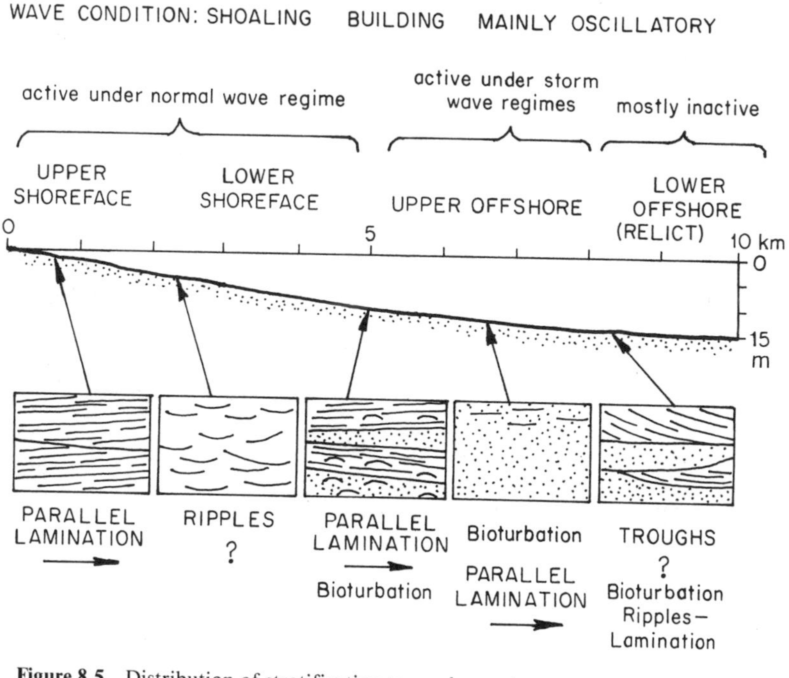

Figure 8.5 Distribution of stratification types along a low wave-energy coast (after Howard and Reineck, 1972).

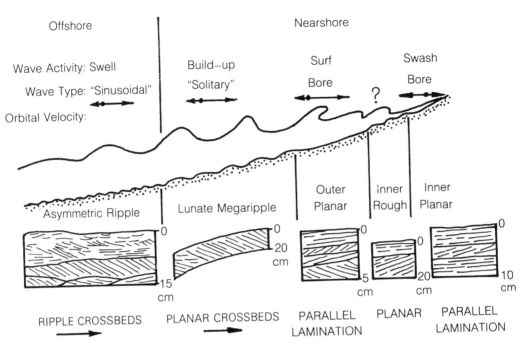

Figure 8.6 Wave dynamics along a nonbarred, high wave-energy coast showing change in wave shape and resultant currents, and distribution of bedforms (after Clifton and others, 1971).

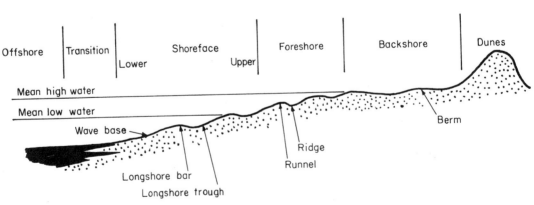

Figure 8.7 Profile of a mesotidal wave-dominated coastline showing the main morphologic elements in relation to water level (after Reineck and Singh, 1980).

progressively shoreward acts as a beach face, where swash processes form parallel laminae dipping gently seaward. Eventually, waves break over the ridge crest and sand is carried over the crest, where it is deposited on high-angle, landward-dipping foresets that prograde into the runnel landward of each ridge (Wunderlich, 1972).

Water flowing over the crest is collected in runnels, forming currents that commonly are sufficient to transport sand in ripples. A typical sequence formed by a migrating ridge and runnel system consists of a coarse storm lag at the base that is overlain in turn by ripple-bedded sands of the runnel, high-angle, landward-dipping foresets of the ridge, and low-angle, seaward-dipping laminae formed after the ridge has welded itself to the beach (R. A. Davis and others, 1972) (Fig. 8.8).

Beyond the limit of wave run-up on wave-dominated coasts is a backshore inudated only during storms or spring tides; it is usually separated from the foreshore by a low ridge or berm. Storm waves may wash over the backshore and build ridges, with a sequence of structures similar to those of ridge and runnel systems (Fraser and Hester, 1977). During subsequent fair-weather periods, sand in the backshore dries and is reworked by the wind. On prograding coasts, the storm ridges collect this windblown sand and act as cores of coast-parallel dune ridges (Fig. 8.9). Internal structures of dunes in the backshore vary in character. Steeply dipping foresets occur in dunes where strongly developed wind resultants occur (Bigarella and others, 1969). Where weak resultants prevail, however, or where the baffling effect of vegetation is a major factor in dune aggradation, sets of gently dipping, diffuse, parallel laminae may form (Goldsmith, 1973; Fraser and Hester, 1977).

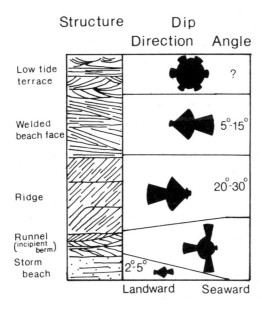

Figure 8.8 Sequence developed by aggradation of a ridge and runnel coastline (after R. A. Davis and others, 1972).

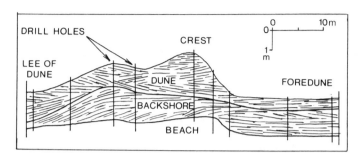

Figure 8.9 Internal structures of a coastal dune in a region lacking strong wind resultants (after Fraser and Hester, 1977).

Depositional Sequences of Wave-dominated Coasts

Although differences may occur among beaches where wave, climate, and tide vary, the sequences they deposit are broadly similar. In general, prograding sequences coarsen upward both by an increase in sand size as well as sand percent. Preservation of primary sedimentary structures also increases upward in response to increased frequency of bedload transport in shallower water and a decreasing effectiveness of infauna to rework the sediment. And, in general, the change in primary structures upward reflects increasing shear stress and decreasing water depth in shallower water. Specific sequences may vary, however, not only in response to variations in physical and biological processes, but also to eustatic trends in sea level and to the gross geomorphic setting in which the beaches occur.

The fair-weather wave base is relatively shallow along coasts with relatively low wave energy, and shoreface deposits are correspondingly thin. On the Georgia coast, for instance, the outer limit of the lower shoreface is at a depth of only 2 m below mean low water, where it grades into bioturbated sand and mud of the offshore zone (Howard and Reineck, 1972).

Nearby terraces of Pleistocene age consist of beach sediments deposited in an environment similar to that of the present-day Georgia coast (Howard and Scott, 1983). The total thickness exposed is about 12 m, and of that 2 m of extensively bioturbated sands probably were deposited on the shoreface. Shoreface sediments are underlain by a coarsening-upward sequence of bioturbated muds and laminated sands deposited in the offshore and transition zones, and they are overlain by sands in plane beds and structures characteristically formed by ridge and runnel systems of the foreshore (Fig. 8.10). Capping the sequence are laminated and bioturbated backshore sands overlain by root-mottled dune sands with traces of high-angle crossbeds.

The degree of bioturbation of the shoreface sands is atypical of that depositional setting which is subject to frequent periods of bedload transport. The sands were bioturbated interstratally, possibly, by *Callianassa* burrowing through foreshore sediments after they had prograded over the shoreface sands.

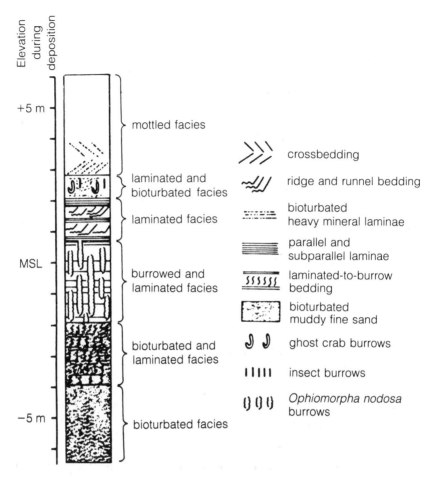

Figure 8.10 Shoreline sediments of Pleistocene-age deposited on a low wave-energy coastline (after Howard and Scott, 1983).

The lower shoreface extends to greater depths on coasts experiencing a more severe wave climate. Off the Ventura–Oxnard coast, for example, the lower shoreface/offshore transition occurs 17 m below mean low water (Howard and Reineck, 1981). The offshore zone consists mainly of bioturbated sandy silt, but occasional beds of sand in plane beds or ripple laminations record episodes of storm-generated wave and current action to depths of at least 22 m (Fig. 8.11).

Sediments of the lower shoreface are deposited in response to periods of storm surge alternating with periods of slow deposition. Storms generate beds of plane-bedded and ripple-bedded sand resting on lag surfaces, and fair-weather periods are represented by bioturbated silty sands. Little evidence of bioturbation exists in upper shoreface sediments, which consist of fine- to medium-grained sand in plane beds, ripple beds, and trough sets.

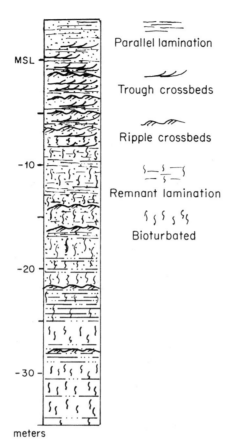

Parallel lamination

Trough crossbeds

Ripple crossbeds

Remnant lamination

Bioturbated

Figure 8.11 Hypothetical sequence that would develop by progradation of a high wave–energy coastline (after McCubbin, 1982, from data by Howard and Reineck, 1981).

Sands of the foreshore are fine- to medium-grained and occur in plane beds formed by wave swash in the intertidal zone. The sequence of structures is not greatly dissimilar to that found on shores of the southeast United States, but the sequence is much thicker on the Oxnard–Ventura coast because of the greater depth of effective wave base (Howard and Reineck, 1981).

Wave-dominated Shoreline Sequences in the Rock Record

The Gallup Sandstone of New Mexico is typical of the many shoreline sandstones of Cretaceous age in the Rocky Mountains (McCubbin, 1972). Bioturbated siltstones and shales of the offshore zone are overlain by lower shoreface sediments consisting of bioturbated sandstones in beds up to 1 m thick that may record substrate response to passing storms (Fig. 8.12). Each bed shows an erosional base and a bioturbated or ripple-bedded top, and within each bed the sandstones occur in hummocky crossbeds or subhorizontal plane beds.

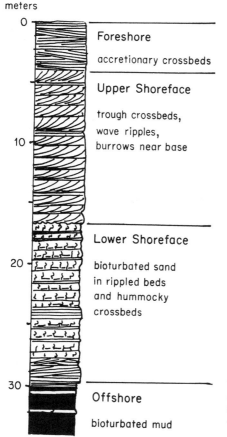

Figure 8.12 Vertical sequence in part of the Gallup Sandstone of New Mexico showing textures and structures similar to those that would form on a high wave–energy coastline (after McCubbin, 1972).

Upper shoreface and foreshore deposits consist mainly of sandstone in trough cross sets. Minor structures include horizontal plane beds, ripple crossbeds, and occasional siltstone or shale beds. Ripple crests are oriented parallel to shore, suggesting onshore transport during fair-weather periods, but orientation of trough sets indicates longshore transport during periods of higher wave energy. The uppermost unit in the sequence consists of low-angle, seaward-dipping laminae probably deposited in the foreshore, and the rooted sands capping the sequence may have accumulated in the backshore.

The Ordovician Starved Rock Member (St. Peter Sandstone) of the upper Great Lakes region, USA, probably was also deposited on a prograding beach (Fraser, 1976). The basal unit of the member consists of ripple-bedded and bioturbated sand that was probably deposited in the lower shoreface (Fig. 8.13). The degree of bioturbation increases downward in the unit, which displays a gradational basal contact with the underlying bioturbated silty sands of the Tonti Member of the St. Peter.

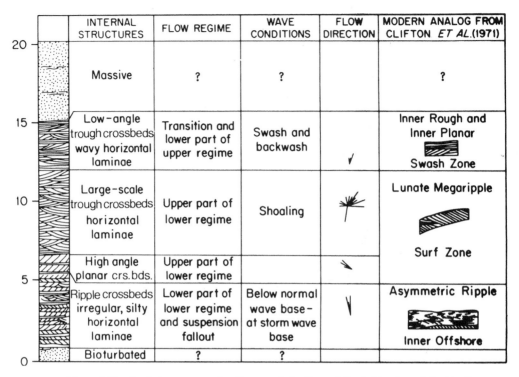

Figure 8.13 Vertical sequence of sediments of the Starved Rock Member of the St. Peter Sandstone showing a distribution of stratification types similar to that which would form by progradation of a nonbarred, high–energy coast.

Upper shoreface deposits consist of sand in large-scale planar and trough cross sets. These, in turn, are overlain by low-angle, small-scale trough and planar cross sets in beds separated by horizontal laminae that were probably deposited on the foreshore.

The uppermost unit consists of apparently massive, medium- to fine-grained, very well sorted sand. The stratigraphic position suggests that it was deposited in the backshore dune environment, and its apparently massive character may be due to lack of heavy minerals in the St. Peter that normally outline bedding.

Predominant direction of transport in the Starved Rock Member was onshore, and the sequence of sedimentary structures suggests that the unit was deposited on a nonbarred coast similar to that described by Clifton and others (1971). Interspersed throughout the sequence (except in the upper unit), however, are relatively thin beds of thoroughly bioturbated silty sands. These beds probably were bioturbated during calm conditions when no bedload transport occurred, and their presence throughout much of the Starved Rock Member suggests that the shoreline was only periodically subject to high wave energy conditions.

Systems on Wave-dominated Coastlines

Introduction

Prograding shoreline sequences can occur in sand bodies deposited by barrier islands or strandplains. Barrier islands and strandplains may be considered end members of an evolutionary sequence (Swift and others, 1985). Most barriers apparently are created during transgressions when beaches are inundated and detached from mainlands as lagoons form in back-beach swales. If a transgression continues and the shoreline retreats, the barrier migrates landward over a platform composed of its own washover fans and flood tidal deltas, and the landward margin of the lagoon encroaches on the mainland. Thus, the relative positions of the barrier, lagoon, and mainland shorelines are maintained.

If, however, sea-level rise slows or sediment yield increases (an uncommon but not unknown circumstance), the barrier may begin to prograde. The oceanward beach can prograde rapidly, but the lagoon is frozen in place, and as it gradually fills, inlets are abandoned. When this occurs, the barrier has evolved into a strandplain.

Despite the genetic connection, however, it is useful to discuss these two geomorphic end members separately. Because of the presence of lagoons and inlets along barrier coastlines and their absence on strandplains, the facies associations that exist in the two are quite different, and there are even fundamental differences in the vertical arrangement of those facies they have in common.

Barrier Island/Lagoon Systems

A barrier island is a low, elongate body of sand that parallels the coast and is separated from the mainland by a lagoon. The barrier forms part of a system that includes, in addition to the barrier and lagoon, an inlet that provides communication between the lagoon and basin waters (Fig. 8.14).

Barrier island.   Sediments of the barrier include all those normally found on wave-dominated coasts. Lower shoreface deposits consisting of intercalated sand and mud occur at the toe of the beach face. During fair-weather periods, active sedimentation occurs only intermittently, and infauna have the opportunity to thoroughly bioturbate the sediments (Fig. 8.15). Upper shoreface sediments consist of relatively clean sands. The degree of biogenic reworking varies, although it tends to be somewhat small, and primary structures, consisting of plane beds dipping offshore at low angles, high-angle foresets dipping onshore, and trough and ripple cross sets oriented along shore, are commonly preserved (Fig. 8.15).

Foreshore deposits consist of sand (and possibly gravel, depending on availability) in seaward-dipping plane beds, structures formed by ridge and runnel systems, and in seaward-dipping high-angle foresets formed in rip channels. Backshore sediments accumulate on dunes, storm berms, or washovers. Sands are somewhat

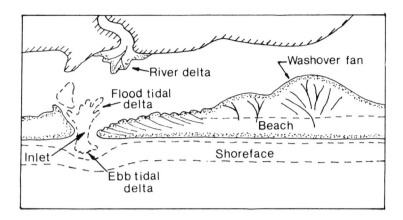

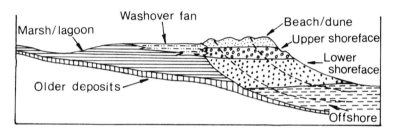

Figure 8.14 Diagrammatic views of a barrier system showing morphologic elements in plan view and distribution of sediment types in cross section (after McCubbin, 1982).

finer-grained than those of the foreshore, and they occur in seaward-dipping accretionary foresets of the beach, landward-dipping foresets of beach ridges and washovers, and planar crossbeds of variable dip direction and angle formed on dunes.

Ebb-tidal deltas form at the mouths of tidal inlets on the ocean side of barriers where flow expansion takes place during ebb tide (Fig. 8.16). The sedimentological characteristics of ebb deltas are variable because they reflect the complex interaction of tide and waves. They are poorly developed on coasts with high wave energy, but on coasts experiencing low wave energy they may form shoals extending several kilometers offshore (Hubbard and others, 1979).

Ebb-dominated currents transport sand in large-scale sand waves in the axis of the channel(s) that cut through the delta. Smaller-scale, flood-oriented bedforms occur on the margins of the channels, and on the ebb platform a complex assemblage of plane beds, sand waves with multidirectional orientations, swash bars directed onshore, and megaripples and ripples reflecting both onshore and along shore transport are found (Hine, 1976) (Fig. 8.16).

A progradational sequence would consist of large-scale ebb-oriented cross sets of the channel axis overlain by flood-oriented crossbeds in thinner sets of the channel margin. Platform deposits capping the sequence would consist of a varied assem-

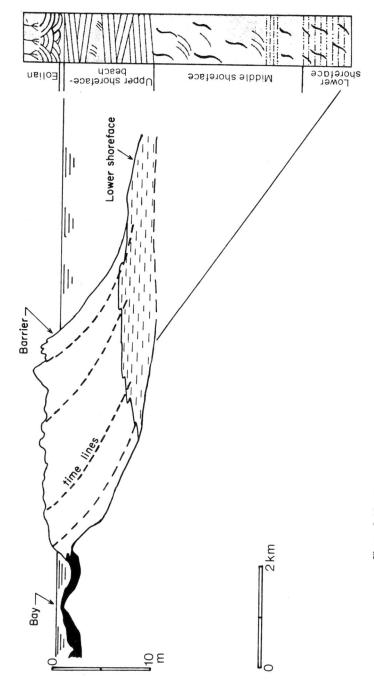

Figure 8.15 Cross section through Galveston Island showing the vertical sequence of sediments developed during progradation of the barrier (after Dickinson and others, 1972).

171

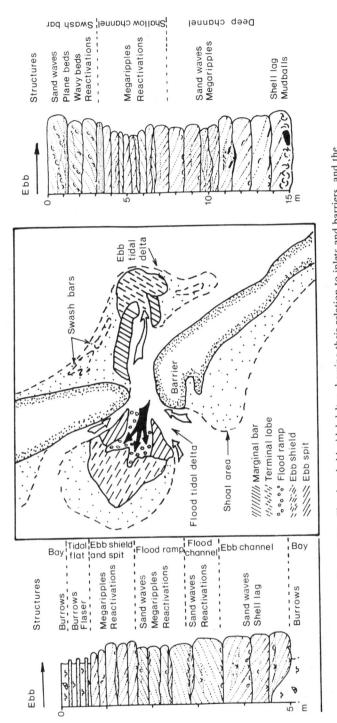

Figure 8.16 Flood and ebb tidal deltas showing their relation to inlets and barriers, and the sedimentary sequences that might develop during their aggradation (based on Reinson, 1977; Hayes, 1976).

172

blage of plane beds and cross sets of varying scale and dip azimuths reflecting the interacting tide- and wave-generated currents (Barwis and Hayes, 1979) (Fig. 8.16).

Lagoons.   Lagoons are narrow, frequently shallow bodies of water separated from the open ocean by the barrier. They accumulate mainly fine-grained sediments out of suspension, but they also possess a coarser-grained component introduced into the lagoon by streams, through inlets in the barrier, and by waves washing over it.

Muds accumulate in the deeper, central portions of the lagoon and on fringing intertidal and supratidal mudflats. The muds may be finely laminated, but in most cases primary structures are destroyed by burrowing. A substantial portion of the lagoonal sediments is organic detritus brought in by rivers draining vegetated hinterlands and produced in situ in fringing swamps and mudflats. Even in arid or semiarid climates, where aquatic marsh flora cannot be established, algal mats may form on tidal flats and add organic material to the fine-grained sediment.

Coarser-grained sediments occur in river deltas, tidal flood deltas, washover fans and aprons, and intertidal flats. In addition, in mesotidal lagoons, considerable amounts of water must be exchanged during a tidal cycle, and bedload transport of sand takes place in channels and on point bars of tidal creeks where ebb and flood flows are concentrated.

In relatively humid regions, rivers entering lagoons may add significant amounts of sediment and form deltas that fill the lagoon on the landward side. Because they are protected from waves, these deltas most often are river dominated. They deposit coarsening-upward sequences consisting of prodelta and interdistributary bay muds at the base and distributary mouth bar and channel sands flanked by levee and overbank deposits at the top (Fig. 8.17). The laminae in the lagoonal muds in front of river deltas are commonly preserved because sedimentation rates are large enough to reduce the impact of burrowing organisms.

Currents in tidal creeks may be sufficient to move sand in medium- or large-scale bedforms. They are active only during flood and ebb flow, however, and during the slack-water period after flood tide, the bedforms may be draped by mud settling out of suspension. Coarsest sediments in the lagoon commonly are found as lags in channel axes of tidal creeks.

Tidal creeks may meander and produce point bars that morphologically are similar to fluvial counterparts, and even straight reaches may shift laterally over considerable distances. Sediments deposited on the side opposite the cutbank consist of layers of mud and sand that dip into the axis of the creek perpendicular to the flow direction of the creek. Within layers, however, other directional indicators such as dips on foresets give true flow direction. The ability of tidal creeks to migrate in lagoons suggests the possibility that lagoonal sequences may be capped by sands in a variety of medium- and large-scale cross sets, with a wide dispersion of dip directions reflecting the complex drainage patterns that the creeks commonly display. Carter (1978), for instance, interpreted deposits displaying these characteristics in

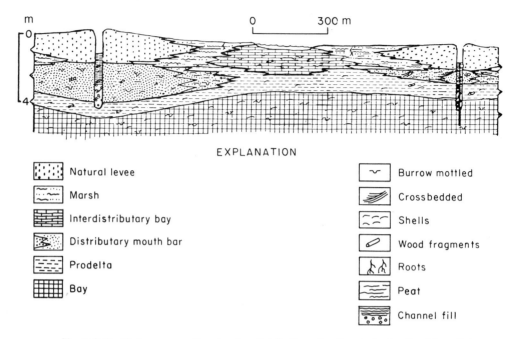

EXPLANATION

Natural levee	Burrow mottled
Marsh	Crossbedded
Interdistributary bay	Shells
Distributary mouth bar	Wood fragments
Prodelta	Roots
Bay	Peat
	Channel fill

Figure 8.17 Strike cross section showing distribution of sediments in the Guadalupe Delta prograding into Galveston Bay (after Donaldson and others, 1970).

the Cohansey Sand (Tertiary) of New Jersey to be tidal creek sands deposited in a barrier-protected environment.

Intertidal flats can occupy large portions of a mesotidal lagoon, and if they border channels where ebb and flood flow are concentrated, they may consist of sandflats covered by ripples superimposed on larger-scale bedforms. Dips on foresets of the large-scale bedforms may show wide dispersion because of the complex interaction of ebb and flood flows, and dip directions on ripple cross sets may further complicate the overall flow pattern because ripples are late-stage bedforms that appear when flow is being directed by local topography.

Tidal deltas are similar to fluvial deltas in that they form where flow expansion takes place, in this case at either ends of tidal inlets. During rising tide, flow through the inlet expands at the mouth on the lagoon side, and sediment carried by the flow is rapidly deposited as a flood tidal delta.

A flood tidal delta consists of three main parts. The flood ramp is a broad platform that extends lagoonward from the mouth of the inlet (Fig. 8.16). Its surface is covered normally with sand waves, and it may be cut by one or more flood channels. Sand spills over the edge of the ramp and accumulates as lobes at the mouths of flood channels forming the ebb shield. The ebb shield acts to deflect ebb flow around the flood ramp and into ebb channels that usually occur at the margin of the ramp. Thus, the morphology of the flood delta acts to segregate flood and ebb flows.

An aggradational sequence developed by a flood tidal delta might consist of a basal unit of sand in seaward-dipping, large-scale foresets deposited in ebb channels (Fig. 8.16). These would be overlain by sand in landward-dipping, large-scale foresets deposited in flood channels and on the flood ramp, and these would, in turn, be overlain by sands deposited on the ebb shield in smaller-scale, seaward-dipping cross sets. Organic muds capping the sequence would be the deposits of intertidal flats and supratidal marshes established on the delta after its abandonment.

Washover fans are formed when sand is eroded from the beach face, carried over the barrier, and deposited in the lagoon. This process is normally restricted to storm periods when high-amplitude waves break on the barrier shoreline and either wash over the barrier in a sheet or wash through the barrier in channels cut during the storm.

Expansion occurs in either case as the flow enters the lagoon and the bedload is rapidly deposited. Each washover event deposits a thin sheet of sand, and successive events cause a washover fan to aggrade upward and prograde into the lagoon. Primary structures within a fan consist of sets of planar laminae dipping lagoonward at low angles and sets of higher-angle foresets also dipping into the lagoon (Schwartz, 1975) (Fig. 8.18). These structures are not normally preserved, however, because the sand is rapidly reworked by infauna soon after active sedimentation ceases (Frey and Mayou, 1971).

Relatively small washover fans develop on mesotidal coasts where a large number of relatively deep tidal inlets apparently can accommodate the exchange of water during storm surge (Deery and Howard, 1977) (Fig. 8.1). Deposits of such fans would occur as isolated lobes embedded within finer-grained lagoonal deposits. On microtidal coasts, however, inlets are spaced farther apart, and surging water during storms washes over the barrier or through numerous small storm-cut inlets into the lagoon (Fig. 8.1). Large composite fans are characteristic of such barriers, and they may merge together to form a continuous apron of sediments fringing the lagoon side of the barrier (Hayes, 1976) (Fig. 8.1). The apron consists of coalescing fans, each made up of packages of sediments deposited during separate storm events and separated by lag surfaces, windblown sand, or lagoonal deposits (Andrews, 1970; Schwartz, 1975).

Tidal inlets. The third major component of barrier systems is tidal inlets, which are the major pathways for the exchange of lagoonal and open marine waters. Three facies may be found within an inlet. In the deep channel, the currents are the strongest and gravel lags may develop on the floor. Sands deposited in the deep channel occur in large-scale trough or planar cross sets that may show a dominant transport direction produced during ebb or flood flow (Kumar and Sanders, 1974) or a bidirectional transport direction indicative of current transport reversals during a tidal cycle (Hubbard and Barwis, 1976) (Fig. 8.19). In shallower parts of the channel, sands are deposited in small- or medium-scale trough sets, plane beds, and washed-out ripples in response to lower flow velocities and shallower water depths.

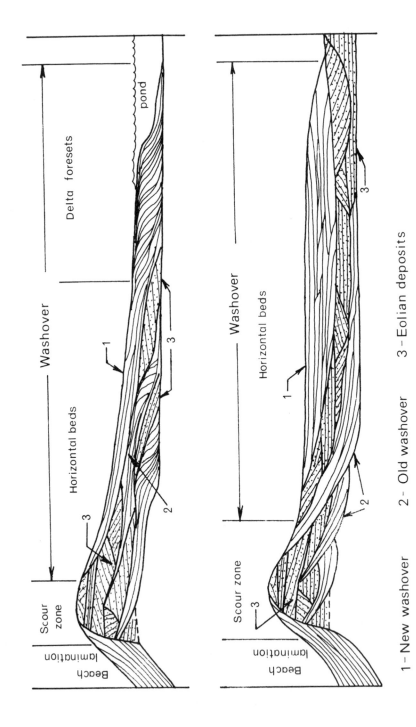

Figure 8.18 Cross sections showing internal structures of washover fans prograding into wet (upper) and dry (lower) back-beach areas (after Schwartz, 1975).

1 – New washover   2 – Old washover   3 – Eolian deposits

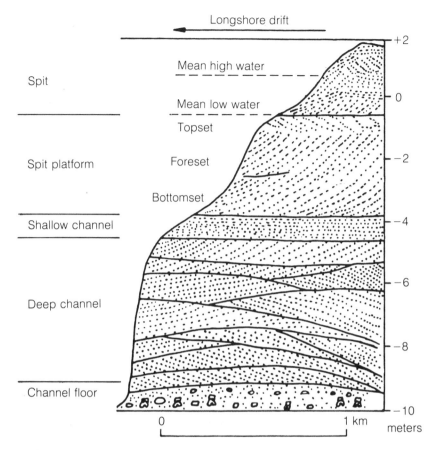

Figure 8.19 Sequence of sediments deposited by a migrating tidal inlet (after Kumar and Sanders, 1974).

Most tidal inlets migrate in the direction of longshore drift. As they migrate, sand is eroded on the downdrift side of the inlet and is deposited on the updrift side. A spit platform is built and progrades into the channel, and a spit is constructed on top of the platform. The dominant features of platform deposits are the large-scale planar cross sets deposited at the platform margin as it progrades into the channel. Smaller-scale structures with bidirectional orientations imposed by the interaction of reversing flows are also present in the spit platform.

The spit forms the leading edge of the barrier island as it progrades into the inlet and consists of sand in structures formed dominantly by wave processes. These structures consist mainly of sets of low-angle planar laminae formed by swash processes and are analogous to the beach laminations of the barrier. Dip azimuths of these sets, however, may be highly variable due to the changing orientation of the beach as it curves around the end of the spit.

Origin of barrier islands.   The origin of barrier islands has long been a topic of debate, and it is important for stratigraphers to understand their evolution because the mode of evolution of a barrier island complex determines, to a large degree, the nature of the material on which it rests. Four hypotheses have been advanced to account for the occurrence of barrier islands, including engulfment of mainland ridges, shoreface retreat, progradation and segmentation of spits, and aggradation of shoals (Fig. 8.20). A considerable body of evidence exists that suggests that all these mechanisms can operate along modern coasts, but that shoreface retreat and engulfment of mainland ridges are, by far, the most important ones.

Hoyt (1967) proposed, from his work on the Georgia coast, that barriers might result from engulfment of mainland beaches, especially those representing former barriers. Such a mechanism would be especially important on a broad, low-relief coastal plain, such as that of the southeast United States, where even a small rise in sea level could inundate large areas of the plain, leaving coast-parallel sand ridges surrounded by water (Fig. 8.20).

Transgressions of coasts with more irregular topography form indented coasts with prominent headlands and deep embayments. If a sufficient amount of sand is available and if significant longshore transport takes place, a spit can prograde across the mouth of an embayment from an adjacent headland (Fig. 8.20). By this process a spit can evolve into a barrier island (Field and Duane, 1976), and in places where it is occurring, the process can act very rapidly. Along the mid-Atlantic coast of the United States, for instance, rates of spit progradation up to 6 km/century have been measured (Kumar and Sanders, 1974).

From morphology alone it is difficult to demonstrate the origin of a barrier by shoal emergence, but with the added information contributed by detailed stratigraphic studies, it was possible to demonstrate that such a process may occur (Otvos, 1970, 1978). The process appears to operate best in a zone of significant longshore sediment transport where a newly emergent bar may become a core for an island that can evolve into a barrier as it enlarges. The process has been shown to operate in a few instances (Otvos, 1977, 1978), but even in these cases, the shoals emerged because the rapid post-Pleistocene transgression produced, in some places, shoreface profiles in disequilibrium with present-day hydraulic factors. Most, in fact, are probably the result of accretion on preexisting shoals or disintegrating barriers (Swift and Moslow, 1982).

Sediments deposited in barrier complexes formed by engulfment of mainland ridges accumulate over soils developed on older coastal plain sediments. These underlying sediments can consist of fluvial or fluviodeltaic deposits, but commonly they consist of barrier island sediments deposited during previous high sea levels. The new lagoon forms over the deposits of the old lagoon as the old barrier is engulfed, and new barrier sediments are welded onto the old (Fig. 8.20).

One important implication of this process is that preexisting barrier islands may become the preferred sites for subsequent barriers during successive sea-level fluctuations. The Holocene barriers of the Florida and Georgia coasts (USA), for instance, are welded on to barriers formed during late Pleistocene high sea levels

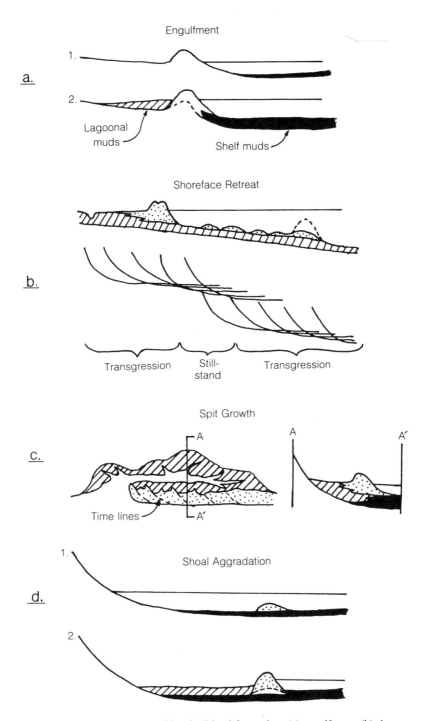

Figure 8.20 Four methods of barrier island formation: (a) engulfment, (b) shoreface retreat, (c) spit growth, and (d) shoal aggradation (after Hoyt, 1967; Swift, 1975).

(Oertel, 1979). In some cases, the Holocene barrier sands rest directly on sands of the previous barriers, but in other cases the sands are attached to delta deposits that prograded out in front of the older barriers (Fig. 8.21).

Barriers formed by spit progradation evolve penecontemporaneously with the substrate over which they are migrating. Cape Henlopen, for example, is presently prograding into Delaware Bay (Kraft and others, 1979). Estuarine sediments in the bay overlie pre-Holocene fluvial deposits, but the barrier overlies and interfingers with estuarine and shallow marine sediments. Similarly, barriers that were initially emergent shoals might be expected to interfinger with lagoonal deposits on the landward side and marine sediments on the seaward side. But both barrier and lagoonal sediments would overlie shallow marine deposits (Otvos, 1979).

It should be remembered that present-day barriers owe their origin, ultimately, to the rapid post-Pleistocene sea-level rise. Field and Duane (1976) present compelling evidence that the barriers along modern coasts probably originated far out on the shelf and have migrated to their present positions along with rising sea level. The association of barriers with transgressive shorelines is almost completely without exception, and it suggests that barriers are only one stage in the development of stratigraphic sequences deposited in the coastal transition zone (see, for example, Ryer, 1977).

Clearly, however, the same barrier that existed on the shelf is not the same as the barrier on the present-day coast (Field and Duane, 1976). Indeed, the variation in barrier morphology seen on modern coasts should be a clue to the ways barriers evolve in response to changes in pretransgressive topography and availability of sediment through time. It is more than likely that, as the barriers retrograded across an irregular substrate, barriers, spits, strandplains, and shoals coexisted on different parts of the coast and succeeded one another as the conditions governing their formation and evolution changed.

The process by which this may have occurred was outlined by Halsey (1979) in a study of the evolution of the Atlantic coast of the Delmarva Peninsula (USA). In the northern part of the study area, the pretransgressive topography consisted of a relatively dense network of deeply incised stream valleys separated by high interfluves (Fig. 8.22). The estuaries grew as sea level continued to rise and evolved into lagoons as small barriers prograded across bay mouths and linked adjacent headlands.

In the central part of the study area, the drainage network was initially less dense and less deeply incised. Again, during initial stages of inundation, beaches formed on headlands and estuaries formed in valleys; but because interfluves were lower, a point was eventually reached where flooding of low areas in the interfluves occurred and formed a lagoon. The initial deposits of the lagoon accumulated in marshes and on tidal flats, but as sea level continued to rise, the lagoon deepened and expanded. At the same time, the barrier migrated landward and upward until the interfluves were completely covered and the barrier became almost completely detached from land (Fig. 8.22).

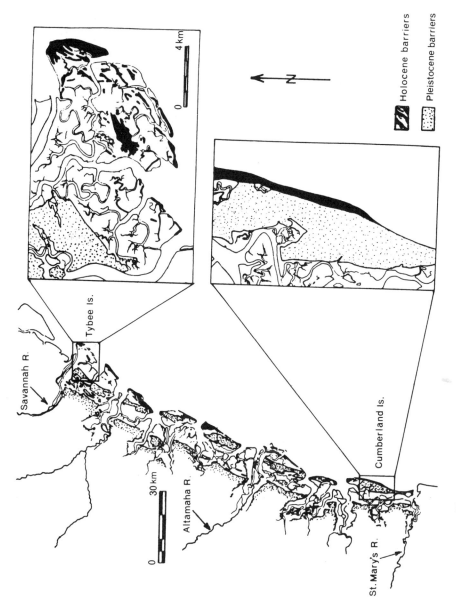

Figure 8.21 Sequential development of barriers and deltas along the southeast coast of the United States. Barriers are developed during high sea-level stands, and deltas develop in front of major rivers during lower stands. Where major rivers do not reach the coast, series of barriers are welded together (after De Pratter and Howard, 1977).

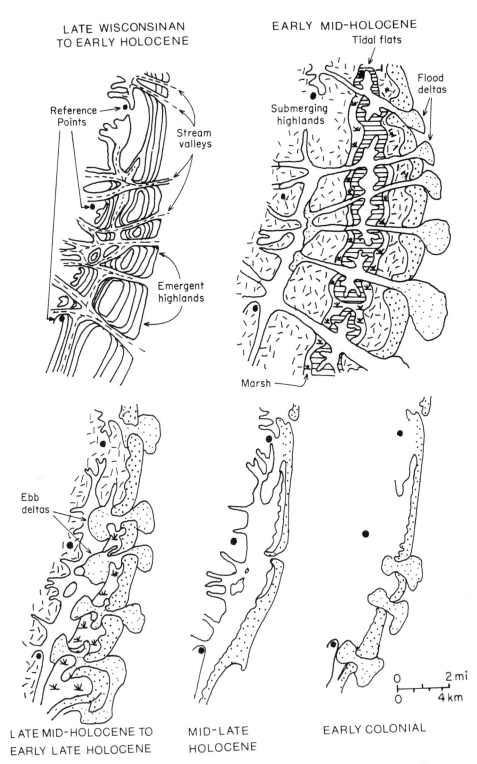

Figure 8.22 Development of a barrier coast during transgression as a function of preexisting topography (after Halsey, 1979).

This model combines aspects of both coastal inundation and spit progradation in explaining the present-day aspects of these barriers. In some areas, barriers were initiated as beaches attached to mainland ridges and lagoons developed by drowning of lowland areas. In other areas, spit growth and segmentation functioned to form lagoons out of estuaries. It is also conceivable that emergent shoals might also evolve into barriers where inundation of low-relief coastal plains might result in formation of bathymetric highs that may have been too low or discontinuous to initially serve as barriers. They might, however, serve as collecting points for sand being transported along the coast and eventually emerge as subaerial features.

Intrinsic and extrinsic controls on the evolution of barrier island/lagoon systems.   At any given instant, a barrier island system may be retrograding or aggrading in place, but once it begins to prograde, it rapidly evolves into a strand-plain. The stratigraphic sequence in the process of being deposited by a barrier system at any particular time may be relatively simple. It should be stressed, however, that although either of these conditions may apply over a long period of time it is probably more often the case that they cyclically succeed one another through time as the system responds to changes in hydrographic regime, sediment yield, and sea level. The resulting stratigraphic sequence will be correspondingly complex.

Transgressive sequences develop when barrier systems retrograde. On present-day coasts, this has occurred where the sediment yield has not been sufficient to overcome the effects of the post-Pleistocene sea-level rise, and the barrier islands have encroached on the coastline. During this process, sediments of the foreshore are eroded, especially during storm periods, and added to the shoreface, where they are redistributed in order to bring the offshore profile into equilibrium with the rising water levels (Brunn, 1962). In addition, sediment from the beach face and backshore is eroded and added to the lagoon by storm washover. The typical transgressive sequence consists of lagoonal sediments in various states of preservation, overlying a variety of substrates, and in turn overlain by back barrier ridge sediments (Kraft, 1978) (Fig. 8.23). Facies boundaries dip basinward in sequences deposited by retrograding barrier systems. Time lines cross facies boundaries and are concordant with the depositional surface on the landward side of the barrier. They reverse their dip on the seaward side, however, and converge on the surface of the transgressive sand sheet (Fig. 8.23).

If the rates of sediment influx and sea-level rise approach a balance, the shoreline continues to retrograde and the facies boundaries continue to dip seaward. Boundaries dip at increasingly high angles as the rates come more nearly into balance, however, and an increasing proportion of the sequence can potentially be preserved.

In detail, the depositional sequence may be quite varied. Backshore sediments may consist of washover fan or flood tidal delta deposits, lagoonal sediments on intertidal flats, or supratidal marshes on the margins of the lagoon (Morton and Donaldson, 1973). Lagoonal deposits on mesotidal coasts might even consist, to a substantial degree, of sediments deposited by migrating tidal creeks.

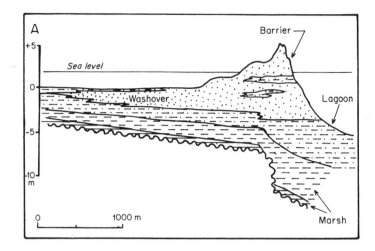

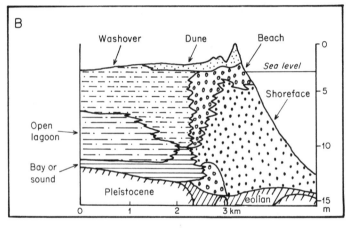

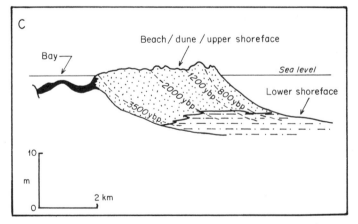

Figure 8.23 Sequences of barrier island sediments developed during (A) transgression, (B) upward aggradation, and (C) regression [after Kraft, 1971 (A); McKee and Sterrett, 1961 (B); and Bernard and others, 1962 (C)].

Actively migrating tidal inlets can significantly alter this sequence. Along the Georgia coast, for instance, deep tidal inlets cut completely through the deposits of the Holocene barrier and lagoon into the pre-Holocene substrate (Hoyt and Henry, 1967). The stratigraphic record left by that retrograding system might consist, to a large degree, of basal tidal channel sediments (Fig. 8.24).

The degree of inlet migration, however, depends to a great degree on the magnitude of the longshore sediment flux. If little sediment is available to it, the updrift margin of the inlet may not prograde at all. Pierce and Colquhoun (1970), for instance, have shown that inlets on the North Carolina barrier system have remained stable during transgression (Fig. 8.25). Even where sediments are available, inlet migration may not occur if the sediment being transported downdrift bypasses the inlet in the form of swash bars migrating across the ebb delta (Fitzgerald and others, 1984).

Tye (1984) showed that there is an intrinsic control on the migration of some tidal inlets because they become hydraulically inefficient as they migrate. Eventually, a threshold is reached, and the inlet breaches its tidal delta to establish a new, hydraulically efficient channel updrift near its original position. Some inlets, therefore, may migrate only within fairly narrow zones on a barrier.

If the Bruun theory works without exception, very little of the shoreface and backshore of a transgressive barrier system can be preserved in the face of a rising sea level; but Kraft (1971) proposed that if the rise is rapid enough, total retention may be attained (Fig. 8.26). The degree to which the Bruun rule applies determines to a large extent the basic mechanism of barrier migration during transgression.

There are three possible ways that barriers can retreat across shelves. A barrier can aggrade during a period of relative sea-level stability and then drown in place during renewed sea-level rise. By this process, the surf zone oversteps the old barrier ridge and forms a new barrier against the landward edge of the former lagoon (Sanders and Kumar, 1975; Rampino and Sanders, 1980, 1981). It is important to recognize that the surf zone does not pass continuously over the back-barrier area during stepwise retreat. Instead, the surf zone jumps the lagoon, allowing considerable preservation of the back-barrier sequence.

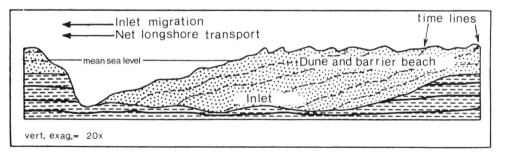

Figure 8.24 Longitudinal cross section through a barrier island showing the distribution of inlet, dune, and barrier beach facies with respect to the substrate under a barrier with an actively migrating inlet (after Hoyt and Henry, 1967).

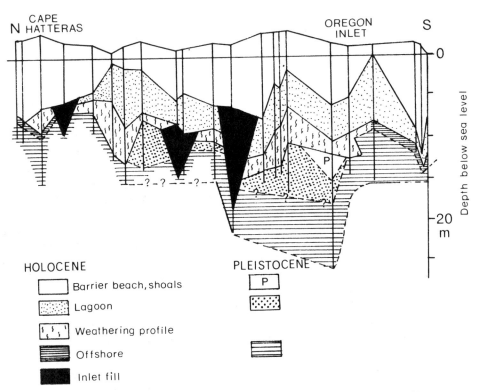

Figure 8.25 Longitudinal cross section through a barrier island showing distribution of facies with respect to the substrate in a barrier with stable inlets (after Pierce and Colquhoun, 1970).

However, the dynamic response time of barriers to normal coastal processes is rapid compared to the potential rate of sea-level rise (Swift and Moslow, 1982), and if the Bruun rule applies, barriers should migrate continuously across the shelf. The surf zone should traverse the entire barrier ridge, level the barrier, and erode most or all of the back-barrier sequence.

The Bruun rule, however, would not apply if sediment yield to the system increased, and in such a case, or if the rate of sea-level rise slowed, a barrier might migrate discontinuously across the shelf (Swift, 1975; Swift and Moslow, 1982; Panageotou and Leatherman, 1986). Upward aggradation would occur at the expense of landward translation, and the surf zone would ride upward on the aggrading barrier. If the rate of sea-level rise increases or if the rate of sediment influx slows, the retreat would resume, but the surf zone would traverse only the upper part of the overthickened barrier ridge, leading to preservation of a substantial part of the barrier ridge sequence.

These methods of barrier migration produce significantly different stratigraphic sequences during transgression (Rampino and Sanders, 1982). Inlet depos-

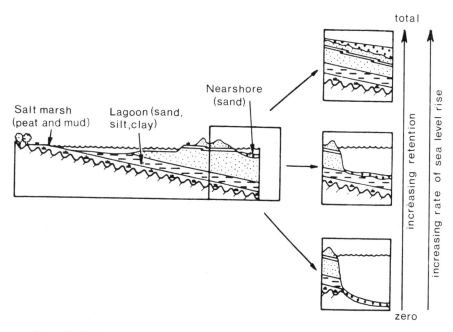

Figure 8.26   Preservation potential of barrier island and lagoon sequences in the face of different rates of sea–level rise (after Kraft, 1971).

its, for example, would be confined to the area of the barrier ridge in the case of stepwise retreat, but would underlie the entire area of the retreat track in the case of continuous or discontinuous shoreface retreat.

Thick deposits of barrier ridge and lagoonal sediments would be preserved during stepwise or discontinuous barrier retreat, but would stand little chance of preservation during episodes of continuous retreat. Instead, these sediments would be spread as a transgressive sand sheet over the retreat path. The sand sheet is commonly remolded into shoreface-attached sand ridges during transgression. These would form over the entire retreat track during continuous retreat, but they would develop only in front of stranded barrier ridges, where retreat occurs in stepwise fashion.

A barrier system begins to aggrade more or less in place when the rates of sediment influx and sea-level rise are in balance. Aggradation produces essentially vertical facies boundaries, with time lines coincident with the depositional surface.

Padre Island has been advanced as an example of an aggrading barrier system (Dickinson and others, 1972). The island itself has remained stationary through most of its history. The landward side of the island, however, has migrated up to a kilometer into Laguna Madre in recent times, although the seaward margin has not retrograded similarly. Barrier sands, exclusive of their dune cover, are nearly 20 m thick, an accumulation of barrier sands much thicker than would be expected given the offshore profile of the shelf (Fig. 8.23).

Barriers cannot continue to grow upward indefinitely, however, because as a barrier aggrades it must continue to extend the toe of its shoreface profile into continually deeper water in order to maintain an equilibrium profile (Dillon, 1970) (Fig. 8.27). Before this can occur, the inner shelf must aggrade to a depth sufficiently shallow for the shoreface to be established on it, and the amount of sediment needed to accomplish this must increase at a geometric rate in order for aggradation to take place. If the sediment yield does not increase, the rate of growth will slow until an intrinsic threshold is reached and retrogradation begins once more. The uppermost part of the sequence, down to the level of the existing shoreface, may be eroded, but the bulk of the sequence previously deposited during early stages of aggradation has the potential of being preserved.

Volumetric considerations, therefore, mandate that for progradation to occur sea level must stabilize or fall, or else progradation will be limited by an intrinsic threshold based on the available sediment. If these conditions are met, a barrier can prograde; and even though it will quickly evolve into a strandplain, it is useful to consider the facies distributions established during the early phases of this evolution.

Most of the barriers cited in the rock record are relatively narrow features, and probably only in interdeltaic areas, such as the Frio Barrier System of the Tertiary Gulf Coast, can sufficient sediment be provided to maintain active growth over long periods of time (Boyd and Dyer, 1964). Even these may be amalgamated barriers, rather than single depositional units, however. Ryer (1977), for instance, showed in the Cretaceous of Utah that many of the units traceable over long distances are actually repetitive sequences formed by cyclic progradation and retrogradation of barriers superimposed on major offlap and onlap sequences (Fig. 8.28).

Prograding barrier islands produce regressive sequences consistent with those produced by any prograding sandy shoreline, with exception of the potential inclusion of inlet sequences within the record. Where inlet sediments are not included, however, sequences typically coarsen upward, with a concomitant increase in the preservation of primary sedimentary structures and a decrease in infaunal traces (Fig. 8.15).

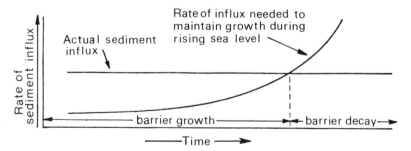

Figure 8.27 Diagrammatic representation of the relationship between the actual rate of sediment yield to a barrier and that rate needed to sustain growth during rising sea level (after Dillon, 1970, reprinted with permission of the *Journal of Geology,* University of Chicago Press).

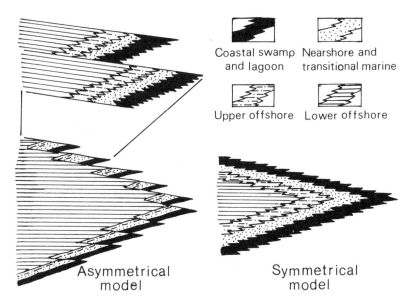

Coastal swamp   Nearshore and
and lagoon   transitional marine

Upper offshore   Lower offshore

Asymmetrical   Symmetrical
model   model

Figure 8.28   Development of sheetlike deposits of barrier island sediments by minor regressive pulses superimposed on a major transgressive event (after Ryer, 1977).

Galveston Island is one of the relatively few well-studied examples of a prograding barrier system. The shoreline of the barrier has prograded seaward about 3 km in 3500 years and has deposited a sequence consisting of bioturbated muddy sands of the lower shoreface, bioturbated sand of the middle shoreface, and planar-bedded and trough-crossbedded sand of the upper shoreface and beach. Dune sands in trough crossbeds cap the sequence (Bernard and others, 1963) (Fig. 8.15). Galveston Island is on a microtidal coast, and tidal inlets are neither deep nor abundant and apparently do not figure greatly in the stratigraphic sequence. Hurricanes are common along the coast, however, and one might expect channels cut by storm surges to form important portions of the foreshore and beach facies (Hayes, 1967).

In a progradational sequence, lagoonal sediments should form the uppermost unit. In fact, it appears the reverse is true, with backbarrier washover fans overlying lagoonal sediments on both Padre (aggradational) and Galveston (progradational) islands (Fig. 8.23). In most cases, lagoonal sediments are confined to the original basin in which they form, and the prograding barrier itself evolves into a strandplain consisting of a broad sand body with only local occurrences of fine-grained sediments in low areas. It is probable that lagoons are ephemeral features that occupy only the most landward part of a barrier sequence.

It has been proposed that lagoons may enlarge themselves as a barrier progrades if migrating tidal creeks are able to erode the lagoonward side of the barrier. By this process, it is possible that lagoonal sediments may form the topmost unit of a regressive barrier sequence at the expense of beach and dune sediments. The mechanism presupposes that a freely migrating creek can move continuously only in one direction for an extended period of time. It also demands that washover processes will be unable to fill the creeks adjacent to the island. Tidal creeks were able to rework parts of Kiawah Island along the South Carolina coast (Duc and Tye, 1987). These creeks were established to drain swales between digitate beach ridges on the prograding end of the island. The creeks migrated rapidly, reworked the upper part of the barrier island sequence, and redeposited a fining-upward sequence of channel lag and point-bar sediments overlain by salt-marsh deposits. These creeks are part of the barrier rather than the lagoon, however, and their deposits are only a thin veneer that would not be confused with true lagoonal sequences.

Although the process of tidal creek migration can operate locally along the barrier over short periods, the ability of the mechanism to operate over the whole barrier for an extended period must be questioned. The most likely place for it to occur would be on a mesotidal coast with few storms where currents in tidal creeks would be sufficient to erode back-barrier sands. It may be necessary to invoke the process to explain the widespread occurrence of lagoonal deposits overlying regressional barrier island sediments, such as in the Cretaceous Fox Hills Sandstone of Wyoming (Land, 1972) (Fig. 8.29).

Large portions of barrier island sequences might consist solely of sediments deposited in actively migrating tidal inlets. Regressive sequences formed by such barrier systems would consist of tidal channel deposits at the base, and spit and spit platform deposits at the top. Deposits of ebb tidal deltas might replace normal shoreface and foreshore sands, and, where lagoons are present, flood tidal deltas may form a coalescing apron on the backshore.

It has been shown that the deposits of tidal inlets may form appreciable portions of transgressive barrier sequences (Kumar and Sanders, 1974; Hoyt and Henry, 1967), but an assessment of their importance in regressive sequences has not yet been made. Lagoons are such effective sediment traps that they have the potential to aggrade to effective wave base faster than sea level can rise (Swift and Moslow, 1982). As a lagoon is filled, the volume of water flowing through the inlet during a tidal cycle declines and the ability of the inlet to erode and migrate declines. Once the lagoon is filled, the inlet should close, but the lagoon might be kept open if the lagoonward shore was continuously eroded by tidal creeks. Objections to this mechanism acting as a regular feature of barrier progradation, however, have already been stated.

Barrier island sequences in the rock record.   Much of the late Tertiary and Quaternary coastal plain of the eastern United States consists of fluviodeltaic, estuarine, and barrier island sediments. The late Miocene Cohansey Sand of New Jersey is only one of the coastal plain units deposited in a barrier complex not unlike the systems presently forming on the North Sea coast of Holland (Carter, 1978).

The Cohansey Sand conformably overlies the marine shales of the Kirkwood Formation, and the gradational contact between the two units represents the transition from offshore to lower shoreface sediments. The sediments overlying the

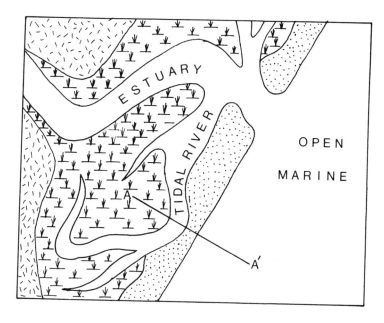

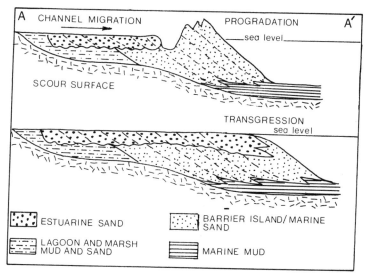

Figure 8.29 Possible mechanism for enlarging a lagoon at the expense of dune and backshore sediments during progradation of a barrier island (after Land, 1972).

transition occur in two types of sequences. One is interpreted as a barrier sequence consisting of shoreface, foreshore, backshore/dune, and lagoonal sediments. Trough-crossbedded sand and granules were deposited by longshore currents in the surf zone. These are overlain by plane-bedded sand of the foreshore and bioturbated sand of the backshore. The sequence is capped by peat and laminated clays deposited in barrier-fringing marshes (Fig. 8.30).

The basal unit of the second sequence consists dominantly of series of thinning-upward sets of tabular crossbeds that occur in channels up to 20 m wide and 5 m deep. Accessory features include smaller-scale trough sets, reactivation surfaces, flaser bedding, and rare burrows. These sediments are interpreted as the deposits of subtidal channels, and they are overlain by intensely bioturbated sand with remnants of trough cross stratification and flaser beds that may have been deposited in an intertidal flat setting. Interbedded sand and clay occur locally in channels cut into these sediments and may represent deposits of tidal creeks that were only intermittently active.

Carter (1978) interpreted the Cohansey Sand to represent a coastal environment consisting of a barrier island and an extensive back-barrier area dominated by tidal channels and tidal creeks (Fig. 8.30). The lack of mud in the sequence, unusual in back-barrier settings, is attributed to the ability of migrating tidal creeks to erode mud and flush it out of the system.

The La Ventana Tongue of the Cliff House Formation is one of the numerous examples of coastal sand bodies occurring in the Cretaceous of the Rocky Mountains. It consists of a series of stacked sandstone units that apparently were deposited in barrier systems (McCubbin, 1982).

Each unit consists of an upward-shoaling sequence of shoreface and beach sediments deposited along a northeastward prograding shoreline. The sandstones thin rapidly to the northeast in an offshore direction, where they pinch out in marine shales, and to the southwest they are replaced by dark shales and coals deposited in lagoons (Fig. 8.31). Thin sandstones extending from the main sandstone into the lagoonal sediments were probably deposited as washover fans.

Sandstones are as much as 60 m thick along the axis of the barrier. This anomalous thickness may have resulted from an overall balance between sediment yield and basin subsidence; but because several units exist within the sandstone, each consisting of a shoaling-upward sequence, it is more likely that there are several progradational episodes represented within the La Ventana Tongue. The progradational episodes were separated by minor transgressions within a framework of coastal onlap, and deposition on the barrier system ceased as the complex was overstepped by marine mudstones of the Lewis Shale.

Strandplain Systems

Sedimentary processes and products.   Strandplains are broad accumulations of sand that extend along the coast for a significant distance, and whose surfaces are commonly marked by closely spaced, coast-parallel ridges. The ridges of some strandplains may be strikingly parallel, but most often the ridges are grouped into sets of parallel ridges oriented at slight angles to each other (Fig. 8.32). Changes in orientation of beach ridges represent realignment of the coastline caused by

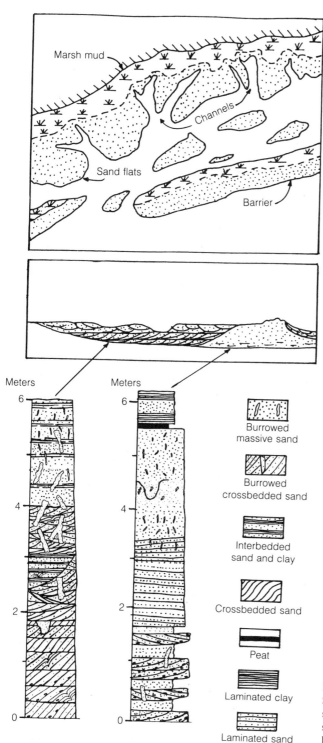

Figure 8.30 Sediments of the Cohansey Sand of the New Jersey coastal plain showing the different stratigraphic sequences developed on the barrier and behind the barrier (after Carter, 1978).

Marsh mud

Channels

Sand flats

Barrier

Meters

Meters

Burrowed massive sand

Burrowed crossbedded sand

Interbedded sand and clay

Crossbedded sand

Peat

Laminated clay

Laminated sand

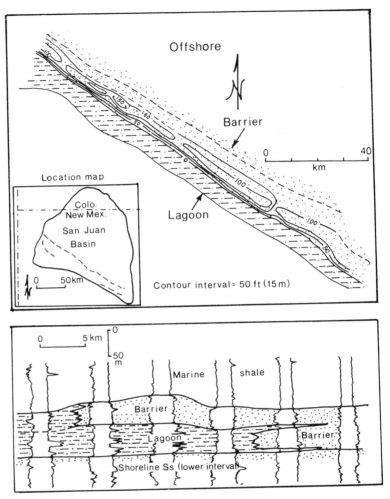

Figure 8.31 Barrier island and lagoonal sediments of the La Ventana tongue of the Cliff House Formation (after McCubbin, 1982).

changes in oceanographic conditions or sediment supply (Curray and Moore, 1964; Curray and others, 1969).

Curray and others (1969) suggested that ridges form when longshore bars aggrade above sea level during periods of reduced wave activity and large sediment influx. Most other workers, however, attribute them to storm berms built above the level of fair-weather wave action where they may be stabilized by vegetation. For the most part, the ridges form only a core for eolian sands that constitute the bulk of the ridge (Fraser and Hester, 1977) (Fig. 8.33).

Ridges are separated by swales. In continuously humid climates, swales are almost always moist and frequently contain standing water. In contrast to the ridges,

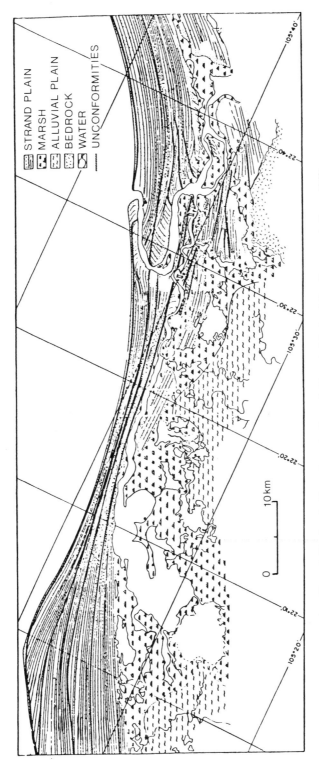

Figure 8.32 Beach ridges along the Costa de Nayarit, Mexico (after Curray and others, 1969).

STRAND PLAIN
MARSH
ALLUVIAL PLAIN
BEDROCK
WATER
UNCONFORMITIES

10 km

0

105°40'
22°40'
22°30'
105°30'
22°20'
22°10'
105°20'

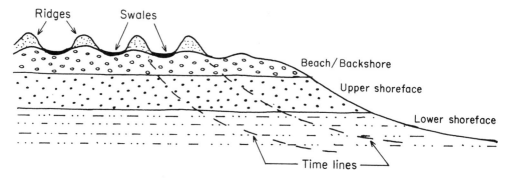

Figure 8.33 Diagrammatic cross section through a prograding strandplain.

swales are commonly vegetated and are effective sediment traps for windblown sediments. In dry climates or in regions that experience extended dry periods, swales can be dry or they may contain hypersaline water (Curray and Moore, 1964).

Where rivers cross the strandplain, swales may accumulate alluvium during flood stages. During early stages of progradation, only swales may be sites of such deposition, but as swales aggrade in older parts of the plain, ridges may be buried and the strandplain sequence may be capped by overbank muds (Curray and Moore, 1964).

Extrinsic controls on strandplain systems.   The shorelines of strandplains are nearly unbroken along the coasts they occupy. A progradational sequence, therefore, would consist of a relatively simple upward progression from offshore through shoreface and into foreshore and backshore sediments. Thickness of each facies, types of sedimentary structures present, and the degree of bioturbation would vary according to the conditions at the time of deposition. Normally, however, one would expect a coarsening-upward sequence concomitant with a decrease of infaunal traces and an increase in evidence of wave-induced bottom shear. A prograding strandline would produce a sheet sand with horizontal facies boundaries and time lines that are concordant with the shore profile at the time of deposition (Fig. 8.33).

The most intensely studied Holocene strandplain is on the Costa de Nayarit, Mexico. The plain extends for over 200 km along the coast and is as much as 15 km wide. Extensive borings through the plain reveal that the sand body consists of two parts (Curray and Moore, 1964; Curray and others, 1969). The basal part consists of sands deposited during the early Holocene transgression. They are the erosional remnants of a barrier island because they, in turn, overlie Holocene lagoonal sediments. The upper part of the sand body was deposited during a regressional phase. Landward, the regressive sands overlie the transgressive sheet sands directly, but the part of the sand body near the coast is currently prograding over inner shelf muds (Fig. 8.34).

Strandplain sequences in the rock record.   Numerous examples of strandplain deposits have been interpreted from the Cretaceous of the Rocky Mountains.

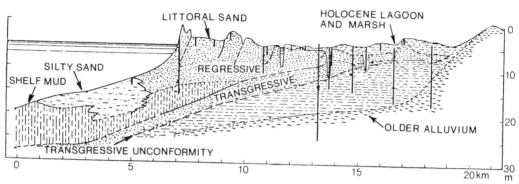

Figure 8.34 Cross section through the strandplain of the Costa de Nayarit, Mexico (after Curray and others, 1969).

The Parkman Sandstone of Wyoming, for instance, consists of a series of strand-plain or barrier island sandstones (Asquith, 1970). Each sand body within the Parkman thins downdip to the southwest, and the unit as a whole intertongues downdip with marine siltstones and shales (Fig. 8.35). Sandstones at the base of each sand body are fine-grained and thin-bedded, and they have a gradational lower contact with the marine siltstones and shales. Massive sandstones occur at the top of the unit, which is overlain with coals, mudstones, and thin sandstones. These nonmarine units overlie the massive sands with a sharp contact to the northwest, but to the southeast, in the direction of progradation, relatively thick sand bodies occur near the base of the nonmarine sequence, which eventually pinches out against the massive sandstones (Fig. 8.35).

The Lower Cretaceous Muddy Sandstone of the Bell Creek Field in Montana has also been interpreted as a barrier island deposit that closely resembles the progradational sequence within Galveston Island (Davies and others, 1971) (Fig. 8.36). It thus probably represents the initial stages of evolution of barriers into strandplains. Bioturbated mudstones of the lower shoreface grade upward into bioturbated sandstones of the middle shoreface, and upper shoreface/beach deposits consist of sandstone in low-angle planar crossbeds. The sequence is capped by massive sandstone with vertical root traces suggestive of back-barrier settings.

The total sequence here is less than 10 m thick and was probably deposited during a single progradational episode. The barrier did not prograde very far, however. The sands of the barrier thin gradually to the northwest into marine mudstones, and to the southwest they are flanked by lagoonal sediments. The lagoonal sediments are variable, but, in general, fine-grained sediments occur at the base of the sequence, and sandier sediments representing deposition on marginal areas of the lagoon occur near the top. The sequence is most commonly capped by root-mottled claystones that may have been deposited in a marsh established after the lagoon was filled in.

The lower part of the Gallup Sandstone of northwest New Mexico has been interpreted by McCubbin (1972) as a strandplain/delta system. The lower interval

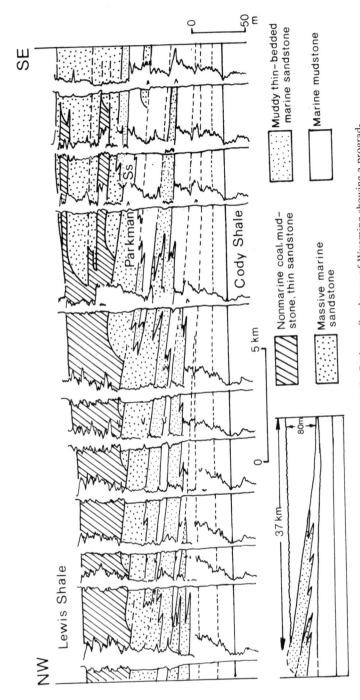

Figure 8.35 Cross section through the Parkman Sandstone of Wyoming showing a prograding sequence of coastal sandstones deposited in an offlapping sequence into a shallow sea (after Asquith, 1970).

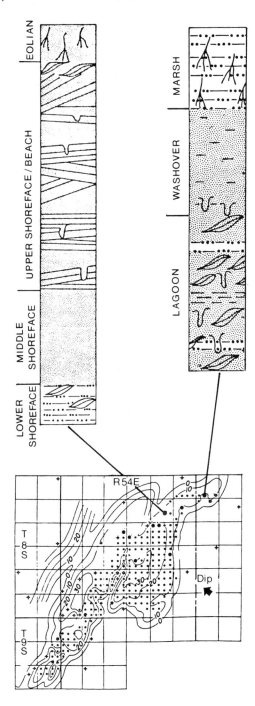

Figure 8.36 Barrier island and lagoonal sediments of the Muddy Sandstone in the Bell Creek Field of Wyoming (after Davies and others, 1971).

rests with a gradational contact on the offshore marine Mancos Shale, and it is overlain by barrier island and lagoonal deposits of the upper part of the Gallup Sandstone (Fig. 8.37).

The strandplain sediments form a progradational sequence that coarsens upward from offshore bioturbated mudstones through shoreface and foreshore sandstones. The sequence then fines upward through beach and backshore deposits. The sandstones thin gradually in the offshore direction, but landward they pinch out abruptly against delta plain sediments. Lagoonal deposits are rare in the sequence,

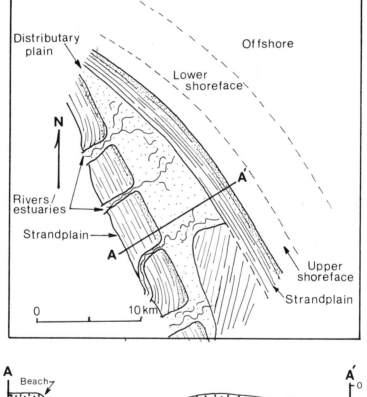

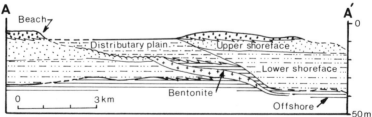

Figure 8.37 Prograding strandplain deposits of the lower Gallup Formation. The sandstones occur in two bodies separated by deltaic sediments (after McCubbin, 1972; McCubbin, 1982).

and this part of the Gallup Sandstone probably formed as a strandplain attached to the seaward edge of an abandoned delta (Fig. 8.37). The presence of another strand-plain sequence landward of the delta deposits suggests that episodes of reduced sediment influx alternated with periods of greater influx, resulting in cycles of deltaic and strandplain deposition along a continuously prograding coast. A similar arrangement of depositional systems can be seen in the Pleistocene terraces of South Carolina (Colquhoun and others, 1972) and Georgia (Hoyt, 1972), although these formed during eustatic lowering of sea level during periods of glacial advance.

9

Deltaic Coastlines

Introduction

Deltas form where rivers enter standing bodies of water that are incapable of completely redistributing the incoming sediment. Flow expansion takes place at the river mouth, and the carrying capacity of the river is rapidly attenuated. Sand-sized sediment is deposited almost immediately, but basinal processes may redistribute it to a greater or lesser extent. The suspended load of the stream is transported into the basin as a buoyant plume.

Basinal processes, however, are not necessarily confined to the river mouth. Channel and floodplain facies of the delta plain component may both respond to a high tidal range, and the delta plain may also feel the impact of strong storm surges. Even where deltas prograde into basins with weak tides and low wave energy, the effects of salt wedges moving up distributary channels may be felt far from the delta margin.

Deltas are probably the single most complex depositional system because they develop through the interaction of a large number of variables. In addition, the complexity is magnified in three dimensions because all the variables, or any combination of variables acting together, are subject to change through time.

In the broad sense, these variables may be divided into those affecting the sediment yield and those affecting the basin into which the delta is building (Fig. 9.1). Source-area variables primarily affect the quantity and caliber of the sediment provided to the delta and the timing of its delivery. Large basins with integrated drainage nets can supply large volumes of sediment that can potentially build extensive deltas. Mature drainage basins also tend to be drained by trunk streams with

202

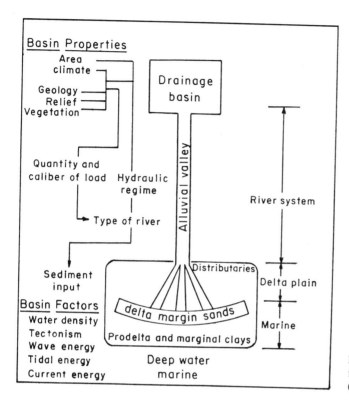

Figure 9.1 Flow chart showing the factors involved in delta construction (after Saxena, 1976).

large suspended loads. These build permanent channels confined within levees on the delta plain (Kolb, 1963), and they also can build broad mud platforms that favor rapid progradation of subaerial delta components (Galloway, 1975). Under such conditions, fluvial-dominated deltas form, whereas rivers that drain immature basins tend to be bedload rivers that build deltas dominated by basin processes (Fisher and others, 1969).

It should be observed that this relationship is not an invariable rule. The Amazon River, for instance, is the world's largest river in terms of discharge, but it posesses only a poorly developed delta. It drains an extensive basin and carries a large suspended load, but it builds into a basin where ocean currents sweep the mouth of the river and carry the suspended load downdrift, where it is deposited on the mud beaches of the Guiana and Surinam coast (Wells and Coleman, 1981).

Immature drainage basins or those subject to severe seasonal climatic fluctuations (Mediteranean or monsoon-type climates) characteristically show periods of large discharge that can produce two effects on deltas. First, episodic discharge is often associated with bedload streams carrying coarse sediments. Distributary channels carrying coarse sediments are unstable, and their load is deposited rapidly at river mouths, where it is reworked into sandy beaches and barriers. Under these conditions, the sand may be distributed evenly along an arcuate delta margin, rather than being concentrated along a narrow front.

The timing of maximum sediment yield also affects delta shape. If the period of maximum input is in phase with the period of maximum energy imposed on the delta by basin processes, sediment may be redistributed continuously along the delta margin. If, on the other hand, maxima are out of phase, there may ensue alternating periods of rapid progradation, followed by reworking and retrogradation (Coleman and Wright, 1975).

The character of both basin and river waters also controls, to some extent, the shape of the delta. Bates (1953) modeled the interaction of river and basin waters in terms of jet-flow theory. If river waters are denser than basin waters, either because of temperature or suspended load, a two-dimensional jet forms and density-driven underflows carry sediment into the basin. If river and basin waters are nearly equal in density, a three-dimensional jet forms. Bedload is deposited at the river mouth, and suspended load is carried a short distance basinward. Three-dimensional mixing is rapid in such a case, however, and the flow decelerates rapidly.

In most marine basins, river waters are less dense than basin waters because of salinity differences. The river water becomes buoyant at the river mouth, and a two-dimensional jet enters the basin at the surface. Bedload is deposited at the river mouth because bottom shear is no longer maintained, but suspended load may be carried far into the basin, because only two-dimensional mixing acts to decelerate the flow.

Once sediment is introduced to the delta front, basin processes begin to redistribute it. The single most important factor in this process is the wave regime (Coleman and Wright, 1975). The configuration of deltas may range from those affected very little by wave action (Mississippi delta) to those dominated by wave activity (Senegal or Grijalva deltas).

Wave activity is small in enclosed basins or those with broad shallow shelves. A river can extend its mouth seaward with little interference in such a setting, and a lobate delta results. Rapid progradation of the river mouth decreases the longitudinal profile of the stream, resulting in an unstable situation that can be relieved only through avulsion upstream. The process may occur frequently in fluvially dominated deltas, which may consist of numerous overlapping delta lobes.

Wave activity tends to be large in basins with steep, concave margins (Coleman and Wright, 1975). Where deltas prograde into such basins, channel patterns tend to be more stable, and they advance along a broad front consisting of reworked sediments originally deposited at river mouths.

Tidal processes cause a major impact on delta shapes, and they affect both delta plains and delta margins (Wright and others, 1973). At the river mouth, sediments are reworked into linear shoals that parallel tidal currents. If the currents are particularly strong, they may totally overwhelm fluvial processes in distributaries, causing upstream movement of bedload. Fluctuating current velocities accelerate aggradation of channels, causing them to be choked with sandy shoals and augmenting processes of lateral accretion as well.

Where the tidal range is exceptionally high, the delta plain may display features more characteristic of tidal flats (Coleman and Wright, 1975). During flood peaks, for instance, water levels may overflow channel margins, and during ebb they drain through sinuous tidal creeks. On many deltas, the level of the lower delta plain is adjusted to mean sea level, but on tide-dominated deltas, the delta plain aggrades to the level of mean flood tide in a manner similar to that of tidal marshes.

Oceanic currents driven by wind, tide, or deep oceanic circulation may affect deltas by redistributing sediments parallel or subparallel to depositional strike (Coleman and Wright, 1975). The Guiana Current, for instance, entrains much of the sediment provided to the coast by the Amazon River and carries it westward to the muddy coast of Guiana and Surinam (Wells and Coleman, 1981). In the Malacca Straits, strong tidal currents transport bedload away from river mouths and build sand ridges parallel to the axis of the straits (Keller and Richards, 1967).

The basin geometry also exerts a considerable control on the ultimate character of a delta. Where shelves are wide and their slopes gentle, wave energy may be dampened considerably by frictional forces. Rapid progradation of delta fronts occurs (Coleman and Wright, 1975) as a series of overlapping lobes because of rapid switching of distributary channels. Persistent high wave energy occurs where deltas prograde out onto steeper slopes. Distributaries break off at common points upstream, and channel paths tend to be more stable. Distributaries are simultaneously active, and the delta advances along an arcuate front.

The slope of the shelf also exerts a control on the shape of the delta. On broad, shallow shelves, rates of progradation can be rapid because the volume of sediment needed to aggrade the delta profile above sea level is less compared to that needed on steeply sloping shelves. With slower rates of progradation, the delta margin is subjected to increasingly longer periods of exposure to basin processes, thereby increasing the ability of the processes to redistribute incoming sediments. Through a feedback loop, even a delta prograding onto a shallowly sloping shelf may evolve from a rapidly prograding elongate system to an arcuate one, simply because it is increasing its volume geometrically, although its sediment influx remains constant.

Several attempts to classify deltas were proposed on the basis of the relative importance of the various factors discussed above. The most recent of these by Coleman and Wright (1975) and Galloway (1975) classify deltas according to the manner and degree in which basin processes redistribute incoming sediment.

Coleman and Wright (1975) proposed six models intended to be representative of the variation that can be expected where fluvial processes interact with waves and tides. Galloway (1975) defines fluvial-, wave-, and tide-dominated deltas as end members of a ternary relationship, where the deltas may be positioned in the diagram according to the relative importance of the three processes (Fig. 9.2). Galloway's classification emphasizes the gradational character among delta types, rather than categorizing discrete "model" deltas.

The different components of a delta system may also react differently to the imposed stresses (Elliott, 1978b). A delta plain, for instance, is normally dominated by its distributaries, even though the delta margin may be thoroughly reshaped by

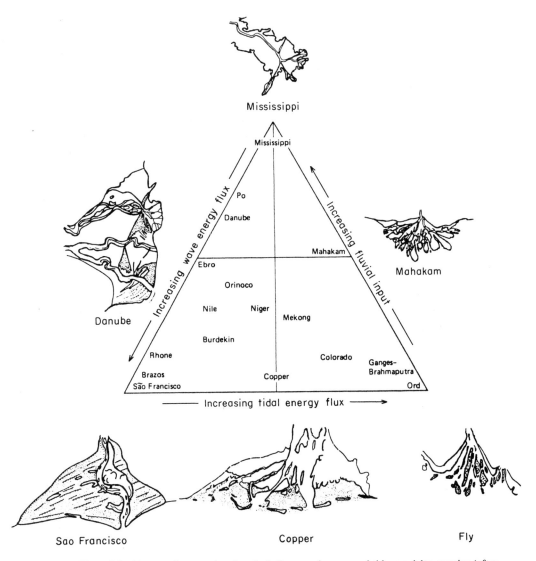

Figure 9.2 Ternary diagram showing the influence of waves and tides on delta margins (after Galloway, 1975).

waves. Deltas in basins with a high tidal range may possess a river-dominated upper delta plain, a tide-dominated lower delta plain, and a tide/wave-dominated delta margin, as in the case of the Niger Delta. It is apparent, therefore, that any classification system must account not only for the variability of processes, but also for the lateral variation of responses within a delta system.

Sedimentary Processes and Products

Delta Plain

River-dominated delta plains are characterized by unstable distributary channels with unidirectional flow separated by interdistributary areas occupied by marshes, lakes, or shallow bays (Fig. 9.3).

Fining-upward sequences develop in distributary channels but, unlike other rivers, they develop in response to channel abandonment rather than lateral migration. They also differ from other rivers in that, even in basins of lesser energy flux, episodes of abnormally large wave or tide activity can effectively stop discharge, causing some of the suspended load to be deposited in the channel as clay drapes on sandy bedforms.

Channel-margin facies develop in response to periodic flooding during large discharge. Flooding may top channel margins as a sheet flow that deposits coarse-

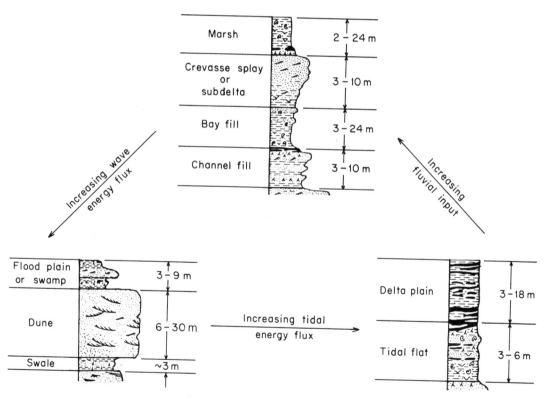

Figure 9.3 Ternary diagram showing typical delta plain sequences developed in river-, wave-, and tide-dominated deltas (after Galloway, 1975; Coleman and Wright, 1975).

fine couplets in response to rapid deceleration of the flow as it leaves the channel. A levee forms and its deposits become finer grained and thinner away from the channel, and a coarsening-upward sequence develops as the levee expands into interdistributary areas.

Flow may also leave the channel through crevasses in the margin. Small crevasses may take the form of levee-bounded channels leading into interdistributary areas, or they may form lobate masses of coarse sediment formed by successive flows that are separated by muddy sediments deposited during slack water. Large splays may prograde into interdistributaries as minor deltas complete with distributary bars where semipermanent crevasses occur.

Splays also form coarsening-upward sequences as they prograde. Interdistributary muds, commonly with a high organic content, grade upward into silt and then sands that display all the features coincident with highly episodic flow, including coarse–fine couplets, reactivation surfaces, and fining-upward sequences deposited by density flows (Fig. 9.3).

Except near the channel, interdistributary facies consist of muds, often organic and highly bioturbated, that accumulate in an environment only episodically disturbed by flood currents or wind-generated waves. The character of the interdistributary facies is dependent, to a great extent, on the climate. Organic productivity is high in tropical areas, and the bulk of the sediment may consist of peat. Productivity is low in arctic regions, but the processes of biological degradation are also low, and sedge and moss peats dominate the sediment (Coleman and Wright, 1975). Fine-grained clastics dominate interdistributary fillings in temperate regions where peats are thinner and intermittent.

Evaporative processes are important in interdistributary areas in arid climates. Chemical precipitates can form, and their composition may range from calcrete and gypsum to halite, depending on the amount of rainfall and its seasonal distribution.

Tide-dominated delta plains occur when the tidal range is sufficiently high to enter distributaries during flood and spill over channel banks into interdistributary areas. Tidal distributaries tend to be straight and funnel shaped, and they are subject to short-period variations in flow velocity and direction.

Clean sands are deposited in the central parts of channels, whereas, on the margins, muddy sands accumulate because flow velocities are insufficient to remove suspension deposits that settle during slack water. At some upstream point, flow becomes impounded by tides during flood, and bedload is deposited as a series of sand bars and mudflats that have been termed "inner deltas" (Elliott, 1978b). Linear sand ridges form in the lower delta plain. They are oriented parallel to channel margins, and they form in response to reversing tidal currents (Fig. 9.4).

Tidal channels build fining-upward sequences that may dominate the delta plain sequence in response to pronounced channel migration (Oomkens, 1974). Scoured bases are overlain by crossbedded sands that grade upward into muddy bioturbated sands. These sands may display reversed crossbedding, flaser bedding, and other features indicative of reversing flow of short-term period. The sequence is capped by interdistributary sediments (Fig. 9.3).

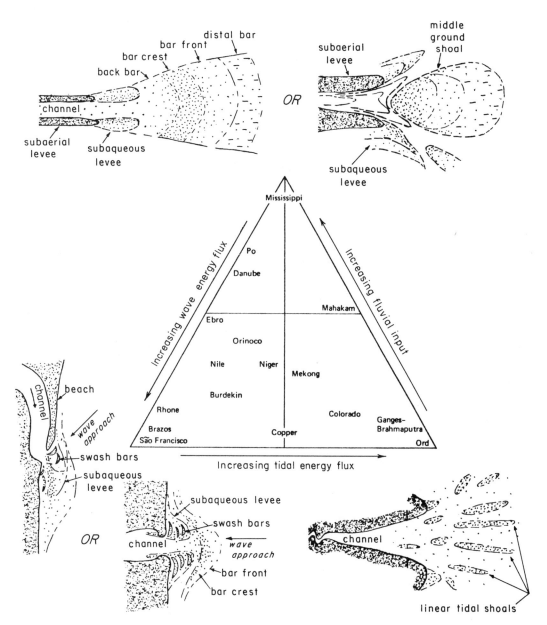

Figure 9.4 Ternary diagram showing the morphology at the distributary mouth of river-, wave-, and tide-dominated deltas (after Galloway, 1975; Wright, 1977).

Interdistributary areas of tide-dominated deltas are strikingly different from those of river-dominated ones. The entire delta plain may be inundated during tidal flood and drained during ebb. The plain may be intricately laced with sinuous tidal creeks, each with its own set of tributaries. These creeks also tend to migrate and deposit sand in channels and on point bars. The surface of interdistributary areas builds to the level of the spring tide, but during neap tide the surface may be exposed. Swamps with high organic productivity form in humid tropical areas, whereas salt flats form in arid ones.

Two types of delta plains occur in wave-dominated deltas. Where deep-water waves approach normal to the coast, the plain consists of arcuate beach ridges arranged symmetrically about the river mouth (Fig. 9.4). The high wave energy produces exceptionally clean sand beaches capped by dunes, and the high elevation of the ridges prevents inundation either by river or tidal flood waters. Over much of the plain, therefore, fine sediments can accumulate only in the swales between dune ridges. The resultant sediments may consist almost entirely of sands reworked by marine processes (Coleman and Wright, 1975).

A pronounced littoral drift is produced when deep-water waves approach the shore obliquely. Sand deposited at the river mouth is transported downdrift, producing a sandy beach barrier that protects the river channel from wave attack and deflects the river mouth from the delta apex. As the river lengthens its course, it becomes unstable, and eventually it breaks through the barrier near the apex and the process begins once more.

The resulting delta plain consists of a series of broad barriers, oriented more or less parallel to the coast, that are separated from each other by lowlands representing the former river courses. These lowlands are much broader than the swales separating beach ridges, and they may fill with muddy organic sands that accumulate after channel abandonment. These deposits are likely to be preserved because of the protection offered by the barrier (Fig. 9.3).

Delta Front

The Mississippi Delta represents an extreme where very large amounts of sediment are provided to a basin with negligible tide or wave energy flux. The delta front is characterized by elongate lobes consisting of a channel flanked by levees, splays, and subdeltas and fronted by a distributary mouth bar (Fig. 9.5). The bar consists of a back bar, a crest, and a bar front that slopes offshore into the prodelta. It progrades as the back bar and bar crest are eroded, and the products of the erosion are added to the bar front.

Progradation produces a coarsening-upward sequence consisting of sand, silt, and clay laminae at the base and becoming siltier and sandier upward (Fig. 9.5). These are bar front sediments that accumulate in response to alternating periods of suspension deposition from the buoyant sediment plume and periods of wave reworking of the sediment.

The sand becomes coarser and less muddy upward in the sequence, and it may

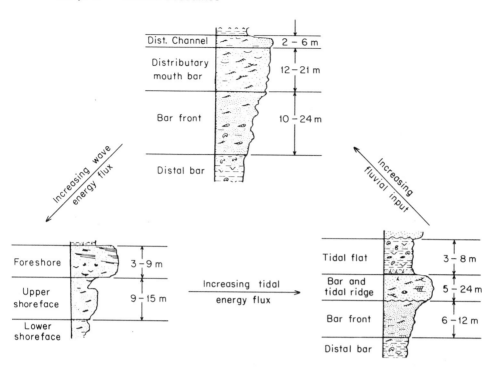

Figure 9.5 Ternary diagram showing typical sequences developed at margins of river-, wave-, and tide-dominated deltas (after Galloway, 1975; Coleman and Wright, 1975).

display small-scale primary structures. These sediments represent the bar crest where wave attack is concentrated. The distributary may cut through the bar crest in one or two places, and upon abandonment these channels can be filled either with fining-upward sequences or with mud. The back-bar sediments stand relatively little chance of preservation because they are eroded by the channel as distributary bars prograde.

   Most commonly, river-dominated deltas experience a far greater wave energy flux than that that attacks the Mississippi Delta. They characteristically exhibit a relatively smooth arcuate margin, with minor protuberances produced by subdued distributary mouth bars. Distributary mouth bars may prograde much like those of elongate deltas, but interdistributary areas are less pronounced; the delta front may be composed of beach ridges, and wave-formed sand bars may occur on the surface of the distributary mouth bar (Nelson, 1970).

   The delta front of wave-dominated deltas consists of broad beaches composed of well-sorted sand. Where waves attack frontally, river flow decelerates rapidly, and bedload is deposited a short distance from the river mouth. Steep velocity gradients also occur at the lateral margins of the flow, resulting in rapid deposition of bedload and the formation of subaqueous levees (Wright, 1977). As they are deposited, however, they are being reworked continuously into wave-formed shoals and swash bars, and together with sandy beaches they may form a delta front that is entirely marine in character (Fig. 9.4).

An exception occurs in those cases where a strong littoral drift transports sand down the coast from the delta apex, forming a barrier that can protect the distributary mouth bar from wave attack. In the Senegal Delta, for example, the bar consists of interbedded clay, silt, and ripple-laminated sand (Coleman and Wright, 1975) (Fig. 9.5).

The mouths of distributaries of tide-dominated deltas display a broad funnel shape and contain linear sand shoals oriented parallel to the channel (Fig. 9.4). The tidal equivalent of a distributary mouth bar may consist of a broad shoal area consisting of river-borne sediments that have been reworked by tidal currents into a complex of sand ridges and flats. Tidal ridge sands are well-sorted and commonly occur in crossbeds with bidirectional dips. The sands may grade landward into channel-fill deposits consisting of crossbedded sands with occasional thin clay layers and lenses of mud-filled scours. The channel fills rest on sharp scoured bases, and the infilling sands commonly occur in landward-dipping crossbeds (Fig. 9.5).

Downdrift from the mouth of the distributaries are broad tidal flats. Near the river mouths, where rates of sedimentation are large, the tidal flats may be nearly devoid of infauna; but farther away from the river mouth, they may be thoroughly bioturbated like other tidal flats.

Active migration of tidal channels, however, reduces the preservation potential of tidal flats. Instead, delta-front sequences may consist entirely of shoal and tidal sand ridge sediments.

Prodelta

In front of deltas are the prodeltaic deposits consisting of muds that are often bioturbated and rich in organic debris. They are the result of suspension deposition of those fine-grained sediments carried basinward by a buoyant plume from the river mouth. These sediments are occasionally reworked by storm waves into coarse–fine couplets, and density currents carry sands from the river mouth into the prodelta area.

Coarsening-upward sequences consisting of silts and clays at the base and silts and sands at the top are the rule, but variations in thickness and composition occur among the three major delta types. Buoyant forces carry fine-grained sediments far out from distributary mouths of river-dominated deltas (Wright, 1977). These sediments can be carried great distances in suspension, and they may settle out in relatively deep water. There they can accumulate to great thicknesses and form broad platforms over which delta-front sediments prograde (Fig. 9.6).

Tidal mixing in front of distributary mouths of tide-dominated deltas destroys vertical density gradients, reducing the effect of buoyant forces (Wright, 1977), and strong tidal currents are invariably present in offshore areas. The combined impact of these two processes reduces the thickness of prodelta deposits and can produce bedload transport of the coarse fraction. The resulting deposits consist of inter-

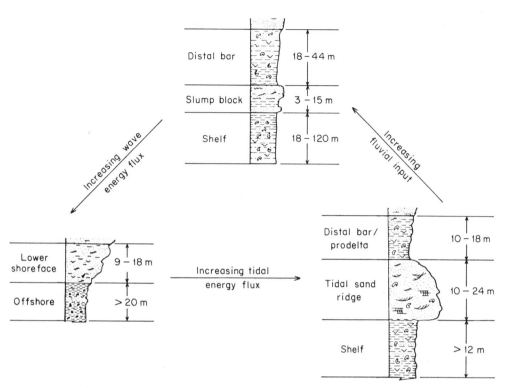

Figure 9.6 Ternary diagram showing typical sequences developed in prodelta aras of river-, wave-, and tide-dominated deltas (after Galloway, 1975; Coleman and Wright, 1975).

bedded muds and ripple-bedded sands that, in some cases, may enclose thick lenses of crossbedded sands deposited as linear tidal ridges (Fig. 9.6).

River flow is also rapidly attenuated at distributary mouths of wave-dominated deltas. Sediment transport capacity is lost within short distances of the mouth, and buoyant effluences are effectively precluded (Wright, 1977). The sediments in front of wave-dominated deltas are also relatively thin and coarse-grained, because finer-grained material may bypass the prodelta area when wave-induced turbulence keeps it in suspension. Frequent reworking by storm waves reduces the impact of infauna, and primary structures are commonly preserved. The base of the sequence may consist of muds with abundant crossbedded silt and sand layers that increase in thickness and abundance upward (Fig. 9.6).

Intrinsic and Extrinsic Controls on the Evolution of Delta Sequences

Simplistically, delta sequences consist of sediments deposited during a period of progradation that are capped by sediments deposited during a period of delta abandonment when retrogradation takes place. Progradation occurs in response to conditions where basinal processes are unable to redistribute incoming sediments, and in a similar manner retrogradation occurs when basinal processes begin to overwhelm the sediment being transported to the delta margin; normally, this occurs when the rate of influx declines. Sediment influx may decline because of intrinsic processes such as distributary abandonment or because of extrinsic factors such as a change in climate.

Walther's principle may be applied to facies distributions in modern deltas so that, during phases of progradation, prodelta sediments are progressively overlain by delta margin and then delta plain sediments. In fact, Coleman and Wright (1975) constructed idealized progradational sequences by applying Walther's principle to lateral facies distributions.

This technique, however, may produce some misconceptions, because it ignores three basic factors involved in the formation of deltaic successions. The first is the most obvious: not all facies show an equal opportunity for preservation. Mangrove swamps, for instance, make up the majority of the delta plain, and beach ridges make up the majority of the delta front in the Niger Delta. Yet, in vertical sequence, these two facies comprise only a small part of the column, which instead is occupied mainly by sediments deposited in rapidly migrating tidal channels (Oomkens, 1974) (Fig. 9.7).

Second, not all facies may be present either parallel or perpendicular to depositional strike. Elongate deltas, in particular, present an extremely complex internal architecture along strike. Prodelta muds, for instance, may form a broad platform that underlies the whole complex. On the platform, however, distributary mouth bar sands flanked by associated channel margin facies are deposited, and these are enclosed by interdistributary muds locally interbedded with crevasse splay sands. Certainly, no single vertical sequence can be representative of these sediments. Even wave-dominated deltas, which show a more uniform internal distribution of sediments along strike, are not uniform perpendicular to strike, and tide-dominated deltas can display considerable complexity in both directions.

Third, delta morphologies change through time, and the facies distribution present during initial stages of construction may not be present during later phases. Earlier, the influence of shelf slope angle on the rate of delta progradation and on the ability of basinal processes to redistribute incoming sediment was cited. If the amount of sediment provided to the delta does not increase through time, the rate of progradation must slow as the amount of sediment needed to construct the prodelta platform increases at a geometric rate. This, in turn, triggers a feedback loop, because as the rate of progradation slows, basin processes are allotted more time to redistribute sediments at the delta margin, further decreasing the rate of progradation.

The Niger Delta, for instance, is wave- and tide-dominated, and the delta front is occupied by beach ridges except where the margin is cut by tidal channels. Using Walther's principle, one could predict that the delta-front sands extend updip into the body of the delta below the delta plain muds.

In the Niger Delta, however, a prism of sand is present at the present delta

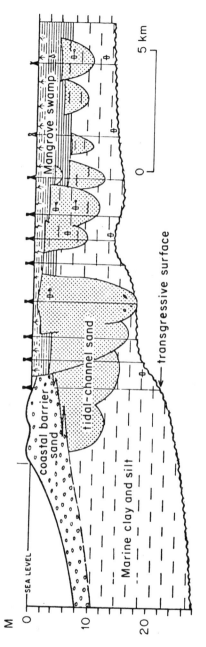

Figure 9.7 Cross section of Holocene deposits of the Niger Delta showing the dominance of tidal channel deposits throughout the delta (after Oomkens, 1974).

front, but not under the delta plain (Fig. 9.7). The initial stages of delta progradation apparently were too rapid to allow even the relatively high wave energy flux in
front of the Niger Delta to extensively rework sands delivered to the distributary
mouth into a broad arcuate delta front. As the volume of the delta increased, however, the rate of progradation may have slowed sufficiently for redistribution of the
sediments to take place.

The process described above is the result of the intrinsic factors involved in
delta building. A similar effect might be created, however, where climate changed
and sediment yield was reduced or where it was reduced in response to changes in
the feeder system described previously for alluvial fans (Schumm, 1977).

Progradational sequences may be bounded by thinner retrogradational sequences formed during phases of delta lobe abandonment initiated by distributary
channel shifting. When a channel shifts, one lobe is abandoned while another begins
a growth phase. Because underlying prodelta muds are water-saturated, abandoned
lobes founder as the muds compact, and the delta plain of the lobe is eventually
transgressed. The modern Mississippi Delta, for instance, consists of seven overlapping lobes that have formed and been abandoned during the last 6000 years (Kolb
and van Lopik, 1966).

During abandonment, the thickest accumulation of sediment occurs at the former delta margin, where shoreline sands and distributary mouth bars are reworked
into barrier islands. The islands separate the open marine environment from a shallow, sometimes brackish water sound that is developed over the foundered delta
plain. Sediments in the sound may consist of clayey silts and sands, with local concentrations of shell material (Coleman and Gagliano, 1964).

Abandoned elongate deltas form discontinuous barrier chains because delta-
front sands are restricted to distributary mouths. The Chandeleur Islands of the
Mississippi Delta are an example of such a barrier chain. Where sands are more
evenly distributed, however, the barrier may be more complete. More even distribution may result from wave reworking, as in the case of Cape Hatteras, which sepa-
rates the Atlantic from Pamlico Sound (Swift and Sears, 1974), or it may result
from transgression of a river-dominated delta, where the distributaries are closely
spaced and form a continuous ring of distributary mouth bars. The abandoned La-
fourche Complex of the Mississippi Delta was such a lobe, and on abandonment it
formed a delta-front sheet sand.

As transgression continues, the delta margin islands must migrate landward.
But as they cross the relatively sand-poor delta plain, they are separated from their
sediment source, and eventually they lose their definition and spread out to form
an extensive, but thin, sheet sand. As the upper delta plain approaches mean sea
level, it may become a wetland covered by a peat blanket (Elliott, 1978), and as
transgression continues the entire delta plain may become submerged. The Teche
and Maringouin complexes are the oldest of the Recent lobes of the Mississippi
Delta, and they remain only as broad, submerged shoals (Kolb and van Lopik,
1966).

Elliott (1978) suggested that because delta lobe abandonment is initiated by avulsion, the process may be restricted to river-dominated or at least river and wave-dominated deltas where avulsion is most common. Other factors affect the amount of sediment reaching the delta front, however, and these are not specific to any delta type. Even small sea-level rises may also profoundly affect a delta, and where sediment input ceases or sea-level rises, basinal processes may be able to impose a phase of retrogradation on wave- or tide-dominated deltas.

A hiatal surface is the result of delta lobe abandonment. Chemical sediments such as glauconite and phosphate may form when the influx of terrigenous clastic sediment ceases, or shell material may accumulate. Relict sediments become extensively bioturbated, and abundant fecal matter may be added to the sediment where it is commonly glauconitized.

The hiatal surface is oldest in the prodelta and becomes progressively younger landward onto the delta plain. If the abandoned delta is reoccupied, construction of a new delta occurs in a stepwise fashion across the underlying sediments of the abandoned delta lobe (Frazier, 1974) (Fig. 9.8).

More significant interruptions of deltaic progradation result from regional tectonism, isostatic readjustment, or significant eustatic sea-level rise. When such events occur, long-term hiatuses form second-order hiatal surfaces that bound depositional complexes consisting of overlapping and overstepping lobes separated by first-order hiatal surfaces (Frazier, 1974) (Fig. 9.8).

During stillstand, delta progradation occurs in a stepwise fashion across the shelf as delta lobes form and are abandoned. If sea level drops because of deepening of the basin or uplift of the margin, stream entrenchment occurs. Relatively coarse grained sediments are introduced beyond the delta margin, progradation is accelerated, and the former delta plain becomes a relict surface that is no longer being actively provided with sediment and is, in fact, subject to erosion (Fig. 9.8).

Eroded valleys become estuaries if transgression occurs, and they act effectively as sediment sinks that prevent sediment from reaching shelf or slope areas. Major hiatal surfaces are thus formed in these offshore areas, and they merge updip into the hiatal surfaces developed on the delta plain during low stands of sea level (Fig. 9.8).

Formation of a new depositional complex begins as soon as transgression ceases. New deltas develop and migrate across both subaerial and subaqueous hiatal surfaces, and if transgression is not renewed, these sequences eventually prograde beyond the margin of the previous complex, resulting in an offlapping configuration (Fig. 9.8).

The stratigraphic sequence that would develop as a result of such a series of events is based on the sequence deposited by the Mississippi Delta as it responded to changes in sea level caused by Pleistocene glaciations. Minor cycles caused by yearly floods are apparent in coarse–fine couplets in delta plain, levee, and splay deposits and in the episodes of rapid progradation of distributary mouth bars. These cycles are superimposed on longer-period cycles formed in response to major changes in channel margin morphology that occur during the formation of large splays or subdeltas, where relatively thick coarsening-upward sequences develop.

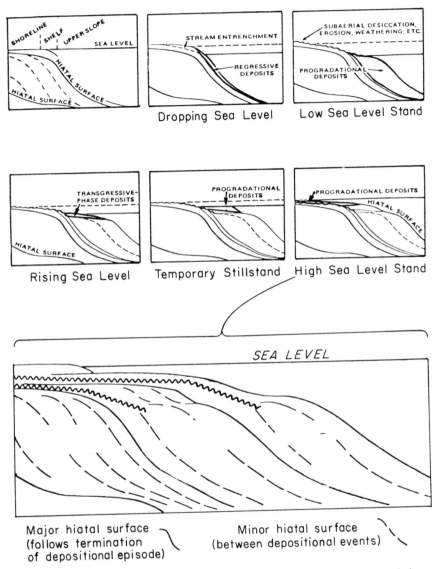

Figure 9.8 Diagrammatic representations of stages in delta growth during a period of large-magnitude, eustatic sea-level changes (after Frazier, 1974).

These, in turn, are superimposed on even longer-period cycles where very thick coarsening-upward sequences record the progradation of individual delta lobes.

Seasonal flooding may impose minor growth cycles on wave- or tide-dominated deltas, but lobe abandonment is probably not an important process, and major cycles form in response to relative sea-level changes. Regressive sequences of the Niger Delta, for instance, consist of bioturbated muds with thin sand layers at the base that grade upward into delta-margin barrier or tidal channel sands, or delta plain swamp deposits depending on position in the delta (Oomkens, 1974) (Fig. 9.9).

In areas away from the coastline, the uppermost regressive sequence overlies strongly bioturbated sandy clays. This thin unit probably represents a time of non-deposition or very slow deposition that occurred following a transgressive phase, and it is probably equivalent to the second-order hiatal surfaces found within the Mississippi Delta.

Lagoon and swamp deposits of the lower delta plain accumulate over upper delta plain sediments during transgression. Sands of the coastal belt transgress muddy lagoonal belt sediments, and because tidal range is high, sands are primarily the result of tidal channel deposition (Fig. 9.9).

In the Niger Delta, both transgressive and regressive sequences grossly coarsen upward. Transgressive sequences, however, develop over soil horizons formed on the deposits of the previous regressive phase, whereas regressive deposits overlie a relict surface formed during a hiatus in deposition at the time of maximum transgression (Oomkens, 1974).

In both cases described above, deltaic sequences were built in response to simple eustatic changes in sea level caused by glaciation, and although the sediment influx may also have varied in response to climatic changes, those fluctuations were far overshadowed by the effect of sea-level changes.

The architecture of deltaic successions depends not only on the ability of marine processes to rework incoming sediment, but also on the ability of the basin to

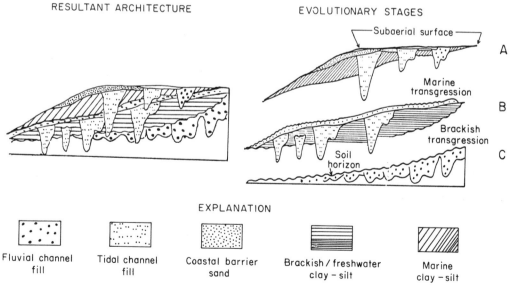

Figure 9.9 Stages in the growth of the Niger Delta during the late Pleistocene rise in sea level (after Oomkens, 1974).

alter its geometry sufficiently to accept it. Where rates of subsidence are small relative to the rate of deposition, deltas build through progradational phases that overstep themselves basinward (Curtis, 1970). Lateral facies relationships are reproduced in vertical sequence, with landward and shallow water facies overlying deeper water ones (Fig. 9.10).

This may be the case in cratonic sequences, but on more active margins the relative rates of subsidence and deposition may be in delicate balance, and the resultant architecture may consist of vertical rather than horizontal facies boundaries (Fig. 9.10). Although it may seem improbable that such a balance could be common or could persist through time, vertically stacked facies distributions can occur where the transfer of sediment from source area to receiving basin leads to subsidence of the basin by loading and to uplift of the source by isostatic rebound. And as suggested previously, such a situation may occur as part of a natural succession of

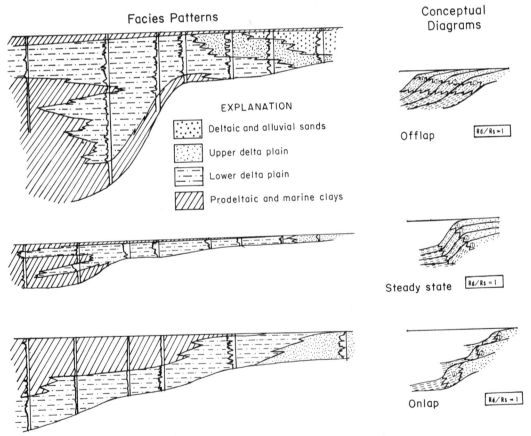

Figure 9.10 Types of deltaic architecture that can develop under varying conditions of basin subsidence and sediment influx (after Curtis, 1970).

deltaic growth when a delta progrades to a point of equilibrium with basinal shape and processes.

An onlapping sequence of progradational wedges is deposited progressively landward where rates of subsidence are in excess of rates of deposition (Fig. 9.10). The lateral facies distributions of progradational wedges are similar to those where rates of deposition are equal to or greater than rates of subsidence; but they are laterally less extensive, and in vertical sequence landward facies are overlain by basinward ones (Fig. 9.10).

These situations occur in regular succession through time in the Miocene deltaic deposits of the Louisiana Gulf Coast (Curtis, 1970). Two megasequences occur in these deposits. During the initial stages of deposition of each megasequence, rates of deposition are far in excess of rates of subsidence, and thick progradational wedges of sediments, consisting of vertically stacked facies sequences, are deposited. Each progradational wedge is bounded vertically, however, by thin transgressive units (Fig. 9.11).

The ratio of deposition to subsidence declined through time, however, so that at some point in the development of the megasequence they were in balance, and lateral facies distributions built vertically. Eventually, the rate of subsidence exceeded that of deposition, an onlapping succession developed, and the megasequence was capped by thick marine shales.

Thus, in the normal evolution of these deltas, the extrinsic factors of sediment influx and basin tectonics control, to a large measure, the three-dimensional patterns of deposition. The patterns, however, do not appear to be functions of chance,

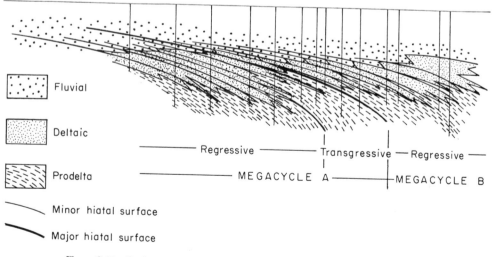

Figure 9.11   Cycles of delta growth and abandonment in the Miocene deltaic deposits of the Louisiana Gulf Coast (after Curtis, 1970).

but develop in response to variations that occur regularly and cyclically through time.

Deltaic Sequences in the Rock Record

Among the most-studied ancient deltaic deposits are the Carboniferous deltas of the Appalachian region (see Ferm and Horne, 1979). Their interpretation has been the subject of much controversy (see Ettensohn, 1980, and Rice, 1985, for discussion), and the deposits, even after decades of study, can be attributed to no single modern counterpart.

The deltaic sequence consists of three major facies associations deposited in beach-barrier and lagoon, lower delta plain, and upper delta plain settings. Quartz arenites deposited on barrier islands separate red and green shales with marine affinities from dark gray coal-bearing shales (Fig. 9.12). Associated with the dark gray shales are graywackes deposited in tidal and fluvial channels.

Quartz arenites are as much as 15 m thick and extend for tens of kilometers along strike. Sedimentary structures consist of seaward-dipping accretionary foresets and trough and planar crossbeds dipping parallel to the axis of the sand body, suggesting a considerable amount of longshore transport (Horne and others, 1978).

Thin sheets of ripple-bedded and burrowed quartz arenites extend from the barriers landward into the gray shales. These quartz arenites probably formed as storm washover deposits and thicker, wedge-shaped bodies of sand that also extend into the gray shales probably originated as flood-dominated tidal deltas (Fig. 9.12). It should be noted that Rice (1985) attributes these sediments to fluvial deposition.

Red and green marine shales occur seaward of the sand bodies. Away from the sand bodies, they are interbedded with thin bioclastic carbonates, whereas in more shoreward positions the mudstones are interbedded with tidal-flat dolomites. Organic shales occur in sequences 10 to 15 m thick and form the principal component of the lagoonal deposits shoreward of the sand body. Sequences are separated by thin coals or burrowed siderite zones, and each consists of claystone at the base grading upward to siltstone or silty sandstone at the top. Shales may be extensively burrowed, but sandstones and siltstones commonly are plane bedded or flaser bedded. These deposits represent shoaling-upward sequences in a lagoonal environment, and the presence of phosphatic brachiopods suggests brackish-water conditions. Thin coals at the top of some sequences indicate that lagoons periodically were filled to the depth where rooted plants could be established.

Tidal channel deposits up to 10 m thick are enclosed within the lagoonal sediments. Channel fills locally consist of fining-upward sands that may extend landward from the quartz arenitic barrier or may be isolated in the lagoonal mudstones. The influence of tidal deposits is indicated by the presence of both landward- and seaward-dipping crossbeds. Channels were abandoned at times and filled with mudstones and coal, and large slump blocks locally constitute an important part of the fill.

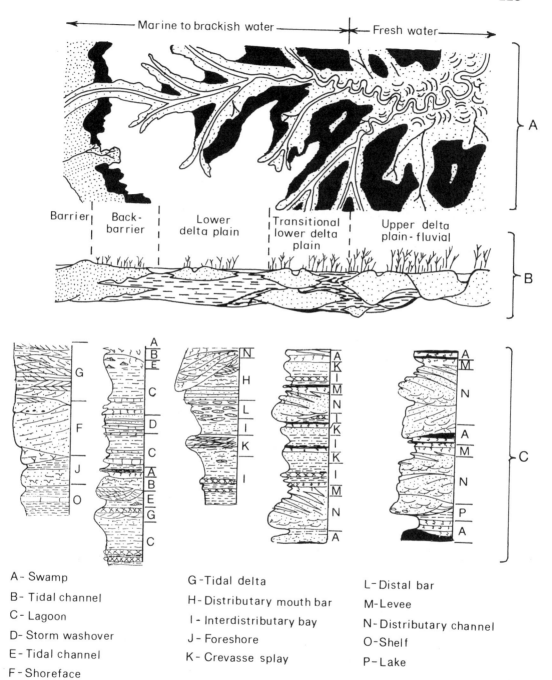

A- Swamp

B- Tidal channel

C- Lagoon

D- Storm washover

E- Tidal channel

F- Shoreface

G -Tidal delta

H- Distributary mouth bar

I- Interdistributary bay

J- Foreshore

K- Crevasse splay

L- Distal bar

M- Levee

N- Distributary channel

O- Shelf

P- Lake

Figure 9.12 Interpretative plan view (A) and cross section (B) of the Pennsylvanian deltaic deposits of the Appalachian area. Also shown are typical vertical sequences (C) that develop in the different components of the delta (after Ferm and others, 1971).

Deposits of the lower delta plain consist of distributary mouth bar, channel, and splay sands along with interdistributary bay sediments that are all comparable to those of the Mississippi Delta. Distributary mouth bar sands are as much as 20 m thick and are laterally equivalent along strike to bay-fill mudstones (Fig. 9.12). Sands are coarsest in the center near the top of the sequences and are finer-grained and muddier along the flanks. Primary structures consist of graded beds, current and oscillation ripples, and contorted bedding, especially on the flanks and front of the bar (Baganz and others, 1975). Continuity of beds, however, is poor because of extensive channeling by distributaries during floods.

Distributary channels are characteristically filled with two types of deposits. Most common are organic mudstones, coal, and thin ripple-bedded sand stringers that fill abandoned channels. Sediments deposited by active channels are relatively few, by comparison, suggesting that lateral accretion was not an important process. Where they occur, they consist of fining-upward sequences of trough crossbedded sand at the base that passes upward into ripple-bedded fine sands and silts. Conglomeratic lags commonly mark the scoured bases of some channel-fill sequences (Fig. 9.12).

Channels are bounded by thin levee deposits consisting of irregularly bedded and root-mottled claystones, siltstones, and sandstones. Splays extend away from levees into bay-fill and backswamp mudstones. They may consist of overlapping lobes where channel splitting occurred, and locally they are cut by channels filled with muddy sediments that accumulated after the splay was closed. Near the channel, the splays consist of coarse-grained, crossbedded sands, and away from the channel, the deposits thin rapidly and consist of finer-grained sands that display ripple drift crossbedding. As the splays grade into interdistributary mudstones, they consist of siltstone containing abundant plant debris (Fig. 9.12).

Backswamp deposits consist of root-mottled claystones and silty mudstones with abundant plant macrofossils. Coals are locally absent where muds accumulated under oxidizing conditions; but for the most part, reducing conditions prevailed in ponded waters of the backswamp where coals accumulated as peat. Scattered thin sands were deposited by flood events that introduced bedload material into the backswamp (Fig. 9.12).

The third major facies association is the upper delta plain where sandstones, mudstones, underclay, and coal were deposited in channel and backswamp settings. Channels are filled with fining-upward sandstone sequences as much as 20 m thick. They are festoon crossbedded at the top, and in some cases the sandstones occur in epsilon-type crossbeds. Basal contacts are erosional, but lateral contacts are gradational where splays interfinger with backswamp deposits. In some cases, channel sand bodies are arranged en echelon in cross section because of avulsion into adjoining backswamps. Backswamp deposits consist of coarsening-upward sequences of shale, with abundant plant fossils at the base overlain in succession by siltstone, sandstone, underclay, and coal. These relatively thin sequences record the filling of swamps and small lakes on the delta plain surface.

On a gross scale, these delta deposits apparently form a single progradational succession (Fig. 9.13). At times, however, progradation was interrupted by marine transgressions. Horne (1979), for instance, described estuarine deposits in the Pocahontas Basin. Estuarine sequences overlie distributary sandstones in deep channels. The sediments are relatively fine grained, often are extensively burrowed, show greatly dispersed transport patterns, and contain faunal remains with marine affinities. The deposits described by Horne (1979) may be comparable to the drowned river valleys of the Mississippi Delta formed by interglacial sea-level rises (Frazier, 1974).

The Eocene Wilcox Group of Texas consists of seven principal depositional sequences. Chief among these is the Rockdale delta system, and although it occurs primarily in the subsurface it is extraordinarily well known because of the close spacing of oil and gas wells drilled into it. The delta system grades updip into the Mt. Pleasant fluvial system and downdip into the South Texas shelf system. Lateral equivalents along strike include barrier bar, strandplain, and lagoon-bay systems (Fisher and McGowen, 1969), but the relationships of these latter three systems with those of the delta will be described later in the discussion of coastal transition zone systems in the rock record. Principal components of the system include delta-plain sandstones, mudstones, and lignite, delta-front sandstones, and prodelta and interlobate mudstones (Fig. 9.14).

Distributary and interdistributary facies occur within the delta plain. Distributary channel deposits of the upper delta plain consist of sandstones deposited in meandering streams that formed multilateral sand bodies containing few nonchannel deposits. Downdip, however, these sandstones merge into narrow sand bodies that are isolated within nonchannel mudstones and lignites. The lower delta plain sand bodies probably accumulated in relatively straight channels that "accreted" in a downward direction through compactional subsidence of underlying mudstones (Fig. 9.14).

Associated with the channel sandstones are interbedded mudstones and sandstones deposited on levees and splays marginal to the channels and in lakes and floodbasins away from the channels in interdistributary areas. Lignites of limited areal extent were deposited in small interdistributary depressions, and lignites of much larger areal extent may represent the landward deposits of destructional episodes.

Thick sequences of well-sorted sands were deposited at the delta front. They occur in lobate or elongate bodies and were deposited on distributary mouth bars. Delta-front sandstones are relatively continuous around the delta margin, however, and their textural characteristics as well as the occurrence of multidirectional ripple crossbedding within them suggest that some redistribution by marine processes took place.

Thick, uniform sequences of dark mudstones were deposited in prodelta areas. Sedimentation must have been rapid because the mudstones retain primary structures and the faunal content is small in numbers and diversity. Mudstones also accu-

226

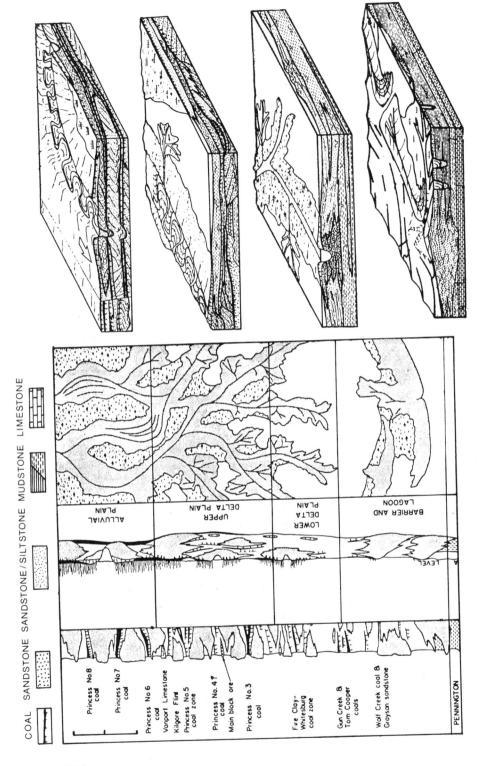

COAL SANDSTONE SANDSTONE / SILTSTONE MUDSTONE LIMESTONE

Princess No 8 coal
Princess No 7 coal
Princess No.6 coal
Vanport Limestone
Kilgore Flint
Princess No.5 coal zone
Princess No.4? coal
Main block ore
Princess No.3 coal
Fire Clay-Whitesburg coal zone
Gun Creek & Tom Cooper coals
Wolf Creek coal & Grayson sandstone

ALLUVIAL PLAIN
UPPER DELTA PLAIN
LOWER DELTA PLAIN
BARRIER AND LAGOON

LEVEL

PENNINGTON

Figure 9.13 Stratigraphy of the Pennsylvanian deltaic deposits of the Appalachian area showing interpretative cross section, plan view, and the three-dimensional architecture that develops in different areas of the delta (after Horne and others, 1978).

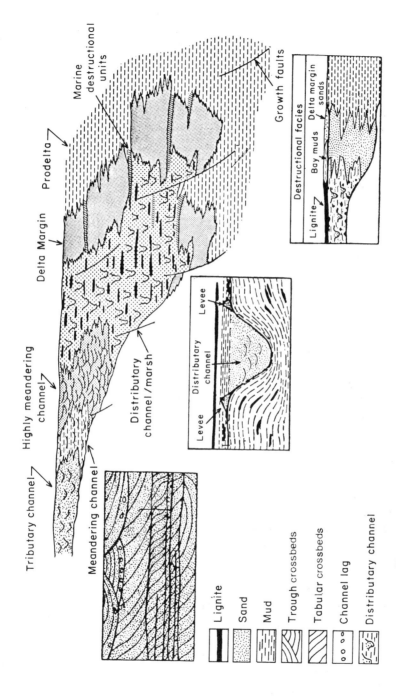

Figure 9.14 Eocene deltaic deposits of the Texas Gulf Coast showing the gross relationships of the various delta components downdip and the detailed facies relationships within components (after Fisher and McGowen 1967, 1969).

mulated in interlobate areas, but those deposits are thinner and are interbedded with fossiliferous or bioturbated siltstones that are atypical of prodelta mudstones.

The three components described thus far are stacked vertically in progradational sequences. A fourth component exists,₀however, that formed during phases of delta destruction. On the margins of deltas, this component consists of glauconitic and bioturbated sandstones, shell debris, and fossiliferous mudstones. Farther up on the delta plain, the equivalent deposits consist of laterally extensive lignites that probably were deposited much the same way that peats blanket abandoned lobes of the Mississippi Delta.

The Rockdale delta system was built in three stages. Thin delta sequences of limited areal extent were constructed during the initial stage, but during the intermediate phase, thick sequences were deposited by elongate deltas that prograded far into the basin (Fig. 9.14). During the culminating phase, thick sequences also accumulated, but they were deposited by lobate deltas that contained large proportions of sand.

The system prograded into a basin made unstable by the loading of large masses of sediment. Rates of subsidence and sediment influx were, for the most part, in balance, because facies patterns aggraded vertically and facies successions are not well developed. Instead, the lateral distribution of facies persists through the succession, and relatively few components may be encountered in the succession at any given locality.

Only by careful analysis of rock sequences in their regional framework is it possible to differentiate wave- or tide-dominated deltas from their nondeltaic counterparts. One well-described example for which this has been done is the lower Paleozoic deltaic deposits of southern Morroco (Vos, 1977).

The deposits consist of stacked progradational lenses bounded vertically by thin shelly beds. During active constructional phases, delta-plain and delta-front components prograded onto a marine shelf without significant redistribution by basinal processes (Fig. 9.15).

A relatively small amount of fine-grained sediment occurs in the delta-plain component, which consists of fining-upward, multilateral and multistoried sandstone bodies. Pebbly sandstones in planar and trough crossbeds occur at the base of the sequences and are overlain in succession by trough crossbedded and plane-bedded sandstones, fine-grained sandstone in ripple drift crossbeds, and root-mottled siltstone. These sequences are interpreted to be deposits of braided stream systems.

Delta-front deposits consist of fine-grained sandstones interbedded with siltstones and silty mudstones. The occurrence of ripple-drift crossbedding, deformed bedding, and vertical escape burrows suggests that sediments were deposited rapidly,

Figure 9.15 Lower Paleozoic delta deposits of southern Morocco showing (A) a dip section with the internal arrangement of facies developed during a single phase of delta growth and abandonment; (B) typical sequences that can develop in different parts of the delta; and (C) three-dimensional representations of different phases of delta evolution (after Vos, 1977).

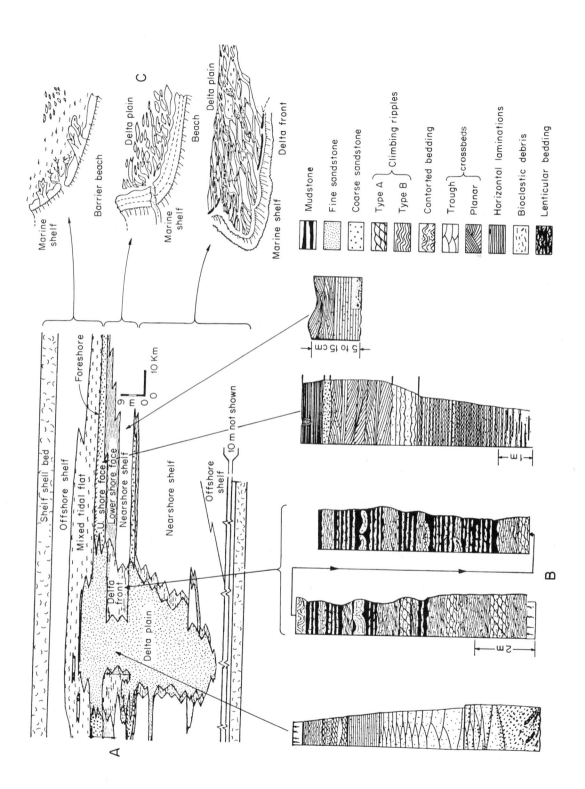

Marine shelf

Barrier beach

Delta plain

C

Beach

Marine shelf

Delta plain

Delta front

Marine shelf

Mudstone
Fine sandstone
Coarse sandstone
Type A ⎤
Type B ⎦ Climbing ripples
Contorted bedding
Trough ⎤
Planar ⎦ crossbeds
Horizontal laminations
Bioclastic debris
Lenticular bedding

Shelf shell bed

Offshore shelf

Foreshore

10 Km

6 m

0

Mixed tidal flat

U. shore face
Lower shore face
Nearshore shelf

Offshore shelf

10 m not shown

Delta front

Delta plain

Nearshore shelf

A

5 to 15 cm

1 m

2 m

B

and the presence of occasional brachiopod molds attests to the marine origin of the deposits.

Laterally equivalent deposits on the shelf include interbedded sandstones and mudstones possessing a characteristic sequence of structures suggestive of deposition by episodic currents. These currents may have been generated by density flows, but the lateral impersistence of the beds is more suggestive of deposition by storm-generated currents.

At some stage during deposition, the fluvial system became less active or basinal processes increased in their ability to rework the incoming sediment, and a wave-dominated delta with a well-developed beach system evolved (Fig. 9.15). On the foreshore of these beach systems, sandstones in low-angle accretionary foresets were deposited. Upper shoreface sediments consist of trough and planar crossbedded sandstones that grade seaward into lower shoreface sandstones and siltstones similar to the nearshore shelf deposits of the actively prograding phase.

The rapid onset of the destructional phase of these deltas resulted in the deposition of tidal-flat sediments directly on top of delta-plain sediments (Fig. 9.15). Tidal-flat deposits consist of flaser-bedded sandstones, siltstones, and mudstones, with occasional crawling trails, ripple crossbedding with bimodal dip orientations, and scattered bioclastic debris. Embedded within tidal-flat sediments are storm washover deposits consisting of plane-bedded or crossbedded sets stacked in units up to 2 m thick. Bioclastic lags and flute and groove casts occur at the base of these units, and their surfaces are marked by a mixture of shell and wood debris and interclasts.

Other components of the destructional phase included delta margin barrier islands, which separated the open marine environment from the mixed tidal flats. On the landward side, the tidal flats passed laterally into delta-plain channel sandstones, and on the back-barrier tidal flat, sedimentation was occasionally interrupted during storm periods by the rapid influx of washover sands.

The destructional phase of each delta sequence reached a climax when a marine shelf setting was established over the delta plain. Offshore shelf deposits consisting of bioturbated mudstones and ripple-bedded fine sandstones were deposited directly over tidal-flat sediments, and the sequence was capped with shell beds probably deposited during a period of nondetrital deposition.

In the three examples just cited, the delta systems evolved from very actively prograding, river-dominated deltas to ones where sediment yield and basinal processes were more in equilibrium. It may normally be the case that the form of a delta changes through time, either by a decrease in sediment influx or an increase in the effectiveness of basinal processes caused by an increase in their strength or an increase in the time available to them to affect the incoming sediments. Only in the case of the end members of Galloway's (1975) ternary diagram might the dominant process so overwhelm the others that the form of the delta can remain constant through time.

10

Estuarine Coasts

Introduction

An estuary is a semienclosed body of water that possesses a free connection with the open sea and within which seawater is measurably diluted with river water (Pritchard, 1967). This definition excludes coastal inlets because they show no significant influx of fresh water, but includes delta distributaries that may experience intrusion of sea water, even in microtidal settings, because of their low gradient.

Four types of estuaries are defined on the basis of the degree of mixing of fresh and salt water (Pritchard and Carter, 1971). Type A estuaries are common on microtidal coasts. Salt wedges move up the estuary and lift the freshwater flow from the river bed; little mixing occurs between the two bodies of water. Mud remains in suspension in the freshwater flow, but bedload is deposited rapidly at the leading edge of the salt wedge.

Partial mixing of the two water bodies occurs in type B estuaries, which commonly occur on mesotidal coasts. They are effective sediment traps because suspended clays brought downriver flocculate in the mixing zone and settle into the upstream flow of the salt layer, rather than being carried seaward by the freshwater flow.

Strong tidal currents mix waters thoroughly in type C estuaries and destroy the vertical salinity gradient, and type D estuaries are both laterally and vertically homogenous. Sediment movement is entirely dominated by tides in type D estuaries, but the lateral salinity gradients in type C estuaries may impose complex circulation patterns on the interacting river and tidal flows.

Drowned river valleys, formed during the post-Pleistocene sea-level rise, are the most common type of estuary on present-day coasts, although they also occur in the distributaries of tide-dominated deltas. A continuum exists between two types of estuaries, which can be differentiated on the basis of tidal influence, size, and pattern of sediment infilling. Protected estuaries generally occur in small rivers on mesotidal coasts where spits prograding from adjacent headlands restrict tidal exchange through a relatively narrow inlet. These are essentially equivalent to the barrier estuaries of Roy and others (1980). Protected lagoons behind the barriers develop coast-parallel marshes laced by tidal creeks that drain laterally into the main channel of the estuary.

Nonprotected estuaries occur in large rivers that are not yet enclosed by prograding spits or on macrotidal coasts where tidal exchange is sufficient to keep estuary mouths free of sand. Because they are not protected by a barrier, wave processes are capable of affecting sediment distribution within these estuaries (Urrien, 1972; Kraft and others, 1979).

Sediment bodies formed in these estuaries are elongate perpendicular to the coast, and because they tend to form in drowned valleys of large rivers, they are filled longitudinally by deltas prograding from the head of the estuary.

Sedimentary Processes and Products

Protected Estuaries

Three major facies occur in protected mesotidal estuaries, including the primary channel, fringing tidal flats, and tidal inlet. These, in turn, can be divided into subfacies on the basis of local hydraulic processes. An overall fining trend exists from subtidal to supratidal settings and from lower parts of the estuary to its upper parts. Thus, the coarsest sediments are found in the primary channel near the inlet, and they become finer-grained upstream in the primary channel and up the secondary channels. Sediments are even finer-grained on the intertidal flats, and the finest-grained sediments commonly occur on supratidal flats and salt marshes.

Fluvial processes are dominant in the upper estuary in the channels. Sand in seaward-dipping, large-scale crossbeds and interbedded muds and ripple-bedded sands are deposited on point bars. Bioturbation is negligible because bedload transport, especially in the channel, is continuous (Dorjes and Howard, 1975; Greer, 1975).

Tidal influence is more pronounced in the middle estuary where bottom sediments may contain abundant shell fragments of marine organisms brought upstream by flood tides. Flow is episodic, and mud is deposited as clay drapes and flaser beds during slack water. Point bars are also muddy in the upper part of the middle estuary, but farther down total mud content declines. Both point-bar and channel deposits may be intensely bioturbated.

The channel of the lower estuary consists of a complex distribution of shifting channels, bars, and sand waves. It tends to be straighter, and the pronounced asymmetry characteristic of the upper reaches is absent. Bars and sand waves consist of sand in large-scale crossbeds and ripple beds and only minor reworking by infauna is evident.

Point bars in the lower estuary may be made up of sand waves with superimposed smaller-scale bedforms. Transport directions can be very complex, with flood and ebb flows segregated into different parts of the bar. Channels cutting across the bar commonly end in flood or ebb-dominated spillover lobes much like flood or ebb tidal deltas, which have been described at length in the discussion of barrier island/lagoon systems in Chapter 8.

Channels in the lower estuary are fringed by tidal flats. Away from the inlet, lower intertidal sandflats are succeeded by mixed flats and upper mudflats. Near inlets, however, tidal flats are exposed to wave action, and their surfaces may be covered by wave-formed ripples. Tidal flats are crossed by tidal creeks that migrate laterally and are capable of replacing much of the tidal-flat sequence with deposits of the tidal creeks.

Inlets are the point of communication between the open marine setting and the protected estuary. On coasts with high wave energy, inlets may be floored by sand waves and bars hundreds of meters long and tens of meters wide. Although they may show steep ebb- or flood-dominated slipfaces, they are apparently composed of amalgamated megaripples that deposit sets of alternately seaward- and landward-dipping crossbeds. On low wave energy coasts, mud can be an important constituent of inlets, where it forms mud interbeds and clay drapes on sandy bedforms (Boothroyd, 1978).

Depositional sequences within protected estuaries could be laterally quite variable, but it must be recognized that, because of the tendency of tidal channels and creeks to migrate, their deposits will tend to make up the bulk of the sequence.

Aggradational sequences in primary channels near inlets consist of sediments that fine upward from erosional bases (Fig. 10.1). Gravel and shell debris at the base of the sequence form a lag that is overlain by sand in large-scale cross sets. Channel sands are overlain by shoal or accretionary bank sediments consisting of crossbedded sands, sometimes with clay drapes and reactivation surfaces. The sands may be interbedded with mud on microtidal, low wave energy coasts, but shoal and bank deposits are sandier along high wave energy coasts with greater tidal range. Channel sequences are capped by bioturbated tidal-flat sands and muds and, in places, with supratidal marsh deposits (Fig. 10.1).

Channel sediments are muddier in tidal creeks away from primary or secondary channels because they are, for the most part, not subject to wave attack, and bedload transport is intermittent in them (Clifton and Phillips, 1981). Channel lags develop on erosional bases, but they are overlain by interbedded sand and mud of the accretionary bank rather than by channel sands. Like the larger channels, deposits of these small creeks are capped by tidal-flat and supratidal marsh deposits (Clifton, 1982) (Fig. 10.1).

Sequences in tidal channel point bars are not greatly dissimilar from aggradational sequences formed by fluvial point bars (Barwis, 1978). Both bedform scale

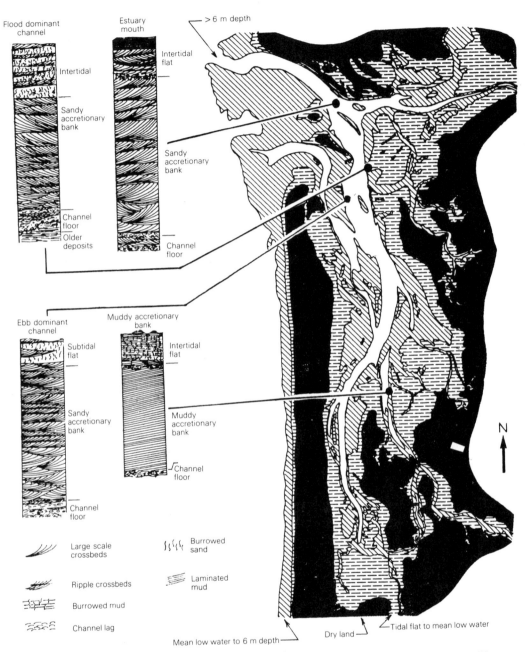

Figure 10.1 Variations in channel-fill sequences in various parts of an estuary (after Clifton and Phillips, 1981).

Tidal creek point bar

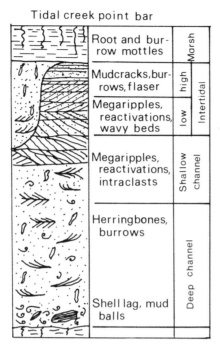

Figure 10.2 Vertical sequence of sediments and sedimentary structures in a tidal creek point bar (after Barwis, 1978).

and grain size decrease upward in sequences that normally rest on lag gravels deposited on an erosional surface (Fig. 10.2). Unlike deposits of fluvial point bars, sediments in tidal channel bars are commonly bioturbated, and they are capped by supratidal marsh deposits.

Unprotected Estuaries

Unprotected estuaries tend to be large. They show a characteristic funnel shape that widens toward the sea and leaves the margin of the estuary open to wave attack. They may also be sufficiently large to generate significant wave energy internally. Sediments may be derived laterally from small tributary streams or from erosion of headlands, but most sediment is provided to the estuary from the main river entering at the head.

The upper reaches of the estuary, therefore, are typically dominated by deltaic processes (Fig. 10.3). Wave action is limited because of distance from the mouth of the estuary, although tides still influence sedimentation. River-mouth bars tend to be sandy, but mud content tends to be large in the rest of the plain, which may consist of tidal flats at the margin and near the channels and supratidal marshes farther inland.

River influence declines gradually beyond the delta plain. Because of lateral variations in salinity, river flow may be confined to one side of the estuary or the

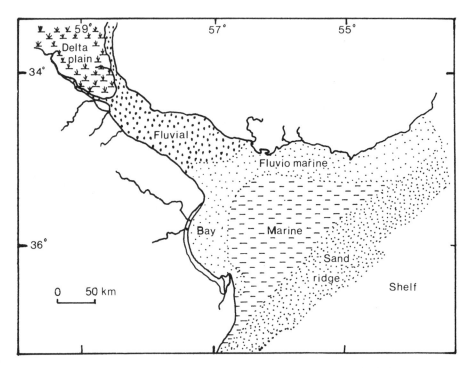

Figure 10.3 Distribution of sediment facies in the estuary of the Rio de la Plata
(after Urrien, 1972).

other, leaving the rest of the estuary to become increasingly affected by the tides.
Sand content decreases gradually through this zone away from the river-mouth bars.

Fluvial discharge, tides, and waves interact in the middle estuary transition
zone. Wave activity in this zone mixes fresh and salt waters, causing clays suspended
in the freshwater flow to flocculate. The turbidity maximum commonly occurs in
this zone, and the bottom sediments tend to be muddy, although sand may be concentrated on beaches along the margin by wave activity.

Spits may form at the mouths of large estuaries, but in most cases they cannot,
or at least have not prograded across the mouth. The volume of sand needed to
accomplish this may exceed the amount available, and the scouring action of the
flow necessary to accomplish the cyclic exchange of large volumes of water may
keep the estuary mouth clear.

Shoals may occur at the mouths of large estuaries, however. These may be
relict barriers formed at lower sea-level stands (Urrien, 1972), palimsest sediments
formed earlier but still in equilibrium with the present hydraulic conditions (Ludwick, 1973), or part of the platform of a prograding spit (Kraft and others, 1979).
In any case, the combined action of tides and waves tends to produce a sand maximum at the mouth that interrupts the seaward-fining trend within the estuary (Fig.
10.3).

Sequences within large estuaries are varied because of the complex and changing interaction of waves, tide, and river flow that occurs within them. The sedimentary sequences being deposited in Delaware Bay, for example, record the changing wave energy away from the mouth of the bay, as well as local differences in sediment supply and tidal energy (Kraft and others, 1979). In addition, stratigraphic sequences vary in response to changes in the underlying topography.

Wave energy is relatively high near the mouth of the estuary and decreases in intensity as the bay narrows toward the mouth of the tidal part of the Delaware River. Sediment is coarsest near the mouth, where it is provided by longshore transport along the Atlantic coast (Fig. 10.4). The Delaware River provides relatively little coarse material, but along the bay coast, local sources of sand include headlands of easily eroded coastal plain sediments that project into the bay.

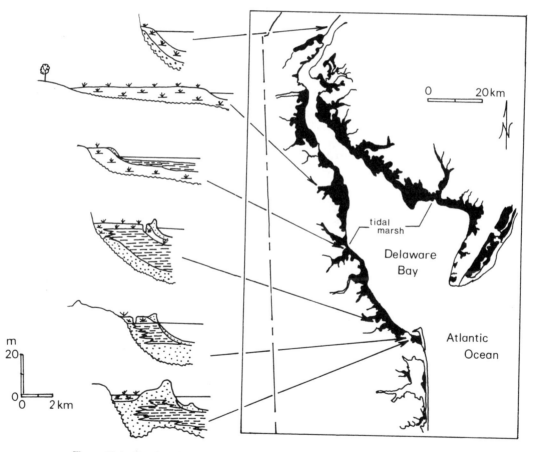

Figure 10.4 Stratigraphic sequences developed in a different parts of Delaware Bay in response to variations in sediment yield and wave activity (after Kraft and others, 1979).

A complex spit system occurs at the mouth of the bay where it is subject to intense wave attack. Sediment is provided by erosion of Atlantic coastal barriers to the south, and the spit is prograding obliquely into the bay (Fig. 10.4). Sand lost to the system by erosion is added to nearby bay-mouth shoals. The spit is prograding, but the overall situation is one of transgression so that sediments of the spit overlie shallow marine–estuarine sediments deposited earlier in the Holocene (Fig. 10.4).

The lower and middle shores of western Delaware Bay include broad coastal salt marshes fronted by thin, narrow barriers or broad tidal flats. Sediments are provided to this part of the shore through erosion of pre-Holocene coastal plain sediments, which crop out at the shoreline.

The whole system is presently migrating landward. Tidal salt marshes are forming at the leading edge of the transgression, and these are followed by the barriers, which migrate as beach faces are eroded and sand is washed over the barrier onto the surface of the salt marsh during storms (Fig. 10.4). Barriers of this part of the bay are taller and wider near the mouth and become thinner and narrower up the estuary.

Broad intertidal flats with exposed relict marsh surfaces occur in the upper part of Delaware Bay. Small pocket beaches form in some areas, and short segments of narrow washover barriers locally protect the marsh; but for the most part waves are insufficient to construct and maintain a continuous barrier (Fig. 10.4).

Intrinsic and Extrinsic Controls on the Evolution of Estuarine Sequence

Estuaries are effective sediment traps that can aggrade quickly to sea level even during temporary stillstands. Under conditions of continuously rising sea level, an estuary may persist for a long time, especially in the case of large ones, and the sequences developed in them can consist of several stacked aggradational sequences deposited during temporary stillstands (Clifton, 1982) (Fig. 10.5). Each unit is tabular except where it fills channels cut in earlier deposits. Primary channels may be capable of cutting through several subjacent units, and particularly active channels may be able to replace the entire underlying unit.

Although estuaries can be filled quite rapidly, the enclosing barrier island may be forced to retrograde across the estuary in the face of continuously rising sea level. Shoreface erosion may expose estuarine sediments to wave attack during this process, and the upper part of the sequence may be planed off, leaving only primary and secondary channel deposits overlain by transgressive sheet sands (Hoyt, 1972).

Subsequent to a high stand, sea level may drop and a new barrier may reoccupy the site of the old one. Tidal channels in the estuary, established behind the newly formed barrier, may be able to rework the surface of the old estuary, producing a subaqueous erosional surface that may resemble a surface scoured during an episode of subaerial erosion. The recognition of oxidized zones, roots, and rhizomes

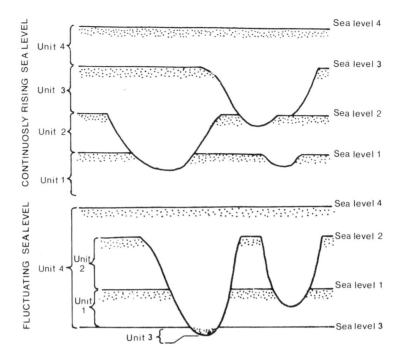

Figure 10.5   Schematic diagram of estuary fill sequences developed during steadily rising sea level and fluctuating sea level (after Clifton, 1982).

in soil horizons may be necessary in order to interpret ancient estuarine sequences (Clifton, 1982).

If sea level drops enough to expose estuarine sediments to subaerial erosion, as it did several times during the Pleistocene, alluvial channels may develop on the exposed plains. Succeeding sea-level rises can cause these valleys initially to aggrade rapidly, but eventually they will evolve into estuaries. Inset stratigraphic relationships will then result as valleys cut into older estuarine sediments fill with channel sands, tidal-flat and marsh muds, and deltaic deposits (Colquhoun and others, 1972).

Sequences in small, unprotected estuaries do not develop with the same diversity that those in large ones do. They are filled mainly by longitudinal progradation of deltas from the head, and lateral progradation from the margins adds only a small volume to the estuarine fill. Lavaca Bay, for example, is a small estuary on the Texas Gulf Coast in a microtidal setting that experiences occasional very high wave energy. It was flooded during the post-Pleistocene sea-level rise, and although spits are currently prograding across the mouth of the estuary, they have not yet closed the bay (Wilkinson and Byrne, 1977).

During the late Wisconsinan low-sea-level stand, the Lavaca River and Garcetas Creek cut a deep valley into underlying Pleistocene sediments. As sea level rose during the Holocene, however, the Lavaca–Garcetas fluviodeltaic system began to aggrade the valley at the same time it retreated up valley in the face of the transgression. Fluvial sands overlain by deltaic sands form the basal units in the estuary, and they record the transgressive event that occurred at this time (Fig. 10.6).

By about 10,000 ybp, sea level had begun to flood the valley, forming open estuarine conditions in the lower part. Muds were deposited in the estuary, and a series of onlapping muddy units were deposited as the transgression flooded progressively more of the valley (Fig. 10.6). Deltaic sands of the Lavaca River currently are prograding into the estuary, and, in a few places, marginal spits are prograding laterally into the bay.

Because sediment influx was large, water depth remained almost constant even as sea level rose. The onlapping sequence of bay muds, therefore, form an almost complete record of the transgression as it moved up valley. The only true regressive deposits are the sands of the delta and the marginal spits.

Tidal deltaic deposits may accumulate at the mouths of small, unprotected estuaries, forming important proportions of estuarine sequences along coasts experiencing a significant tidal range. Port Hacking on the east coast of Australia is being filled from the mouth by subaqueous tidal delta sands that are prograding up-estuary and encroaching on a deep mud basin. Simultaneously, fluviodeltaic sands of the Hacking River are prograding down the estuary into the mud basin (Thom and Roy, 1985).

Estuarine Sequences in the Rock Record

One possible example of an ancient protected estuary occurs in the Upper Almond Formation of southwestern Wyoming (Van Horn, 1979; Weimer and others, 1982). Sedimentation took place in a protected lagoon where freely migrating tidal channels deposited sandstones up to 5 m thick in fining-upward sequences over lag gravels developed on an erosional base. Sandstones display a minor amount of burrowing, occasional slump structures, and carbonaceous laminae draping foresets of crossbeds (Fig. 10.7).

Channel sands are capped by tidal-flat sediments consisting of shale, siltstone, and coal in symmetrical cycles that record minor episodes of transgression and regression. Coal at the base of each cycle was deposited in a freshwater marsh during maximum regression. Salt-marsh deposits form the initial transgressive deposits, and bioturbated lagoonal or tidal-flat sediments mark the culminating phase of transgression. The cycles were completed when salt-marsh and freshwater-marsh sequences were deposited during the regressive phase (Fig. 10.7). The presence of both ebb and flood tidal deltaic sandstones, channel fills of both major channels and minor drainage creeks, and tidal-flat deposits suggests that deposition took place in a protected, mesotidal estuary not unlike those described by Boothroyd (1978).

Recognition of large, unprotected estuaries is hampered by their size. They actually are small-scale sedimentary basins with infilling sequences composed of nu-

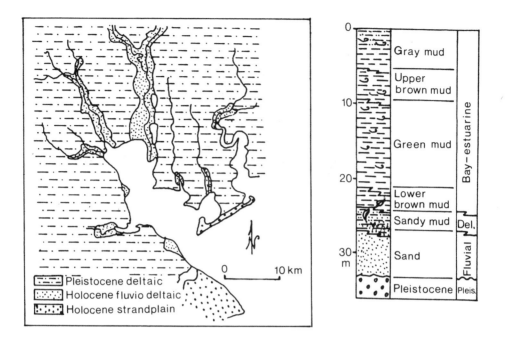

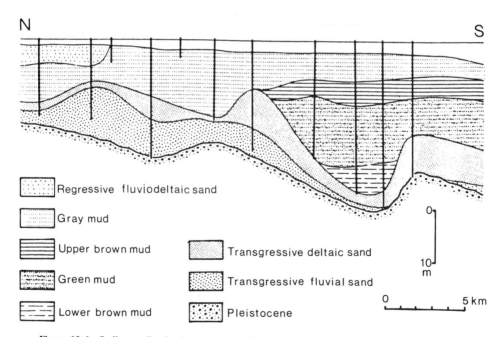

Figure 10.6 Sediment distributions in Lavaca Bay shown in plan view, dip cross section, and typical vertical sequence (after Wilkinson and Byrne, 1977, Figs. 1, 3, and 4, reprinted with permission of the American Association of Petroleum Geologists).

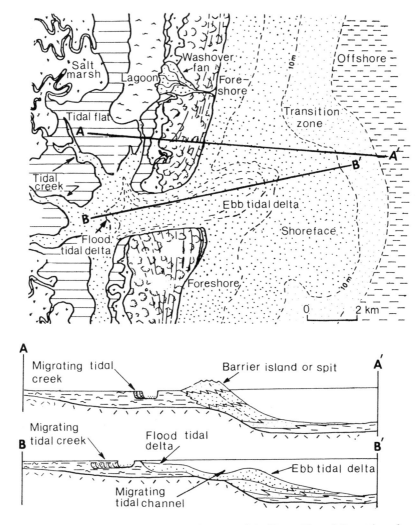

Figure 10.7 Model of the depositional system of the Upper Almond Formation of Wyoming showing lateral distribution of facies and their internal arrangement (after Van Horn, 1979, in Weimer and others, 1982, Fig. 45, reprinted with permission of the American Association of Petroleum Geologists).

merous depositional components. Without good regional control and without reference to paleosalinity indicators, recognition of the separate components, rather than the system as a whole, is the normal result.

Along present-day coasts, however, estuarine deposition took place during the Pleistocene in estuaries that in some cases are still in existence. Where this has occurred, the system is easily recognized, and identification of individual components can be made by comparing sequences to sediments presently being deposited.

The Pleistocene Millerton Formation of Tomales Bay, California, is an example of such an occurrence. Tomales Bay occupies one portion of the San Andreas Fault in northern California. It is approximately 20 km long and less than 2 km wide over much of its length. Its main sediment source is a small river entering at its head, where a delta with fringing tidal flats is currently prograding down the estuary. Other sediment is provided by small creeks draining laterally into the bay and by erosion of headlands composed of Pleistocene-age estuarine deposits that have been uplifted by tectonism along the fault.

The estuarine deposits consist of interbedded marine and nonmarine strata of laterally variable litho- and biofacies (Monteleone and Fraser, 1978) arranged in regressive sequences deposited over transgressive surfaces (Fig. 10.8). Shelly mudstones with open-bay foraminifera occur at the base of each regressive sequence (Fig. 10.9). They overlie bioturbated siltstones or gravel and shell lags that represent the deposits of the transgression as it overrode gravelly deltaic sands deposited during the previous regressive cycle. Estuarine mudstones are overlain by bioturbated tidal-flat mudstones and sandstones, which are locally succeeded by salt-marsh mudstones. These marine strata are overlain by fluviodeltaic sands and gravels deposited predominantly by northward-flowing streams. Locally derived fanglomerates are interbedded in some places with marine mudstones. These can be traced to outcrops of Franciscan melange rocks outcropping along the margins of the bay,

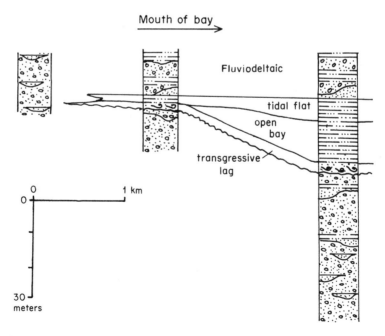

Figure 10.8   Lateral variation in the distribution of estuarine and fluviodeltaic sediments in the Millerton Formation of Tomales Bay.

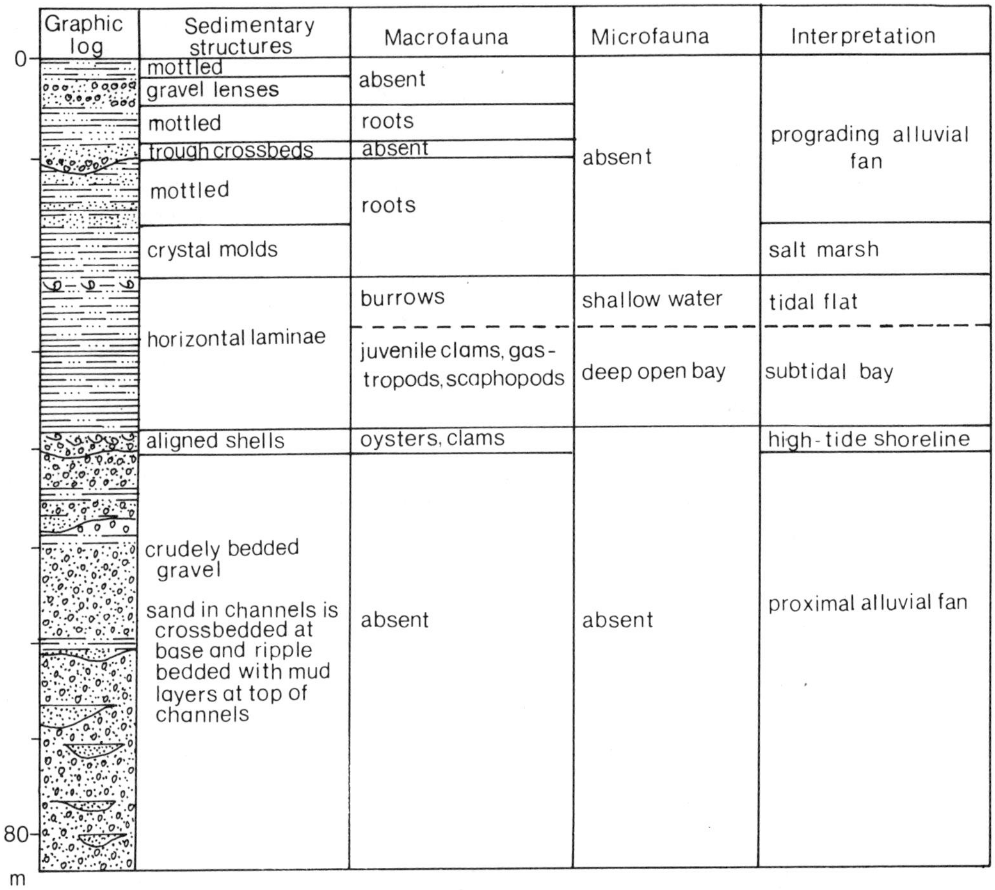

Graphic log	Sedimentary structures	Macrofauna	Microfauna	Interpretation
0 —	mottled	absent		
	gravel lenses			
	mottled	roots		prograding alluvial fan
	trough crossbeds	absent	absent	
	mottled	roots		
	crystal molds			salt marsh
	horizontal laminae	burrows	shallow water	tidal flat
		juvenile clams, gastropods, scaphopods	deep open bay	subtidal bay
	aligned shells	oysters, clams		high-tide shoreline
	crudely bedded gravel sand in channels is crossbedded at base and ripple bedded with mud layers at top of channels	absent	absent	proximal alluvial fan
80 — m				

Figure 10.9 Vertical sequence of aggrading estuarine sediments of the Millerton Formation.

suggesting that lateral sediment sources included alluvial fans that shed debris directly into the bay.

Unlike Lavaca Bay, the sediment yield into Tomales Bay is small, and during repeated transgressions only thin deposits accumulated; the bulk of the estuarine fill was deposited during regressive phases.

Discussion: Stability of Coastal Zone Systems

Thus far, the various systems comprising the transition zone have been discussed separately; yet along modern coasts they commonly occur in lateral juxtaposition, and similar spatial arrangements of these systems can be found in the rock record. The deltaic sediments of the Eocene Wilcox Group of the Texas Gulf Coast have already been discussed, but they are only a part of a complex of coastal and shallow marine sediments that illustrate the interaction of various depositional systems within the continental–shallow marine transition zone (Fig. 10.10) (Fisher and Mc-Gowen, 1969).

The Pendleton Bay/Lagoon System occurred to the northeast of the main area of deltaic sedimentation. Sediments consist mainly of laminated or bioturbated mudstones and massive or crossbedded sandstone. Sandstones contain some glauconite, and lignite is locally common. Marine fossils are present, but the assemblage shows little diversity, suggesting that, although the environment was not extreme,

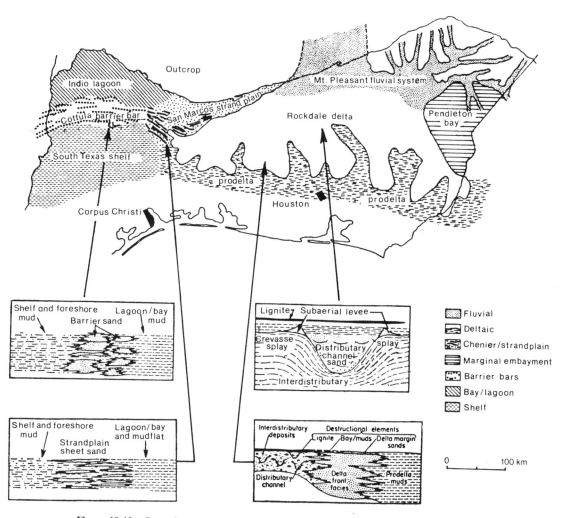

Figure 10.10   Coastal zone sediments of the upper Wilcox Group of the Texas Gulf Coast (after Fisher and McGowen, 1967, 1969).

circulation ranged from open to restricted. Both rock types and their faunal content indicate that sedimentation occurred in lagoons, lakes, bays, and mudflats in an area flanking the main axis of deltaic deposition.

A similar setting can be found to the east of the present-day Mississippi Delta, where shallow, semienclosed basins with fresh to marine waters occur. Mississippi Sound, for example, is a shoal-water area partially enclosed by barriers. Holocene transgression inundated an area underlain by alluvial, estuarine, barrier, and open marine sediments (Otvos, 1978). Open marine conditions initially were established over much of the area, but with the emergence of barrier islands a broad brackish-water bay was formed where muds are being deposited. Swamps, marshes, and mesohaline lakes such as Ponchartrain and Borgne occur north of the bay, and similar depositional settings probably occurred in the Pendleton Bay/Lagoon System.

The structure of the Rockdale Delta System was discussed earlier at length and will only be summarized here. The delta plain, consisting of distributary channels, marshes, and swamps, was fluvial dominated. Sand was concentrated in distributary mouth bars, and during periods of large sediment influx, delta lobes prograded rapidly seaward. During periods of reduced influx, however, lobes were transgressed. Marine sandstones and mudstones accumulated over the lower delta plain, while extensive marsh deposits were forming farther up the delta.

Immediately to the southwest of the delta lay the San Marcos Strandplain/Bay System. Part of the system consists of mudstones similar to those of the Pendleton Bay, and a similar depositional setting is proposed for them. The mudstones grade laterally into interbedded sandstones and mudstones. Sandstones are fine-grained, bioturbated, and slightly glauconitic. They occur either as elongate, strike-parallel bodies or as thin but extensive sheets. Mudstones are pyritic and locally lignitic.

These sediments were probably deposited on a coastline that alternated between periods of chenier plain and strandplain deposition. Muds were transported alongshore during periods of active delta growth and were deposited on a muddy coastline that was periodically reworked to form coast-parallel cheniers. During episodes of delta destruction, large amounts of sand were provided to the system and the shoreline prograded as a sandy strandplain.

The Cotulla Barrier/Indio Lagoon System lay to the southwest of the San Marcos Strandplain. Barriers consisted of sand in elongate, strike-parallel bodies up to 30 m thick and occurred in a belt of stacked sand bodies up to 300 m thick. Northwest of the barrier belt are laminated or bioturbated mudstones that contain some thin lignite beds. Sandstones within this facies are common near the sand belt and were probably deposited as washover fans and small tidal deltas.

The barrier is cut by at least one possible tidal inlet, and isopach patterns on either side of the inlet suggest the presence of both flood- and ebb-tidal deltas. Some underlying control exists on the placement of this inlet for it to have persisted through the deposition of the amalgamated barrier sands. Possibly the position of the inlet was structurally controlled in much the same way as the position of fluvial systems was controlled on the Tertiary Gulf Coastal plain of Texas (Galloway,

1981). Its position may also have been determined by elements of underlying topography, such as drowned river valleys (Price and Parker, 1979).

Because both the Wilcox Group and the Holocene sediments of the northwest Gulf of Mexico were deposited in the same structural basin, it is not surprising that their depositional systems should be similar in size and distribution. Nevertheless, the comparison is quite striking.

Extrinsic controls on the evolution of these systems include intensity of basinal processes, rate of sediment influx to the coastline, and tectonic character of the basin. Stable basins with low depositional gradients promote rapid lateral migration of systems. The Gulf Coast basin, however, was an unstable one, where large amounts of sediment were provided to a rapidly subsiding basin (Fisher and McGowen, 1969). Subsidence occurred in response to depositional loading, and sediment bodies were stacked vertically in zones that were spatially persistent for long periods. By this process, lithosomes were produced that consisted of accumulations of the sediment bodies deposited by the given system that dominated a particular zone.

The zones apparently represented positions of dynamic equilibrium determined by the interaction of the various extrinsic controls. Each system oscillated within these positions in response to variations in these controls, and the maximum extent of advance during progradational phases was determined by the threshold point at which the volume of sediment needed to maintain progradation exceeded the influx. The threshold point was, in part, determined by volumetric considerations discussed earlier and in part by increased erosion of sediments from the coastal systems as they approached the shelf edge and increased their exposure to basinal processes.

These equilibrium positions were remarkably stable. They not only existed during deposition of the Wilcox Group, but also during deposition of other Eocene strata in the Gulf Coast basin of Texas (Ricoy and Brown, 1977). The Eocene section there consists of strata deposited by transition zone systems that alternate with shelf deposits. During the Eocene, offlap sequences were periodically interrupted by transgressional episodes that spread shelf sediments over transition-zone facies. Transgressive episodes, however, were followed by rapid progradation when transition zone systems were reestablished over former equilibrium positions (Fig. 10.11).

These systems migrated in a direction parallel to dip in response to external stress, but they migrated within zones that tended to persist through time. This was, in part, caused by the relative constancy of the tectonic regime, but it was also partly the result of the ability of the adjacent shelf to act as a buffer. Although the shelf migrated landward in response to transgressions, limits of progradation were fixed by the constraints imposed by the amount of sediment delivered to the system, the ability of marine processes to rework them, and the shape, in particular the gradient, of the shelf. As long as these factors did not change drastically, the zone within which the coastal systems migrated remained relatively narrow.

The position of these systems along strike also tended to persist through time. The type of system that can develop at a given point along the coast is determined largely by the intensity of coastal processes, the relative location of point(s) of sedi-

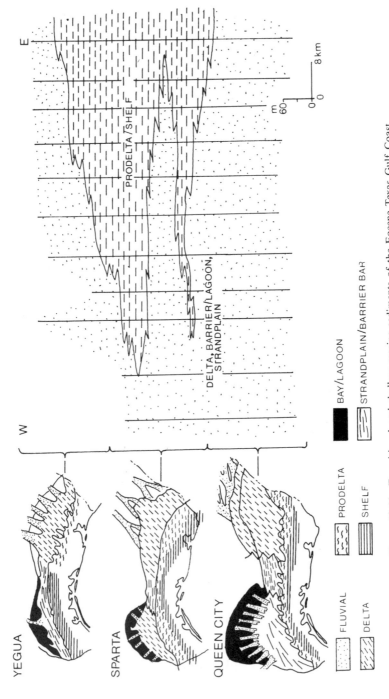

Figure 10.11 Transitional and shallow marine sediments of the Eocene Texas Gulf Coast basin showing positions of the depositional systems through time (after Ricoy and Brown, 1977).

ment delivery, and its magnitude and timing. If these factors vary only slightly, the distribution of coastal systems will remain relatively stable along strike. The magnitude and timing of sediment delivery to the coast is primarily a function of climate, which apparently did not vary greatly, and the intensity of coastal processes, which apparently remained relatively constant. The points of sediment delivery along the coast also remained relatively stable because stream courses were largely structurally controlled.

Part IV

SHALLOW MARINE
ENVIRONMENT

Introduction

The shallow marine environment includes marginal seas or continental shelves and epeiric or interior seas. Sedimentary processes on continental shelves are variable in direction and intensity, and those most effective in transporting sediments tend to be oscillatory and/or episodic. Currents may flow in any direction relative to slope, but those with sufficient velocity to entrain shelf sediments are normally rectified into coast-parallel directions by the interaction of the flow with shelf topography. Nevertheless, local variations in bathymetry and sediment input may produce a complex mosaic of texture and structures in the sediment cover on a shelf.

The partial or complete isolation of epeiric or interior seas must have prevented, to a degree at least, the introduction of wave-, tide-, or climatically induced currents from the deep marine basin into the shallow marine environment. In addition, the shallow depth and accompanying low gradient of epeiric seas probably affected the way even shallow marine processes acted in them.

The types of processes that occur on modern shelves are the same as those that occurred in ancient shallow seas, but modern shelves can only be used as counterparts for ancient shallow seas with care because they have undergone a profound transgression during the last 10,000 years. The sediment cover on modern shelves has formed in response to this transgression, primarily as the result of erosion of retreating shorefaces (Swift and others, 1972). Only on a few shelves have riverborne sediments become an important constituent of the sediment cover. Thus, although modern shelves can be used to directly model the response of the shallow marine environment to transgression, they can only be used indirectly to infer the responses that might result from regression.

11

Continental Shelves

Introduction

Continental shelves are submerged margins of continents. Shelf surfaces slope gently seaward from the coast to the continental slope at depths between 100 and 250 m, where a pronounced increase in slope occurs. Shelves range in width from a few kilometers to several hundred kilometers, and although they are not featureless plains, the relief over large parts of shelves may be on the order of only several meters (Fig. 11.1). Modern shelves have been classified on the basis of dominant transport process, tectonic setting, or stage of development relative to the Holocene sea-level rise.

The dominant transport process on most shelves is either weather-controlled, through wave-drift or wind-forced currents, or tide-dominated. It should be remembered that these are really end members that are interactive on shelves, and the intensity of these processes cannot only change laterally on a given shelf but seasonally as well.

During winter months, for instance, a shelf may be dominated by passing storms producing currents that can entrain bedload and create turbulence that can thoroughly mix the water column. Storms are less frequent and intense during summer months, causing development of a temperature-stratified water column where density-forced currents and oceanic currents can act. Tide-induced currents are more regular in their occurrence, but flow velocities change monthly from spring to neap tide, and when they are reinforced by storm winds, currents of exceptional strength can develop.

The shape of shelves appears to be most controlled by tectonic setting (Emery,

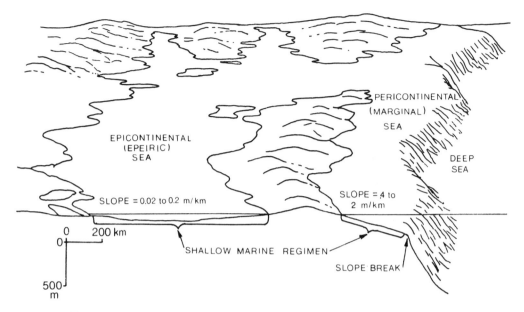

Figure 11.1 Generalized physiography of shallow marine environments in epeiric and marginal seas showing the relationships of the two settings to land masses and deep marine basins (after Heckel, 1972).

1968). Along convergent plate margins, shelves tend to be narrow, where they exist, and they slope steeply seaward to a slope break that commonly is deeper than on other shelf types. Shelves are somewhat wider along transform margins, but they are best developed on trailing edges of plates where their shape is controlled by a deep structure, such as an upfaulted block or reef, that is buried by a seaward-thickening sediment prism (Fig. 11.2).

Some examples of shallow marine environments, such as the Baltic Sea or Hudson Bay, are not shelves sensu strictu. Rather, they are small enclosed or semienclosed seas where the intensity of oceanic processes such as wind- or tide-forced currents is dampened. Their ancient counterparts may include some of the epeiric seas that occupied large parts of continental interiors at times in the past.

Present-day shelves may also be classified according to their relative stage of disequilibrium following Holocene sea-level rise. They include among their sediments varying proportions of relict and palimpsest sediments and normally are covered by only thin veneers of modern detrital sediments locally deposited in areas relatively near the coast. Although the most recent sea-level rise was the product of a rapidly changing climate, other mechanisms may have operated in the past to effect similar, albeit less dramatic, sea-level fluctuations.

Swift (1969a) classified shelves according to their relative proportion of modern sediment cover. Shelves along the east coast of the United States or along the North Sea coast of western Europe are almost entirely underlain by relict or palimp-

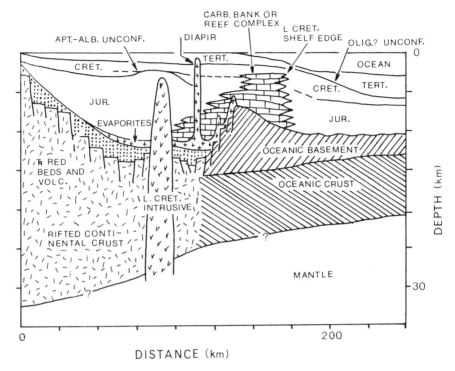

Figure 11.2 Deep structure underlying the continental shelf of the east coast of the United States (after Grow and others, 1979, Fig. 12, reprinted with permission of the American Association of Petroleum Geologists).

sest sediments. These are termed autochthonous shelves or shelves where the sediments were derived from within the basin itself immediately following a period of rapidly rising sea level.

Substantial areas of some shelves are covered by modern detrital sediments derived from outside the basin. Swift (1969a) calls these allochthonous shelves because their sediment is extrabasinal in origin, but it is more important to remember that these shelves are in equilibrium with present-day high sea levels and are in the process of aggrading their surfaces.

In addition to these two basic types, other workers recognize examples in the rock record of very thick successions of shallow marine sediments. These sequences must have developed on shelves where water depth fluctuated within a relatively narrow range over a long period of time so that neither coastal nor deeper marine conditions were ever established. Such shelves have been termed aggradational shelves, and it may be that counterparts can be found on those shelves where a reduced sediment influx is nearly in balance with a slowly rising sea level so that neither pronounced progradation nor retrogradation is taking place. Unfortunately, the time span over which this process could have taken place on modern shelves has been too short to permit significant facies stacking to have occurred and, therefore, one can only infer the mechanics of sediment accumulation from studies of ancient successions.

Swift's (1969a) classification will be used in this section. A tectonic classification can be easily applied because the tectonic environment of most shelf sequences is evident from their structural setting, but the utility of applying such a classification to ancient shelf sequences is limited. The delivery and disposition of sediments on a given shelf will respond in the same way to sea-level fluctuations, even though the rate of such occurrences may vary, and there does not seem to be a significant difference, except perhaps in thickness, among the deposits of convergent, trailing edge, or transform shelf margins.

Modeling experiments by Harbaugh and Bonham-Carter (1977), for instance, have shown that sediments accumulate on shelves until the shelf surface builds to the point where marine processes are able to bypass all incoming sediment to the continental slope. The shape of the isodynamic surface will vary according to the magnitude of the fluid forces expended on it, but all shelves tend to converge to a given equilibrium profile regardless of their initial state, which might be determined by their tectonic setting (Swift and Niederoda, 1985).

It might be argued that Swift's (1969a) classification can only be applied to modern shelves that have undergone a rapid, profound sea-level rise during the Holocene. Ancient shelves, however, may have responded in similar fashion to rises of lesser magnitude. The Miocene shelf sediments of the Caliente Range of southern California, for instance, were deposited on a transform margin (Clifton, 1981a). Climate-induced sea-level changes were the dominant control on facies development, and the shelf responded to changes in sea level in much the same way the shelf of the western North Atlantic did during the Quaternary.

It would be of great utility to understand the dynamic processes that operated on ancient shelves and to differentiate weather- and tide-dominated shelves. Unfortunately, however, processes that are determinable through direct measurement on modern shelves may not be so easily differentiated in ancient deposits because no simple correlation exists between process and product (Swift and Niederoda, 1985). A system of strong tidal flows on the shelf may be thought of as regularly repeating storm events, and storm events can be considered to be random tidal flows, because both interact with shelf topography in much the same way. All the morphological elements described previously have been found on both weather- and tide-dominated shelves, and these elements appear to respond in the same way regardless of the process. Wind-forced currents on shelves, for instance, may be unidirectional for a particular storm or for as long as a season, but depending on location of dominant storm tracks, successive storms may produce reversing currents. The weather-producing factors in the environment may also change seasonally and produce reversing currents, and the sediments may respond in very much the same way that they would to reversing tidal currents.

Tidal currents, in fact, may be unidirectional at a given shelf locality where either ebb or flow are dominant. Ebb and flood currents may even follow mutually exclusive paths that are in close proximity to one another.

At least gross mechanical and morphological similarities exist among sand accumulations on shelves dominated by tides, storms, and ocean currents. As more detailed studies of the morphological elements of shelves, especially of their internal structures, become available, however, it is becoming increasingly possible to confidently assess the relative contribution of various dynamic processes on shelves to the formation of stratigraphic sequences.

Sedimentary Processes and Products

Transport Processes

Currents capable of moving sediments on shelves are produced by long-term oceanic circulation, shorter-term weather patterns, tidal effects, and density differences. Although it might be convenient to think of these components separately, they may act synergistically on shelves, and it is the resultant of the interaction that produces patterns of sediment distribution.

Major ocean currents are driven by large atmospheric circulation cells that maintain relatively permanent positions. They are caused by the earth's effort to transfer heat from tropical to polar regions, and they consist of a shallow component composed of warm, light waters flowing poleward and a dense, cold, deep return flow. Although they tend to lie oceanward of the shelf, they can induce flow on the shelf where shear occurs at the boundary between waters of these currents and those of the shelf. In addition, the current may be displaced shelfward during summer when outer shelf waters may be strongly stratified.

Both direct and induced bottom flow velocities of deep ocean currents tend to be small, although they are sufficient to transport suspended sediment. Some currents, however, may be much stronger. The Gulf Stream, for instance, is apparently capable of entraining fine-grained sediment off Cape Hatteras (Hunt and others, 1977), and the Aghullas Current off the southwest shelf of Africa can re-form available sand into large-scale bedforms (Flemming, 1980) (Fig. 11.3). Such currents dominate less than 5% of modern shelves (Swift and Niederoda, 1985).

Shorter-term weather patterns produce transient currents of varying duration. Passing storms can affect ocean circulation patterns for a few days. When short-term movements of air masses are time averaged, persistent patterns of air movement become evident, and they may produce long-term patterns of water circulation.

Atmospheric circulation induces movement of water by direct shear coupling of wind to ocean surface waters. When wind imparts shear to the water surface, it induces a mass transfer of water. The Coriolis force acts on this unconfined flow and deflects the surface flow 45° from the direction of wind flow. As stress is trans-

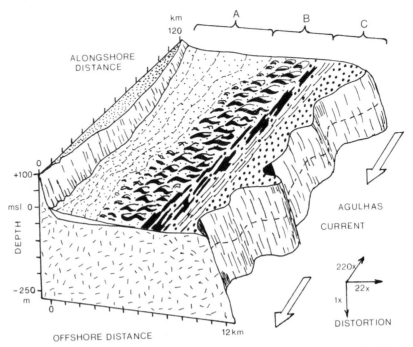

Figure 11.3 Physiography of the northeast African shelf showing the relationship of large-scale bedforms to oceanic currents on the middle and outer shelf (after Flemming, 1980).

mitted down the water column, water layers shear past one another, and the flow is deflected progressively farther from the direction of wind flow with depth. The resultant drift is at 90° to the wind direction, and the resultant flow pattern is called an Eckman spiral. At wind velocities greater than about 10 km/h, Eckman spirals are not stable, and flow takes place in horizontal helical vortices called Langmuir cells that are aligned parallel to mean flow.

Because of Eckman veering, winds blowing parallel to the coast may set up onshore and offshore directed currents. Onshore transport of water by wind may cause an elevation of sea level along the coast. This may be supplemented by low barometric pressure associated with passage of a storm, so that an abnormally high water level can occur on the coast. When this occurs, a pressure gradient is set up within the water column that continues to increase in magnitude until baroclinic forces exceed the frictional forces at the bottom and a return flow is initiated (Fig. 11.4). It flows perpendicular to shore in shallow water, but as it accelerates in deeper water, it is deflected by Coriolis forces. The deflection is relatively small at the bottom where shear at the substrate slows the flow, but it increases upward in the column, inducing a net movement of water parallel to the shoreline (Swift and Niederoda, 1985).

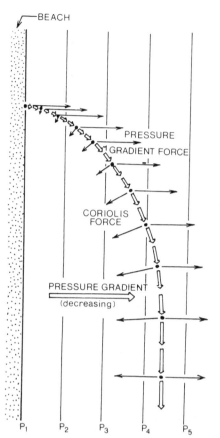

Figure 11.4 Forces involved in forcing geostrophic flow on the continental shelf. Water moving away from the coast in response to a wind-forced pressure head is directed into a coast-parallel stream by the Coriolis force (after Swift, 1976, in D.J. Stanley and D.J.P. Swift, eds., *Marine Sediment Transport and Environmental Management,* John Wiley & Sons, Inc.)

The tidal bulge produced in the open ocean by the gravitational attraction of the sun and moon on ocean waters propagates tidal currents onto the shelf. When the natural oscillatory period of the shelf onto which the tide is propagated is similar to the period of the tidal bulge, resonant amplification takes place, producing higher than average tidal ranges and correspondingly greater tide-induced current velocities. Frictional drag of the wave on the shelf surface causes it to become asymmetrical so that velocities during flood tend to exceed those during ebb, and this asymmetry becomes more pronounced in progressively shallower water.

Coriolis force rotates the tidal flow so that individual water particles follow closed elliptical paths during a tidal cycle. In the open ocean the ellipse is symmetrical, but basin configuration and frictional effects may deform the ellipse so intensely that a nearly rectilinear current is produced. Inequalities between ebb and flood flow under such conditions may produce a residual current capable of inducing significant sediment transport in one direction (Fig. 11.5).

Seaward-directed, density-forced bottom flows can be produced in several dif-

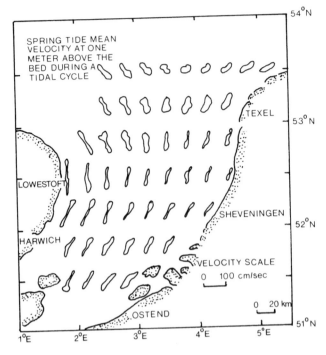

SPRING TIDE MEAN
VELOCITY AT ONE
METER ABOVE THE
BED DURING A
TIDAL CYCLE

TEXEL

LOWESTOFT

SHEVENINGEN

HARWICH

VELOCITY SCALE

0   100 cm/sec

0   20 km

OSTEND

54°N
53°N
52°N
51°N

1°E   2°E   3°E   4°E   5°E

Figure 11.5 Examples of distortion of the tidal ellipse caused by asymmetrical distribution of current velocities measured at various points 1 m above the bed of the North Sea (after McCave, 1971).

ferent ways. Storm waves can entrain bottom sediments, forming a turbid layer that moves downslope toward the shelf edge (see, for example, Nelson, 1982), and density flows can occur when very turbid river water enters the shelf. Large evaporation rates along arid coastlines may produce hypersaline waters that sink at the coast and flow seaward as density flows.

Density-forced currents can also move landward along coasts where significant runoff of fresh water from land occurs. The fresh water flows seaward at the surface, inducing a bottom return flow with a velocity up to a magnitude larger than the outflow velocity of the river water (Leetma, 1976). The process is especially effective in estuaries, but it can also operate on open shelves where the unconfined flow is deflected into a coast-parallel current by the Coriolis force.

The basinward slope of shelves imparts a gravitational acceleration to sediment particles that acts to produce a cross-shelf grading of sediments by diffusive processes. Sediments entrained by storm- or tide-induced currents show a principal direction of movement coincident with the flow. But a downslope component of movement also exists, and as the sediment drifts seaward, it leaves its coarsest fraction behind. This process is repeated until sediment of a particular caliber is deposited at a point where it can no longer be entrained by normal shelf processes. By this process, sediments tend to be arranged in a simple seaward-fining pattern. In reality, of course, sediment distribution patterns are made much more complex by the interaction of advective sediment transport processes with shelf topography, as well as by the amount of sediment provided to the shelf and the rate and manner of its delivery.

Sediment Facies

Sediment sources. Emery (1968) identified five main types of sediments on modern shelves. He included (1) authigenic sediments consisting mainly of glauconite, phosphate, and chamosite; (2) organic sediments consisting mainly of shell debris; (3) residual sediments provided by older rocks exposed at the shelf surface; (4) detrital sediments provided by rivers and eroding coastlines; and (5) relict sediments deposited on the shelf during lower stands of sea level.

Emery (1968) considered relict sediments to be by far the most abundant sediment type, but this conclusion was disputed by later workers. Swift and others (1971) recognized that many so-called relict sediments on the shelf were actually in equilibrium with modern shelf processes, even though they were originally deposited under shallower water conditions. They proposed the term "palimpsest" for these sediments, which can be recognized on most other continental shelves. In addition, gravels, long thought to be relict deposits because of their size, are now considered to be modern deposits on present-day Arctic shelves (Creager and Sternberg, 1972).

Authigenic and residual sediments generally comprise only a small portion of the shelf sediments in any one place, and except in low latitudes, organic sediments are normally only a small fraction of the total shelf sediment cover. On shelves where rates of sediment accumulation are small, however, the relative proportion of these three sediment types increases.

The distribution of the relatively narrow range of sizes of modern detrital sediments on continental shelves is in part a function of the transport processes on the shelf and in part a function of the sediments made available to it. Virtually the only sediment provided to a shelf during a period of rising sea level is derived from coastal erosion by a process known as shoreface bypassing (Swift and others, 1971) (Fig. 11.6). Sediments eroded from the shoreface are added to the offshore surface in an effort to maintain an equilibrium profile (Bruun, 1962), and those that are not eroded are drowned by rising sea level.

This process was described earlier and need not be elaborated on here; but even though these sediments have been transported away from their original depositional setting, they can be reshaped by dynamic processes operating on the shelf.

During periods of rapidly rising sea level, only a relatively small amount of sediment is added to the shelf directly from rivers. Instead, river valleys are transformed into estuaries, which act as sediment sinks, trapping riverine sediments on fringing tidal flats and bayhead deltas. Even sediments being transported along shore in littoral circulation cells are stored on bay-mouth bars and shoals (Fig. 11.6).

Rivers only become important sources of sediment when transgression slows or stops, allowing estuaries to equilibrate with their tidal prisms (Fig. 11.6). When this occurs, the rivers begin to pass sediments out onto the shelf, but during the transport process the sediment load is transformed. Very fine material is trapped on

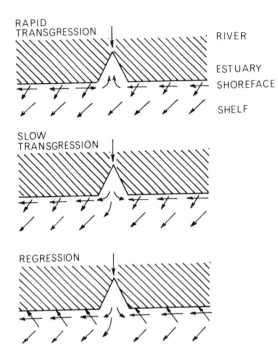

Figure 11.6 Sources and areas of storage of sediment during rapid transgression, slow transgression, and stillstand or regression (after Swift, 1976, in D.J. Stanley and D.J.P. Swift, eds., *Marine Sediment Transport and Environmental Management*, John Wiley & Sons, Inc.)

alluvial plains and interdistributary bays, and coarse material is stored in axes of aggrading distributaries and tidal channels. Only the intermediate sizes are swept out onto the shelf, primarily during periods of river flood.

Shorefaces become sediment sinks when progradation of the coastline begins (Fig. 11.6). Rivers continue to act as partial filters as they trap finer-grained sediment and some coarse sediment, and the remaining coarse sediment is carried to the river mouth, where it is distributed along the coast by littoral processes. Again, it is mainly the intermediate-sized material that can successfully bypass the coastal zone, although coarse shoreface sediments can sometimes be transported out onto the shelf by downwelling currents associated with strong onshore surface flows during intense storms (Swift and others, 1985).

Thus, different sediment types are made available to the shelf at different times during a transgressive–regressive cycle. Sands derived from the shoreface are deposited on the shelf during coastal retreat as sea level rises, but as transgression slows, intermediate grades are delivered directly to the shelf by rivers. In either case, it is possible for both sediment types to be redistributed by dynamic processes once they are delivered to the shelf.

Shelf Sediment Facies

Facies Distribution.   Sediments on the shelf can be broadly grouped into three facies, including sand, mud, and mixed sand and mud. Sand facies include both relict and palimpsest sediments, as well as those on or derived from the nearshore sand prism. Muds are primarily extrabasinal sediments provided by rivers, and mixed sediments appear to be the result of biogenic reworking of the other two facies. The character and distribution of these facies on any shelf reflects not only the transport capability of processes at various positions of the shelf but also the frequency with which transport events occur.

Sand Facies. On some shelves, sand occurs as a broad sheet covered by current- or wave-formed ripples. On other shelves, however, sand has accumulated in low-amplitude bathymetric features with three orders of scale (Swift and others, 1973). First-order features are linear bodies up to 100 km long and 20 km wide called massifs that trend at right angles to depositional strike. These features represent axes of sand accumulation on retreating shorelines, such as capes where littoral drift converges, or coarse sediment sinks such as estuary mouths (Fig. 11.7). Where they can be traced shoreward into estuaries, these massifs are commonly found bordering surface or subsurface channels that may terminate in deltas at shelf edges. Even though the main structure of massifs lies perpendicular to strike, they often show a coast-parallel surface grain that provides some evidence of the ability of contemporaneous shelf processes to reshape the original deposit.

Some of this coast-parallel grain is caused by the superposition of second-order sand bodies on the massifs. Called linear sand ridges, these features are up to 40 m high and tens of kilometers long, and they occur on the massifs as well as on the broad areas of the shelf between massifs. Most ridges appear initially to be constructional features formed on the shoreface by the interaction of tide- and storm-generated currents (Fig. 11.8). Although they were stranded in deeper water during the period of Holocene coastal retreat, these features are capable of being reworked by processes incident to the shelf (Swift, 1976). They normally show a steep side that faces the resultant sediment transport direction, and the presence of inclined strata as revealed by detailed geophysical surveys also indicates that the ridges migrate. Historical records show that some ridges on the shoreface may migrate at rates up to 120 m/year (Swift and Field, 1981), but the activity of ridges in deeper water has not yet been determined.

Transport of sediment on the ridges in both tide- and weather-dominated regimes occurs in the form of smaller-scale bedforms migrating up out of swales and onto flanks, commonly at an oblique angle to ridge crests. The resulting internal structures probably consist of cosets of relatively small scale crossbeds in gently inclined larger-scale beds.

Sand waves are third-order features that are as much as 10 m high and 500 m long. They form in response to tide- and wind-driven currents, and their crests are oriented perpendicular to resultant flow. On tide-dominated shelves, they can be complex features. In shallow water, they may respond to wave-driven currents as well as tides, and surfaces are commonly covered with ripples and dunes in response to short-term variations in flow. In deeper water, however, they respond mainly to tide-driven circulation, and they are simpler in form and lack superimposed bedforms (McCave, 1971). On weather-dominated shelves, however, sand waves may not be present in shallow water where wave action can inhibit growth, and sand

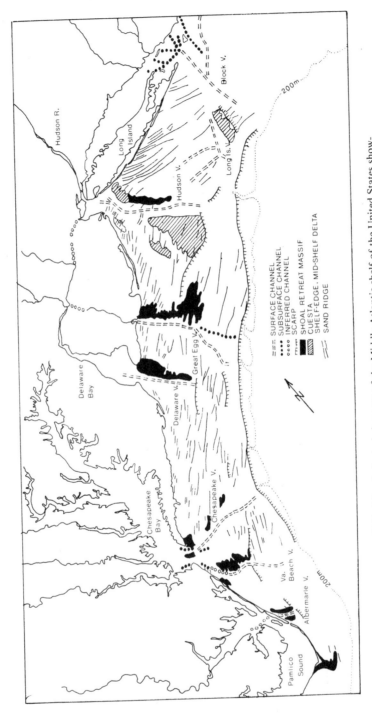

Figure 11.7 Morphological elements of the Middle Atlantic shelf of the United States showing relationships of shelf-edge deltas and shoal retreat massifs to shelf valleys (after Swift, 1975).

SURFACE CHANNEL
SUBSURFACE CHANNEL
INFERRED CHANNEL
SCARP
SHOAL RETREAT MASSIF
CUESTA
SHELF-EDGE, MID-SHELF DELTA
SAND RIDGE

Hudson R.
Long Island
Hudson V.
Long Is. V.
Block V.
200 m
Delaware Bay
Great Egg V.
Delaware V.
Chesapeake Bay
Chesapeake V.
Va. Beach V.
Albemarle V.
Pamlico Sound
200 m

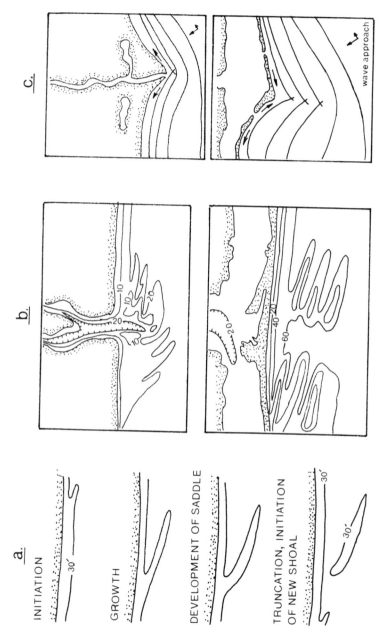

Figure 11.8 Origin of shelf ridges showing (a) detachment of a ridge from a retreating shore-face, (b) organization of ridges into a shoal retreat massif, and (c) a cape retreat massif (after Duane and others, 1972; Swift and others, 1972; and Swift and Sears, 1974).

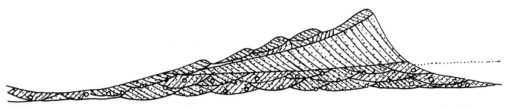

Figure 11.9 Probable internal structure of a sand wave in moderately deep water (after Mc-Cave, 1971).

waves that occur in deeper water may show several smaller orders of bedforms superimposed on them (Field and others, 1981) (Fig. 11.9).

A fourth type of sand deposit, normally associated with tide-dominated shelves, is linear sand bodies called sand ribbons that are oriented parallel to flow direction. They are as much as several hundred meters wide and 15 km long, and their form and internal organization are a function of the velocity of the flow and the amount of sand available to it (Fig. 11.10). They are not greatly dissimilar in size from sand ridges, but they are flow-parallel structures, whereas ridges are flow-oblique, and crestal spacings are much smaller in sand ribbons.

Sand patches are broad expanses of sand lacking large-scale surface morphology. They occur in areas where turbulence is sufficient to keep mud in suspension or to resuspend it episodically after it has been deposited, and where the amount of available sand is small or where current velocities are insufficient to maintain large bedforms. Their surfaces may be covered with ripples, especially following intense

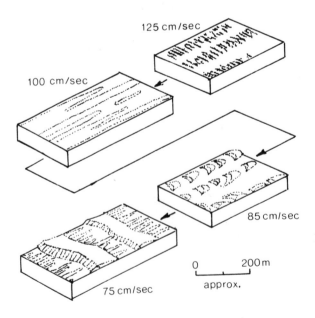

Figure 11.10 Relationship of the morphology of the four main types of tidal sand ribbons to typical near-surface current velocities (after Kenyon, 1970).

storms, but these tend to be degraded with time. Transport events are of short duration, and they are separated by long periods of quiescence during which benthic organisms destroy primary internal structures.

Two other features that appear to be significant in interpreting ancient shelf sequences are hummocky bedding and graded storm layers. Hummocky bedding was first recognized in the rock record, where it was interpreted to be a shelf deposit because of its position within sequences considered to be of shelf origin (for example, Bourgeois, 1980). It was not until 1983, however, that the shelf bedform in which this structure originated was recognized.

Swift and others (1983) recognized hummocky megaripples as flow transverse bedforms with no avalanche faces. They represent sharp-crested megaripples that have been modified by complex storm currents in which the wave orbital current component is equal to or greater than the mean flow component. They have been found to occur in shoreface settings (Greenwood, 1984; Greenwood and Sherman, 1986) and on tidal flats (Bartsch-Winkler and Schmoll, 1984), but their chances of preservation are best on the inner shelf where wave and current activity during fair-weather periods is normally insufficient to remold the substrate.

Graded storm layers are found in somewhat deeper water than hummocky megaripples and occur when the mean flow component is greater than the wave orbital component (Aigner, 1982). Graded storm layers have long been recognized on modern shelves, but they had been identified as turbidites (for example, Hayes, 1967) because they are similar in appearance to Bouma sequences that form from fluid gravity flows. Physical problems, however, exist in initiating and maintaining an autosuspension on a low-gradient shelf (Swift and Niederoda, 1985). It is more probable that graded layers represent the deposits of strong bottom flows generated by barotropic gradients set up by storms that transport sediments put into suspension by cyclic storm wave loading of the substrate (Nelson, 1982).

Mud accumulates on areas of the shelf not subject to large shear stress, such as in deep water or at the end of tidal transport paths, or where rates of influx are so large that episodic shelf transport processes are unable to redistribute all the incoming sediment. This may occur near deltas where mud concentrations are large (in excess of 100 mg/l) and where flocculation increases the suspended weight of clay particles.

Shelf muds are commonly rich in organic debris, especially where they accumulate in upwelling zones. Density stratification may be intense on some shelves during the summer, and when combined with offshore wind-forced surface waters, it may enable cold, nutrient-rich deep waters to intrude onto the shelf. Blooms of single-celled flora occur, and when the organisms die, they settle out of suspension along with mud in outer shelf areas. Decomposition of some of the organic material depletes oxygen in the bottom waters and allows preservation of the remainder, and where denitrification also occurs, authigenic minerals such as glauconite and phosphate can form.

Biogenic reworking of sediments is ubiquitous on shelves, and the degree of bioturbation as well as the type of traces left depend on the type of sediment present, the rate of accumulation, and the frequency of bedload transport episodes. Infauna are few in number and variety on sandy substrates that are subject to frequent episodes of bedload transport, but the numbers of individuals and diversity of species increase in areas of infrequent bedload transport where muds tend to accumulate. The infauna are most commonly deposit feeders that move both horizontally and vertically through the sediment, ingesting it and re-forming it into fecal pellets. In the process, they homogenize the sediment and destroy primary structures, and on some shelves a mixed sand–mud facies is present that is probably the result of vertical transfer of sediment by burrowing organisms (Nittrouer and Sternberg, 1981).

Facies Distribution.   The distribution of these facies on autochthonous and allochthonous shelves is strikingly different. In addition, the manner in which these facies are associated with other shelf sediment types is determined, in large part, by the degree to which a shelf profile has attained equilibrium with respect to its sediment supply.

Only a few modern shelves cannot be considered transgressional shelves because all have suffered the effects of Holocene sea-level rise. However, the North Sea coast of Europe and, especially, the Middle Atlantic coast of the United States represent extreme examples of a transgressional shelf because they are almost barren of modern detrital sediments. Most rivers, in both cases, are not yet in equilibrium with the new sea level, and they act as traps not only for sediments being transported down the river but also for sediments being transported in littoral circulation cells. The two shelves are dominated by radically different transport processes, but the response of the substrate appears to be similar.

The Atlantic coast is storm-dominated (Swift and others, 1972). Major currents are induced by wind shear that forces currents produced both by wind drift and wave drift. The shelf is not tideless, however, and intense tide-induced currents do occur; but they are of very short duration and are a much less important sediment transport process than weather-induced currents (Swift and Niederoda, 1985).

These processes are currently modifying a shelf topography that was formed initially during Holocene sea-level rise. The largest features are sand massifs that represent the migration paths of coastal shoals that formed at capes and estuary mouths during the retreat of the shoreline across the shelf. Second-order features are sand ridges that originate in the shoreface and are detached from it by storm-generated currents (Duane and others, 1972). Although these features originated during the Holocene sea-level rise, they are both maintained and modified by contemporary processes (Stubblefield and Swift, 1976) (Fig. 11.11). Other features of the Atlantic shelf sand facies are ripples, megaripples, sand waves, and current lineations that are all responses of the substrate to modern shelf processes (Field, 1980).

When considering the transgressive sand sheet of the Middle Atlantic coast as a unit in a stratigraphic sequence, it should be remembered that, although some of the features just discussed may be laterally extensive, they rarely exceed a few tens of meters in thickness. In fact, most of the sand sheet on the shelf is much less than

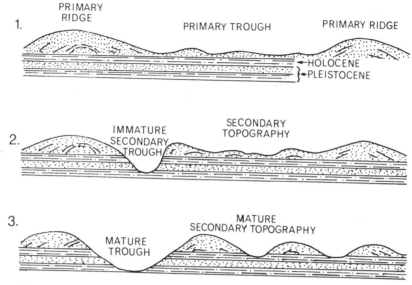

Figure 11.11   Accentuation of ridge topography by shelf processes after initial formation and separation of ridges from the shoreface (after Stubblefield and Swift, 1976).

10 m thick. Sand was provided only through erosion of the shoreface, and as the shoreface retreated, the sediment source was removed. It is apparent that transgressive shelf deposits may occupy only a small part of any stratigraphic sequence.

The tide-dominated shelf of the North Sea off western Europe shows many similarities to that of the Middle Atlantic shelf of the United States. Most major rivers empty into estuaries and most of the coastline is in a state of slow retreat. Extrabasinal muds, however, are being deposited on the shelf in deeper waters below the influence of tide-generated currents and along parts of the German and Dutch coasts, where muds are being provided faster than they can be reentrained and moved.

Most of the sand cover of the North Sea shelf is glacigenic or was derived from shoreface erosion, and the larger morphological features are inherited, in part at least, from shallower water environments. The tide-driven currents, however, are locally more intense and operate more continuously in a given direction than storm-induced currents on the Atlantic shelf, and the substrate, therefore, has been re-shaped to a greater degree to reflect present-day hydraulic conditions.

Sand banks are similar in morphology and are probably analogous in origin to the massifs of the Atlantic shelf. Tidal ridges may also be inherited features, but those, at least, that are in relatively shallow water are now in equilibrium with tidal flow (Kenyon and others, 1981). They are oriented obliquely to the dominant flow, which rotates toward the ridge crest as it moves up the flanks. Frictional forces slow the flow at the crest sufficiently to cause bedload deposition, and as it passes the crest, the flow expands and decelerates still further, causing it to veer toward its original direction (Belderson and others, 1982). Active ridges have sand waves on their flanks, but those in deeper water do not and are presumably no longer active (Kenyon and others, 1981).

Sand also migrates along tidal transport paths in smaller bedforms and sand sheets in areas not occupied by ridges or banks. Sediment accumulates where transport paths converge, and the deposits become thicker and better organized down the transport path in the direction of decreasing midtide velocity until tidal currents decelerate to a threshold below which sand is no longer mobile.

Mud or muddy sand accumulates in two settings on the North Sea shelf. They may accumulate at the end of transport paths where tidal current velocities are insufficient to keep muds in suspension and where storm wave activity is too small to reentrain them (McCave, 1971). Mud may also accumulate in areas where suspended load concentrations are large. Several large rivers enter the North Sea along the German coast, for example, and mud is accumulating at the rate of 15.5 cm/100 years despite the moderate tidal currents in the area (Reineck, 1967). Accumulation of mud is also favored in semienclosed shelf settings like the North Sea, because no pronounced topographic gradient occurs toward a deep-sea basin.

Primary structures tend not to be preserved in muddy sediments because of the mixing ability of infauna in the absence of significant bedload transport. Mud substrates lack bedforms other than ripples, and the most common type of stratification present is caused by rhythmic alternations of coarse and fine layers produced in response to episodic transport induced by passing storms (Reineck and Singh, 1972).

Regressive shelves will occur where coastal sediment sinks have been filled and shorelines have begun to prograde, but they can also occur anywhere rivers deliver significant amounts of sediment to the shelf. The Oregon–Washington coast, for instance, has not yet come into equilibrium with the Holocene sea-level rise, and the mouths of rivers such as the Columbia and Rogue are estuaries where sediments are being trapped. Nevertheless, river-mouth bypassing is taking place, and fine sediments, especially silt, are being added to the shelf.

The dominant transporting mechanism on the Washington–Oregon shelf is driven by wind shear on surface waters. Combined with Eckman veering, wind shear sets up a northward-directed bottom flow. Velocities may exceed 40 cm/s at depths of 80 m and are capable of transporting silt and fine sand as bedload.

Most sediment is delivered to the shelf by the Columbia River, and it enters shelf waters as a surface plume (Kulm and others, 1975) (Fig. 11.12). Maximum density of the plume occurs in the late winter and spring, and most of the suspended sediment consists of silt. The plume is directed north along the coast by the Coriolis force and deposits most of its sediment in a nearly coast-parallel linear band extending to the north from the river mouth (Nittrouer and others, 1979). Southward-directed, wind-forced surface drift in the summer, however, opposes the Coriolis

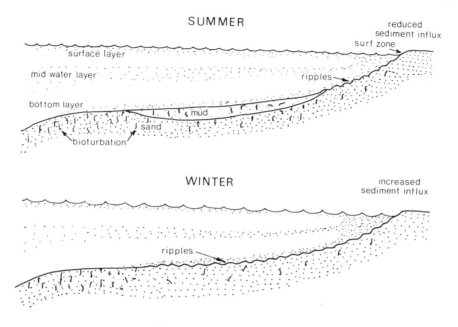

Figure 11.12 Dynamics of supply and distribution of sediments on the Oregon coast (after Kulm and others, 1975, by permission of the *Journal of Geology,* University of Chicago Press).

force and forces the plume to the southeast, where a minor amount of sediment is deposited.

Sediment also enters the shelf as a mid-level turbid layer that occurs at the thermocline. It is relatively compact near the coast, where it receives much of its sediment from river discharge and coastal erosion. In its passage across the shelf, it expands and becomes relatively diffuse at the shelf edge.

Some sediment from the overlying two layers settles into the third layer at the bottom, but most of its sediment is provided to it when storms generate turbulence and flow velocities sufficient to entrain large amounts of bottom sediments. This occurs mainly in the winter so that the dominant drift is northward along the coast, but an offshore component of the drift tends to move sediment obliquely across the shelf to the shelf edge, where it is lost to the system. This mechanism may explain why sedimentation rates are small on the shelf, even though the Columbia River provides up to 8 million tons/year of suspended sediment to it.

Sediments on the shelf are divided into facies that correspond to areas influenced by distinctly different processes. The inner shelf extends from the shoreface to a depth of about 50 m. Sediment transport is dominantly accomplished by shoaling waves and wind-driven currents that are able to winnow out most silt and clay.

The infauna are able to rework the fine sand that remains and, except in the surf zone, only remnants of horizontal laminae and small-scale crossbedding are preserved (Fig. 11.13).

The central shelf is dominated by geostrophic currents produced by density-driven flows and wind forcing. It is covered primarily by silt, with lesser amounts of clay provided to it by the Columbia River. The sediments are transported only intermittently by storm-induced currents, and bioturbation is intense, especially in the summer when transport events are relatively infrequent. Only remnants of primary structures are preserved in this facies (Fig. 11.13).

Silt content decreases into the outer shelf, which is variably influenced by wind-forced currents, shoaling internal waves, and ocean currents. Much of the substrate of the outer shelf consists of relict coarse sands that locally contain in excess of 15% glauconite. Where muds of the central shelf encroach on this facies, infauna mix the two facies, producing a muddy sand facies. The degree of mixing decreases upward away from the sand substrate, however, and where mud accumulates to sufficient thickness, sands are absent in the upper layers.

Most shelf sand bodies are formed during transgressions when sand waves and sand ridges are constructed on the shoreface and then stranded on the shelf as the shoreline retrogrades. Sand bodies, however, can also form on regressive shelves, but in order for them to accumulate in front of prograding shorelines, some mechanism must be invoked that would allow substantial bypassing of the shoreface by sands.

One such mechanism occurs at the Diametta mouth of the Nile Delta, where Coleman and others (1981) have described the detachment and migration of sand plumes from the delta front. These plumes consist of sands deposited at the distributary mouth during flood stage and then swept offshore and downshelf by downwelling storm flows (Fig. 11.14). Separation of the plume from the shoreface is maintained by a back eddy that develops when the coastal flow undergoes boundary layer detachment at the distributary mouth.

Intrinsic and Extrinsic Controls on the Evolution of Shelf Sequences

Third-order shelf sequences form in response to unsteady flows created by storms, which are events that recur with an episodicity measured in days, and by tides, which have a periodic cycle measured in hours. Flow-transverse bedforms respond to shelf flows in a complex manner (Allen, 1980). These are considered to be composed of an unsteady or periodic component and a steady component. If the periodic component is small relative to the unsteady one, a simple sand wave with a single avalanche crest is formed (Fig. 11.15). As the bedform experiences a greater exposure to unsteady flow components, however, the internal organization of the strata becomes much more complex. Reactivation surfaces form as the result of migration of smaller bedforms over the larger ones. As the flow becomes more unsteady, foresets

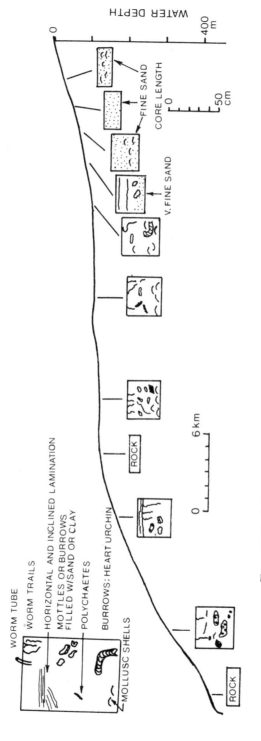

Figure 11.13 Sediments and sedimentary structures of the Oregon coast in relation to position on the shelf and to water depth (after Kulm and others, 1975).

273

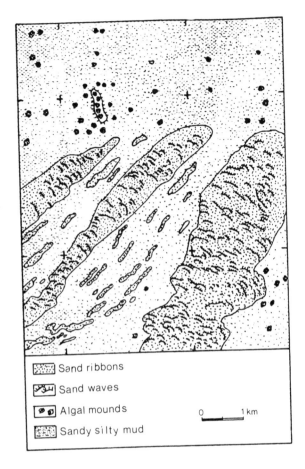

Sand ribbons

Sand waves

Algal mounds   0   1 km

Sandy silty mud

Figure 11.14 Map of sand waves migrating over muddy substrates in front of the Nile Delta (after Coleman and others, 1981).

are separated by sigmoidal reactivation surfaces into bundles that record erosion of sand-wave crests, as well as migration of second-order bedforms. Finally, where flow is greatly unsteady and asymmetric, foresets may consist of a master set of strata inclined in the direction of dominant flow and composed of smaller-scale sets formed by second-order bedforms migrating down lee faces.

Storm flow events produce strata that record nonperiodic events, but tidal flows produce deposits that respond to rhythmic changes in flows that may not only record first-order tidal cycles, but also flow variations produced over a monthly cycle (Visser, 1980). Sand avalanching down the face of the lee slope during flow of the dominant current forms foresets that are partially eroded during the reverse flow of the subordinant current. Bundles of cross strata are thus formed that record first-order tidal cycles. Thickness of the bundles may vary according to second-order cycles because more sand may be transported during peak flows of spring tides than during the lesser flows of neap tides (Fig. 11.16).

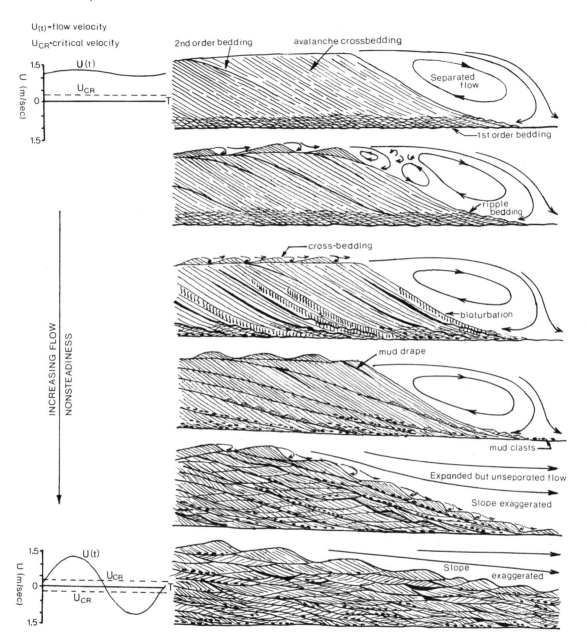

Figure 11.15 Morphology and internal structure of sand waves forming under conditions of variable steadiness of flow (after Allen, 1980).

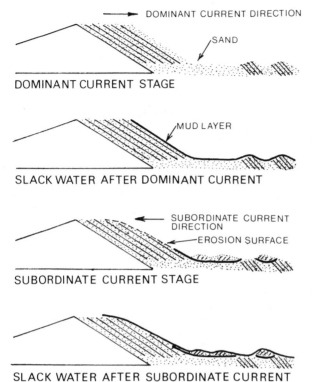

DOMINANT CURRENT DIRECTION

SAND

DOMINANT CURRENT STAGE

MUD LAYER

SLACK WATER AFTER DOMINANT CURRENT

SUBORDINATE CURRENT DIRECTION

EROSION SURFACE

SUBORDINATE CURRENT STAGE

SLACK WATER AFTER SUBORDINATE CURRENT

Figure 11.16 Stages in the formation of tidal bundles (after Visser, 1980).

Hummocky crossbeds are a bedding structure consisting of gently curved laminae without pronounced foresets that are bounded by low-angle discordances (Fig. 11.17). Laminae tend to parallel basal erosional surfaces, and they tend to thicken laterally in a set so that their dip diminishes regularly upward as they fan out laterally. They probably are the result of accretion on hummocky megaripples, and they form under the combined influence of storm flows and wave orbital currents (Swift and others, 1983).

They can occur in sequences with well-defined hierarchical assemblages of bedding structures that reflect the temporal changes in the hydraulic regime that accompany storms (Dott and Bourgeois, 1982) (Fig. 11.18). The hummocky crossbeds fill in scour holes created during peak storm intensity and form soon after storm peaks as the sea floor begins to aggrade. As the storm continues to wane, plane beds and, finally, ripple beds accumulate, and if mud is present, it settles out of suspension and caps the sequence.

Graded storm layers form in somewhat deeper water than do hummocky crossbeds. They also show a hierarchical assemblage of bedding structures that is much like turbidite sequences, and they also form in response to waning flow (Aigner, 1982) (Fig. 11.19). They overlie a scoured surface that may show sole marks,

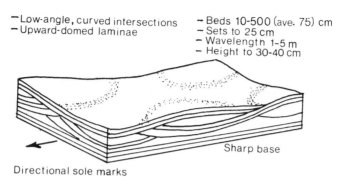

Figure 11.17 Diagrammatic representation of hummocky crossbeds (after Harms and others, 1975).

and they consist of a basal coarse layer with intraclasts that is overlain by an unlaminated fining-upward bed. These beds are overlain by plane beds and wave ripples, and if significant mud occurs in the system, the sequence is capped by a pelitic division.

A shore-parallel zonation of variations of the basic structures of storm-graded layers may exist (Aigner and Reineck, 1982) (Fig. 11.20). They are most common in the offshore–lower shoreface transition zone, where they also most commonly contain crossbedded units. They are coarsest grained shoreward of this point, and they decrease in grain size, bed thickness, and degree of amalgamation seaward. Bioturbation increases seaward, as does the thickness of intervening mud layers. A progradational sequence, therefore, would consist of an upward-coarsening and thickening deposit that shows an accompanying decrease in the degree of bioturbation.

Second-order cycles form in response to climate variations. These climatic variations may take a form that influences the hydraulic regime of the shelf, or they may take a form that alters the magnitude and timing of sediment delivery onto the shelf. In all likelihood, these two effects will be interactive.

Climatic changes that alter the amount or timing of sediment delivery to the shelf result in changes in the degree to which marine processes can redistribute the

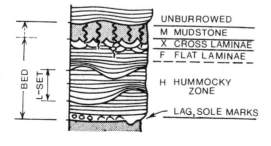

Figure 11.18 Typical vertical sequence found associated with hummocky crossbeds (after Dott and Bourgeois, 1982).

IDEAL SEQUENCE AND
HYDRODYNAMIC INTERPRETATION

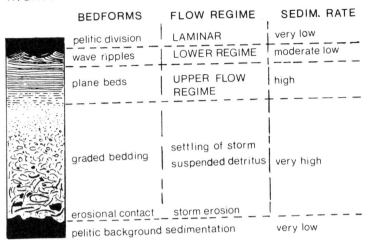

Figure 11.19 Vertical sequence developed in graded storm layers showing vertical change in sedimentary structures and the changes in hydraulic regime accompanying their formation (after Aigner, 1982).

incoming sediment. In 1969, for instance, two severe storms in southern California occurred within one month of each other, resulting in heavy rainfall and record floods in the Ventura River (Drake and others, 1972). A large body of distinctly colored sediments was deposited on the inner shelf as a result of these floods. Nearly two years after the floods, some of the sediments had been redistributed by wave and current activity, but a coherent layer of sediments still existed near the mouth of the Ventura River (Fig. 11.21). These sediments may eventually be completely redistributed, but if additional masses of such sediments were delivered to the shelf during an extended period of frequent intense storms, a layer of unsorted sediment bearing relatively little imprint of the shelf regime might be deposited.

The shoreface may also act as a sediment source that stores sand and then periodically releases it to the shelf. Moody (1964) has shown, for instance, that barriers on the Delaware coast accreted over a long period of time during which the shoreface evolved toward an ideal wave-graded profile. This period was terminated by a single severe storm in 1962, when the shoreface retreated up to 75 m over a three-day period. Some of the eroded sand was transported over the barrier onto the lagoon, but a significant portion was transported onto the inner shelf by rip currents and storm-driven bottom flows.

Both events were out of the ordinary and both delivered amounts of sediment far larger than is normally provided to the shelf. If such events are widely spaced in time, shelf processes may rework them into an equilibrium distribution. Climatic changes, however, may produce extended periods of severe weather that deliver sedi-

FACIES MODEL

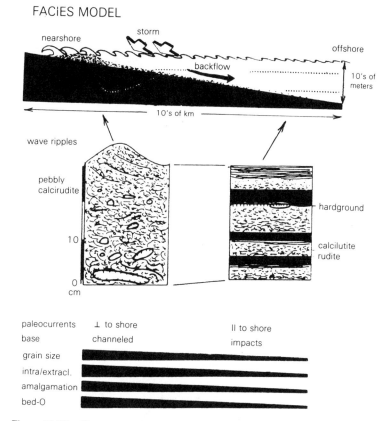

Figure 11.20   Changes in texture and structures within storm-graded layers with distance from shore (after Aigner, 1982).

ment in amounts that overwhelm the ability of the shelf to redistribute them. If the climatic changes are cyclic, the resulting shelf deposits might consist of alternating sequences of strata representing an equilibrium sediment distribution and strata consisting of sediments that were not brought into equilibrium with shelf processes.

First-order cycles form in response to eustatic changes in sea level. The three principal factors affecting deposition of shelf sequences are rate of sediment influx across the shelf, rate of sea-level rise, and rate of subsidence (Pittman, 1978). Assuming that these three factors have never been so delicately balanced that water depth over a shelf has remained constant for an extended period of time, it follows that shelves continuously undergo cyclic change from transgressive to regressive states. A representative shelf sequence, therefore, will consist of sediments deposited during both phases of each cycle.

For the nonmarine and coastal systems, it is possible to propose representative stratigraphic sequences based on successions of Pleistocene- or Holocene-age sedi-

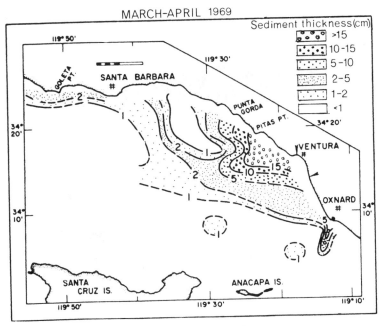

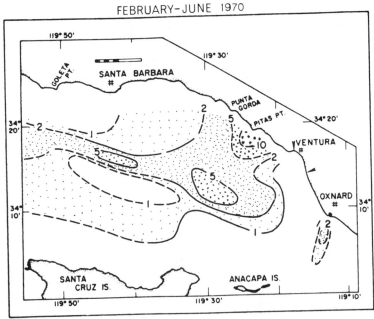

Figure 11.21 Changes over a 1-year period in the distribution of sediments delivered to the Ventura shelf by the flood of 1969 (after Drake and others, 1972).

ments that accumulated within these systems. By doing so, it is possible to assess the effects of known variables on the system as aggradation occurs, and it is also possible to assess the preservation potential of the various components in the systems.

Depositional sequences on modern shelves should provide such information. Unfortunately, modern shelf systems are presently covered only with eroded shoreface sediments deposited during the Holocene sea-level rise. Only a few shelves have been even moderately provided with enough river-derived sediment to initiate a regressive phase.

The Holocene sea-level rise was an event out of the ordinary in terms of its magnitude and the speed at which it occurred. Some evidence suggests, however, that slower transgressions of lesser magnitude may have imposed similar conditions on older shelves. Miocene strata in the Caliente Range of southern California, for instance, consist of about 50 individual progradational units up to a few tens of meters thick deposited in a continuously subsiding basin (Clifton, 1981a). Nearshore and nonmarine components of each unit are separated by lag gravels and offshore sandstones, but transgressive episodes within shelf sediments are marked by the occurrence of bioturbated siltstones and fine-grained sandstones over ripple-bedded micaceous sandstones (Fig. 11.22).

The average length of time that elapsed during each transgressive–regressive episode is similar in scale to long-term climatic cycles produced by periodicity of the earth's orbit. Apparently, even relatively small climatic variations can induce sea-level changes sufficient to radically alter patterns of shelf sedimentation.

Study of modern shelves suggests that they may receive relatively small amounts of extrabasinal sediments during the transgressive phase. Sediments are derived primarily from within the basin itself through shoreface erosion and bypassing, and these are spread as a blanket of sands over the shelf as the sea level rises and the shoreface retreats. On most shelves, these sands are periodically remobilized by shelf processes.

The contact between the transgressive sand blanket and the underlying sediments represents a sequence boundary, and normally it is an erosional disconformity. On the Middle Atlantic shelf, transgressive sands overlie back-barrier sands and lagoonal muds with an erosional contact (Field and Duane, 1976) (Fig. 11.23), and on the North Sea shelf they locally overlie a lag developed on coarse, glacigenic sediments (Houbolt, 1968; Swift, 1976). The sequence boundary may also be marked by a mixture of relict and authigenic sediments. Relict sediments may consist of concentrations of shells and skeletal material, as well as some residual debris, and concentrations of authigenic minerals such as glauconite and phosphate may be indicative of such surfaces.

Sediments of the transgressive part of the cycle may accumulate on broad, nearly level sand sheets covered with ripples and megaripples, hummocky bedforms, and graded storm layers. They may also accumulate in large accretionary forms such as massifs and ridges, or they may be formed by contemporary shelf processes into smaller-scale components such as sand waves.

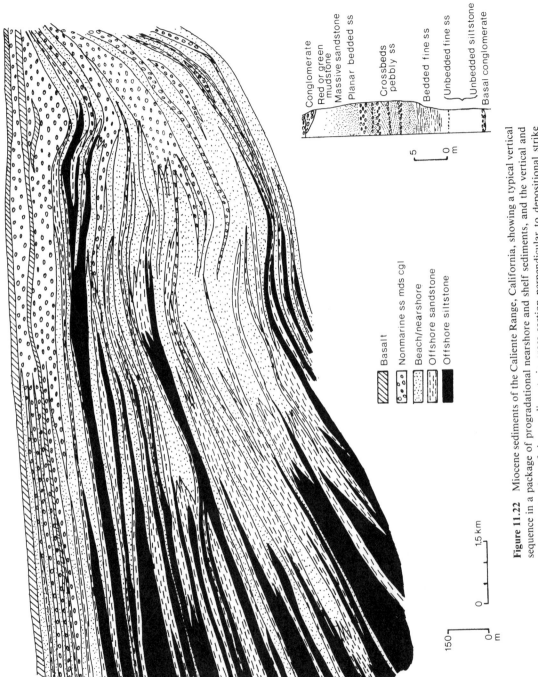

Figure 11.22 Miocene sediments of the Caliente Range, California, showing a typical vertical sequence in a package of progradational nearshore and shelf sediments, and the vertical and lateral distribution of these sediments in cross section perpendicular to depositional strike (after Clifton, 1981a).

Basalt

Nonmarine ss mds cgl

Beach/nearshore

Offshore sandstone

Offshore siltstone

Conglomerate

Red or green mudstone

Massive sandstone

Planar bedded ss

Crossbeds pebbly ss

Bedded fine ss

Unbedded fine ss

Unbedded siltstone

Basal conglomerate

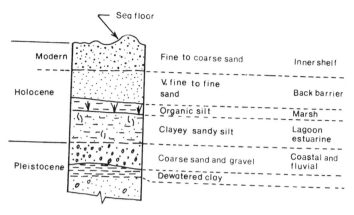

Figure 11.23 Generalized stratigraphy of Pleistocene and Holocene sediments of the Middle Atlantic shelf of the United States (after Field and Duane, 1976).

The preservation of bedding in both accretionary forms and sand sheets is dependent on the rate of accretion, rate of mixing by infauna, and the periodicity of transport episodes (Nittrouer and Sternberg, 1981). In general, biological reworking of the shelf substrate is most intense in the upper 20 cm of sediment, and if the rate of infaunal mixing is small compared to the rate of accumulation, then textural variations within the sediment can be maintained as the sediment is buried beneath the zone of most intense mixing (Guinasso and Schink, 1975).

Sediments may also be reworked by physical processes to such a degree that internal organization is destroyed (J. D. Smith, 1977). If the substrate is being reworked continually by storms and if the rate of accumulation is small, the sediment will consist of the lower parts of graded storm layers that present a massive or at best a subtly laminated appearance (Fig. 11.24). The upper, finer-grained portions of the layers will tend to be preserved, however, if episodes of reworking are infrequent or the depth of erosion is shallow. Where episodes of bedload transport are

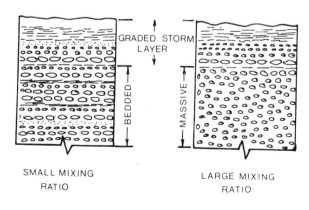

Figure 11.24 The effects on preservation of stratification of physical mixing of the substrate by passing storms (after Smith, 1977).

infrequent, however, smaller-scale features such as ripples may be slowly degraded during intervening periods (Swift and others, 1979).

Once extrabasinal sediments begin to arrive on the shelf in sufficient quantities, the regressive phase of the cycle begins and sediment distribution reflects the contemporary shelf environment. Johnson (1919) envisioned a graded shelf that consisted of a seaward-fining sediment cover deposited in response to a progressive decline in the effectiveness of transport processes across the shelf into deeper water. The competence of shelf processes does tend to decline in deeper water, and the frequency with which events capable of transporting shelf sediments occur also tends to decline. The decrease in transport capacity with depth, combined with diffusive transport of sediments down the shelf gradient, can produce a seaward-fining texture that would result in a coarsening-upward regressive sequence.

The reality is much more complex, however. The distribution of sediments depends not only on the capacity of shelf processes to transport sediment, but also on the shape of the shelf, especially its depth and gradient, the shape of the coastline and its orientation relative to weather-producing elements, and the quantity and caliber of the extrabasinal sediment delivered to the shelf. Because many of these factors are variable both along and across the shelf, it is perhaps not surprising that the sediment distribution on aggrading shelves resembles a complex mosaic pattern, rather than a simple linear one. Diffusive transport provides a framework around which the other factors operate, and it is the interaction of all these factors that determines shelf sediment distribution patterns.

Perhaps the most important interaction is that between sediment influx and sorting ability of shelf processes. Progressive sorting is not a continuous process, but instead is one that occurs during short periods of intense flow, interspersed with long periods of storage. It can only work to completion where the sediment influx does not overwhelm the ability of shelf processes to redistribute and sort it. The process works best on exposed shelves with relatively steep gradients and small to moderate sediment yield from sources distributed evenly along the coast. The Oregon–Washington shelf, for example, is covered with fine sand and relatively coarse silt because storm-induced currents generated primarily during the winter are able to entrain finer-grained sediment and transport it to the shelf edge. A regressive sequence deposited on this shelf or one with similar conditions will contain virtually no clay or fine silt, even at the base of the sequence.

Probably a more typical sequence can be seen in the Gulf of Gaeta, a shallow depression on the Italian Peninsula off the Tyrrhenian Sea. The sea is tideless, but it is affected by summer storms that set up northward-flowing currents on the shelf. The shelf is narrow and slopes steeply down to the deep central part of the semienclosed sea. The sediment delivered to the shelf by the Volturno River consists mainly of silt and clay (Reineck and Singh, 1971). The small amount of sand in the system is confined to the transition zone and shoreface, which extend to a depth of about 15 m. The shelf displays a very pronounced coast-parallel zonation of grain sizes, with the thickness and number of silt layers decreasing continuously away from the coast (Fig. 11.25). A vertical sequence deposited during regression of such a shelf

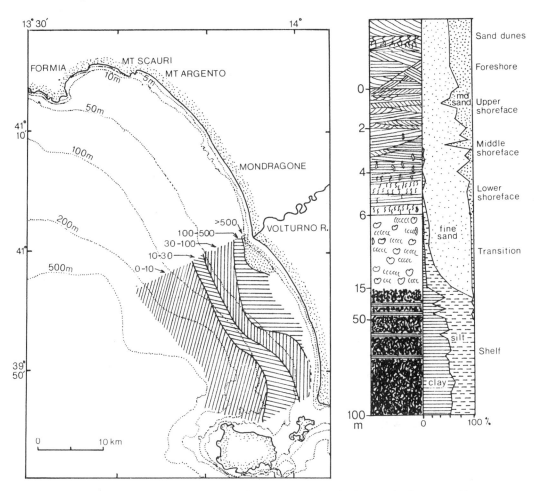

Figure 11.25 Sediments of the Gulf of Gaeta showing the decrease in silt content (shown as the ratio of silt thickness in millimeters per meter of total sediment) and the hypothetical sequence that would develop during aggradation of the shelf and progradation of the shore (after Reineck and Singh, 1971; Reineck and Singh, 1972).

would consist of a very thick section of shelf muds overlain by a relatively much thinner section of sands deposited in the transition zone and shoreface of the pro-grading coast. The shelf muds would coarsen upward, especially in the upper 100 m, by the addition of thicker and more numerous silt layers (Fig. 11.25).

The process may not work as effectively on broad, low-gradient shelves with a large influx of sediment or on protected shelves. On the German North Sea shelf, the gradient is low, the sediment influx is large, and muds are being deposited in water depths as shallow as 15 m (Reineck and Singh, 1972). Storm waves are able to winnow muds from coastal sands, but they collect below fair-weather wave base at rates up to 100 times faster than those accumulating on the Oregon–Washington shelf. Storm waves are able to entrain sediments, but fine-grained sediments that are put into suspension remain in the area where they settle on top of sand layers formed during storm peaks.

The initial deposits of a regressive sequence formed under such conditions would consist of a thick sequence of bioturbated muds interspersed with thin sand layers. The sand layers would increase in thickness, though not necessarily in numbers, near the top of the muddy deposits where the transition into the deposits of the nearshore sand facies begins (Fig. 11.26). The increase in thickness would be relatively abrupt, because it would come primarily in response to the increasing proximity of the prograding shoreline, which is the source of sand (Reineck and others, 1967), and only secondarily because of the progressive increase in the ability of the shelf to winnow fines from the substrate.

The process of progressive down-gradient sorting is complicated, however, when it is combined with other mechanisms of progressive sorting. Sediments do become finer grained in deeper water on tide-dominated shelves as tide-induced currents become less effective transporting agents and the ability of waves to winnow fines is decreased. A tendency also exists for sediments to become finer grained with distance down transport paths, and because these tend to nearly parallel coastlines, they produce a fining trend along shelf gradients that may be superimposed on the normal down-gradient fining across the shelf.

A similar coast-parallel fining trend can be produced where the bulk of the sediments enter the shelf from a single point source, such as on the Oregon–Washington shelf. Sediments enter the shelf as a two-dimensional jet from the Columbia River and are entrained by coast-parallel currents flowing in the mid-shelf. As the plume mixes with shelf waters, sediments begin to settle out of turbulent suspension and are deposited in a northward-fining linear trend that parallels the coast. Because the shelf gradient is steep, however, a strong component of offshore drift enables clays to eventually be transported off the shelf (J. D. Smith and Hopkins, 1972). Because of the incomplete sedimentary record, it is not possible to assess the impact of these other processes on the deposits of aggrading shelves where they may operate.

Despite the operation of some processes in a coast-parallel direction, the degree to which any transport process acts on a shelf is, to a first approximation, controlled by water depth, and sediments are, therefore, distributed according to a depth-dependent zonation. Even where coast-parallel processes are important, they tend to act within zones on the shelf that are also controlled by depth. The hierarchical arrangement of shelf facies according to depth makes it possible, using Walther's principle, to synthesize the vertical successions that would be deposited as facies migrate across the shelf in response to changes in water depth.

Three zones can be defined on the basis of the unique combination of hydraulic factors that operate within them (Fig. 11.27). The nearshore zone is subdivided into the surf zone, where breaking waves induce an along shore transport of sand, and the shoreface where sand transport, at least during fair weather, is dominantly

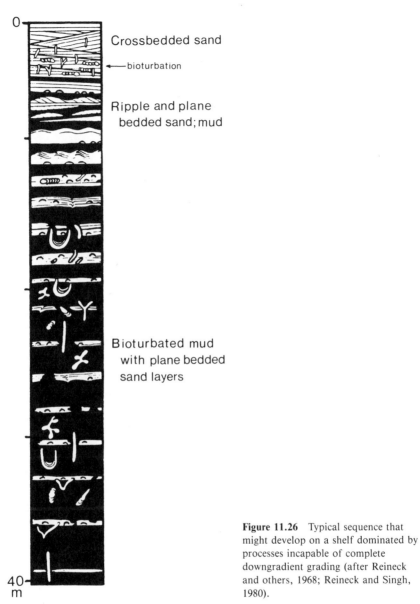

Figure 11.26 Typical sequence that might develop on a shelf dominated by processes incapable of complete downgradient grading (after Reineck and others, 1968; Reineck and Singh, 1980).

onshore. This is accomplished primarily by asymmetric shoaling waves in the upper shoreface and by the interaction of tidal and wind-driven currents with symmetrical wave orbital currents in the lower shoreface. This zone is normally considered an energy fence (Allen, 1970) that prevents much offshore transport of sand. Sand is moved landward by asymmetric shoaling waves to the surf zone, where it is put into

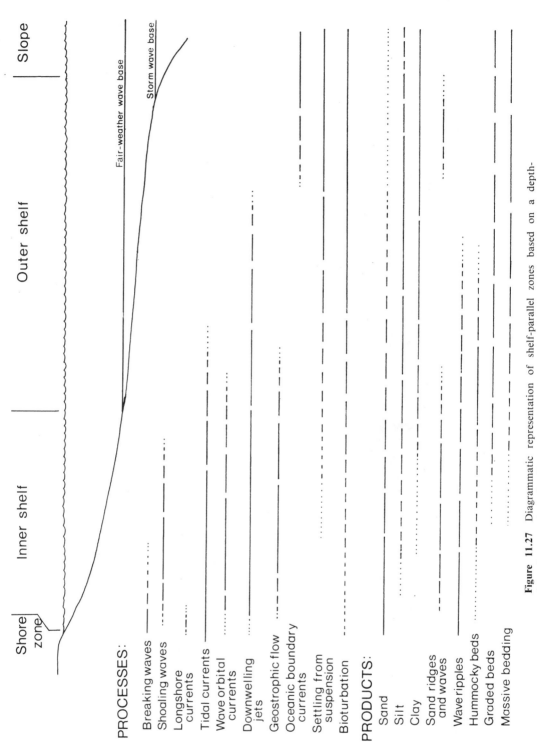

Figure 11.27 Diagrammatic representation of shelf-parallel zones based on a depth-dependent zonation of sedimentary processes.

suspension and returned by rip currents to the shoreface where the process begins again (Cook and Gorsline, 1972).

During storms, however, the transport capabilities of rip currents are greatly intensified, and strong downwelling currents are initiated that act to transport sand beyond the shoreface to the inner shelf where it may be lost to the system (Swift and others, 1985).

Tide-induced currents are most effective on the inner shelf of tide-dominated shelves where episodes of sand transport may occur with short-period cyclicity. The inner part of weather-dominated shelves, on the other hand, is normally active only during storms, and sand transport is episodic but without regular cyclicity.

The inner shelf is dominated by geostrophic flows induced by wind forcing, barotropic gradients, and tide forcing, with dominant transport directions nearly paralleling coastlines. Because of the shelf gradient, however, a small offshore component of transport normally is present, and downwelling currents produced during storms may be particularly effective in transporting sediment away from the shoreface and the inner shelf. Progressive sorting is well-developed, with finer-grained sediments collecting in deeper water where transport episodes are fewer and at the end of tidal transport paths. It is within the inner shelf that large-scale sand bodies, graded storm layers, and hummocky megaripples occur.

The finest-grained sediments accumulate on the outer shelf, where ocean currents, upwelling, and very slow moving, extremely dilute density currents or nepheloid layers are the dominant transport processes. Sediment is introduced onto the outer shelf by slow-moving turbid flows, and the normal process of sediment accumulation is through suspension settling of clay and organic matter.

During sea-level rise or fall, these zones shift back and forth across the shelf in response to changing water depth. The depth range occupied by a given zone on a particular shelf is dependent on the intensity of the various shelf processes, and the shape and sequence of the resulting cyclic deposits is a function of the rate of sediment influx into the shelf and the rate of sea-level change.

A model describing the ways in which these variables interrelate on shelves was proposed by Vail and others (1977a) and Mitchum and others (1977), and later revised by Vail and Todd (1981), Vail and others (1984), and Haq and others (1987). The basic stratigraphic unit defined in this model is the sequence consisting of a relatively conformable succession of genetically related strata deposited during a cyclic rise and fall of sea level and bounded at top and bottom by unconformities or their correlative conformities.

The unconformities are produced when sea level drops. Type 1 unconformities are produced when the rate of eustatic sea-level fall exceeds the rate of subsidence at the shelf edge. The entire shelf is exposed when this occurs so that valleys are incised on the shelf surface and low-stand fans onlap the slope. On the other hand, if the rate of sea-level fall is slower than the rate of subsidence at the shelf edge, withdrawal of the sea is more deliberate and the shelf edge is not exposed. Low-stand fan deposits do not develop in this case, but sediments do accumulate at the shelf margin, and they may prograde over the shelf edge and onto the slope (Fig. 11.28).

As soon as the rate of regional subsidence begins to exceed that of sea-level fall, the relative sea level will rise, initiating deposition of the sequence. Incised valleys are backfilled and coastal deposits onlap the exposed surface, but because initial rates of sea-level rise are slow, sediment influx can exceed the creation of accommodation space, causing shoreline recession and deposition of a low-stand wedge at the shelf edge even as sea level rises.

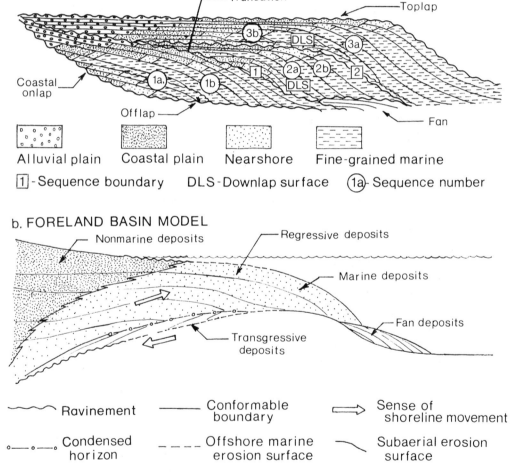

Figure 11.28 Diagrammatic cross sections showing development of sequences, parasequences, and sequence boundaries during a complete cycle of sea level change on (a) a thermally subsiding passive margin (after Vail and others, 1984), and (b) in a foreland basin (after Swift and others, 1987).

Transgression begins only when the rate of relative sea-level rise increases and accommodation space is created faster than incoming sediment can fill it. Coastal sediment traps are created when this occurs, reducing sediment yield to the shelf. The shoreface retreats in the face of rising sea level, and a ravinement is created as the coast encroaches on the subaerial surface and transforms it into a marine one.

Shelf zones migrate landward during the transgression. Sediments deposited on the shelf onlap the ravinement, and as the shelf aggrades, facies boundaries are formed that are inclined upward in a shoreward direction (Fig. 11.29). The angle of

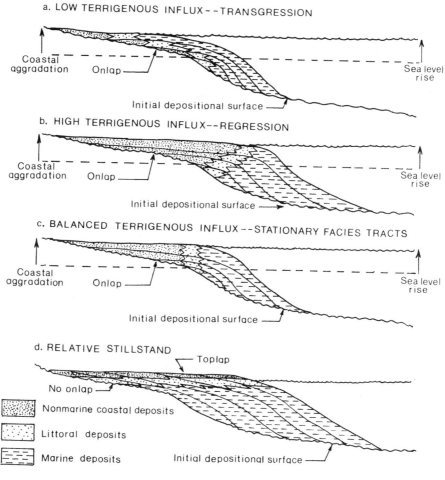

Figure 11.29   Idealized sequences developed during rising sea level on (a) a transgressive shelf where sediment yield is small, (b) a regressive shelf where sediment yield is large, and (c) an aggradational shelf where sediment yield and sea-level rise are balanced; and (d) a sequence developed on a shelf during relative stillstand (after Vail and others, 1977a).

inclination is dependent on the rate of sea-level rise relative to sediment influx. It decreases as sediment influx declines or the rate of sea-level rise increases, and because most incoming sediment is trapped at the coast when the rate of rise is fastest, facies boundaries may nearly parallel the initial depositional surface as the peak of the cycle is approached. At this point, the shelf is starved of sediment and a condensed sequence rich in organic debris and authigenic components is deposited.

The rate of sea-level rise slows as the cycle peaks and eventually a stillstand is achieved. Coastal onlap ceases, coastal traps are filled, and sediment yield to the shelf increases to the point where the shelf begins to aggrade, shifting shelf zones seaward and depositing a downlapping sequence over the condensed horizon. The shoreface progrades and deposits a top-lapping sequence over the aggraded shelf surface. Because accommodation space is not created during stillstand, coastal regression occurs at a rapid rate, giving a characteristic, asymmetric, flat-topped shape to the sequence, causing early versions of the model to assume a rapid fall in sea level following stillstand.

Cycles range from 5 to 10 million years in length. Cycles, in turn, are composed of shorter-term cycles, consisting of rapid sea-level rises without significant fall, ranging in length from 1 to 2 million years. These shorter-term changes in relative sea level produce changes in sedimentation patterns that are reflected in cyclic deposits called parasequences that are meters to tens of meters thick.

The model is most applicable to passive continental margins undergoing thermal subsidence, but it has also been successfully applied to active margins undergoing subsidence due to convergence (May and others, 1983). On such margins, subsidence occurs about a hingeline that lies landward of the shelf edge. The rate of subsidence increases smoothly from near zero at the hingeline to a maximum at the shelf edge (Watts, 1981; Pittman and Golovchenko, 1983), and the sediment wedge that accumulates in such a setting thins onto the hingeline where accommodation space is created slowest. Lithologic boundaries fan out seaward, and erosional sequence boundaries are formed when shelves are subaerially exposed during low sea-level stands.

The model, however, must be modified when applied to foreland basins where subsidence is caused by crustal loading by successive thrust slices. In this case, sediment is poured into a linear half-basin, where the greatest subsidence asymmetrically occurs near the margin (Swift and others, 1987). The sediment wedge that accumulates in such a basin thins seaward away from the hingeline (Fig. 11.28). Lithologic contacts converge seaward and downlap on sequence boundaries. Unlike sediment wedges on passive margins, the proximal part of sediment wedges in foreland basins is the most likely to be preserved, because even a rapid lowering of sea level would normally not exceed the rate of subsidence caused by the crustal loading. Shelf surfaces, therefore, would not be exposed in foreland basins, and sequences would not be bounded by surfaces of subaerial erosion. Instead, outer shelf sequences would be bounded by marine erosional surfaces and inner sequences would be bounded by ravinements (Swift and others, 1987).

Substantive criticisms of the Vail model have been raised, based on the shape of the cyclic sea-level curves, nature of the stratal response to relative sea-level changes, the relative contribution of eustasy and tectonism to local sea-level changes, the supposed global synchroneity of sea-level events, and the methodology used to establish the magnitude of the cycles by seismic techniques.

The model initially was based on identifications of sequence boundaries on seismic records, but Hallam (1984), for instance, questions the initial premise that impedance contrasts invariably follow chronostratigraphic surfaces. Miall (1986) also points out that it is exceedingly difficult to define the contact between nonmarine and littoral facies of coastal onlap strata using seismic records, even though identification of this contact is an essential component of the Vail model. Later versions of the model (for example, Haq and others, 1987) have used outcrop data to overcome these objections; but even where the limits of coastal onlap can be accurately defined, their utility in establishing sea-level maxima may be limited by the complicating effects of tilting, faulting, and folding.

Evidence of sea-level minima in the Vail model includes a downward shift of coastal onlap, erosion of the toplap surface, and development of low stand fans during a relative fall of sea level below the shelf edge. Pittman (1978) and Pittman and Golovchenko (1983), however, have shown that rates of local subsidence at the shelf edge exceed rates of eustasy under most circumstances, so that the shelf surface should be entirely exposed only during exceptionally rapid rates of sea-level fall. Only on old margins undergoing thermal subsidence would the rate of subsidence decay sufficiently to allow exposure of the shelf edge (Watts, 1981).

Even when the potential can be established for drops of sea level in excess of local subsidence, it may be difficult to establish their actual occurrence and magnitude because the evidence can be ambiguous. Miall (1986), for example, points out that submarine erosion of the shelf can be caused by intensified thermally forced circulation during periods of high sea level, and that submarine fans may form onlapping deposits in response to shifts in sedimentation patterns that cannot be directly related to eustatic events. Brown and Fisher (1980) also refer to numerous examples of marine onlap successions that cannot be correlated with a downward shift in coastal onlap.

Despite substantive criticisms of the Vail model, the concept of global sea-level changes as a basic control on shelf and marginal marine sedimentation remains the framework in which interpretations of stratigraphic successions are tested.

Shelf Sequences in the Rock Record

It may be presumptuous at this stage in the study of shelf stratigraphy to propose depositional models. Each intensively studied shelf has displayed a different set of characteristic responses to the almost infinitely variable combinations of shelf processes. An idealized sequence, however, can be proposed as a point of departure for further studies. In this section, shelf sequences deposited in different tectonic settings that were affected by varying combinations of processes will be analyzed in order to assess the potential commonality of the various responses.

Evolution of Sand Waves

Considerable information exists concerning the origin and dynamics of sand ridges and sand waves, but relatively little is known about their evolution and ultimate fate. Fortunately, several studies exist of well-exposed ancient shelf sandstones that provide details about the development of these sand bodies in shelf environments.

Anderton (1976) compared the lateral and vertical facies distributions within the Jura Quartzite Precambrian of Scotland with the pronounced facies changes that occur along tidal transport paths on the North Sea shelf (Fig. 11.30). Sands were deposited in large sand waves under fair-weather conditions at the upstream end of flow paths, producing thick, tabular sets of crossbeds. During moderate storms, however, sand-wave crests were eroded, and sand was transported farther down transport paths, where it was deposited on small dunes, producing trough

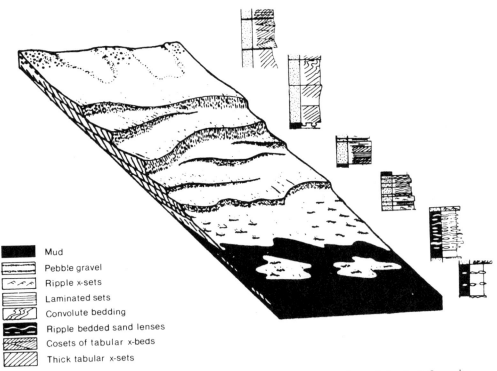

Mud
Pebble gravel
Ripple x-sets
Laminated sets
Convolute bedding
Ripple bedded sand lenses
Cosets of tabular x-beds
Thick tabular x-sets

Figure 11.30 Model for the development of sand waves in the Precambrian Jura Quartzite of Scotland showing downstream change in sedimentary structures and their proposed mode of origin on a tidal sand ribbon (after Anderton, 1976).

cross sets. At the same time, fine-grained sediments were carried to distal ends of transport paths, where they were deposited in thin storm layers.

Sand waves and dunes were largely eroded during severe storms. Winnowed pebble horizons were formed at the upstream ends of flow paths, and the eroded sand was carried downflow, where it was deposited in climbing dunes and then in storm layers. Mud that was put into suspension during the storm settled with its passing and formed a blanket of mud over the sand deposits. Dunes were formed with the reestablishment of fair-weather processes, however, and the mud was partly eroded.

Sandstones of the Precambrian Tanafjord Group of northern Norway were deposited primarily in a shallow marine environment in which repeated episodes of aggradation are represented by a series of coarsening-upward sequences 2 to 10 m thick (Johnson, 1977) (Fig. 11.31). Basal beds of the sequences consist of interbedded siltstones and ripple-bedded sandstones. Siltstones were deposited from suspension during fair-weather periods, and sandstones were deposited at times when wave activity was sufficiently intense to produce symmetrical, low-amplitude ripples.

Fine- to medium-grained sandstones overlie the basal sandstone and siltstone. Fine-grained sandstones are wavy-laminated or ripple-bedded and probably formed under the influence of oscillatory waves. Coarser-grained beds exhibit flat or locally

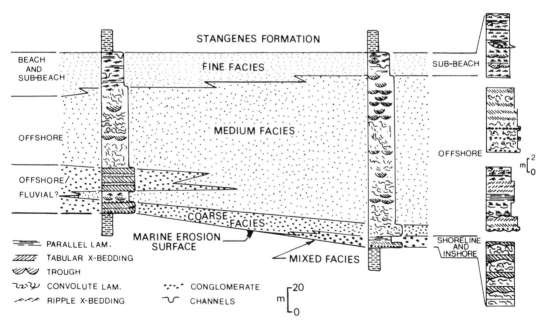

Figure 11.31 Vertical sequence and lateral facies relationships of sediments of the Gronnes Formation at Tanafjord and Varargerfjord (after Johnson, 1977).

erosive bases and are mostly flat-bedded internally. These beds may have formed as storm layers whose surfaces were later reworked by waves during fair-weather periods.

These basal facies occupied areas of the shelf floor between sand waves. Deposits of the sand waves consist of sandstones in trough crossbeds or large-scale, low-angle inclined beds. Trough sets are 15 to 30 cm thick and were deposited by currents capable of forming three-dimensional dunes. Surfaces of the large-scale inclined beds may be covered by symmetrical ripples, and they commonly separate intrasets of trough crossbeds. The large-scale crossbeds probably represent flanks of large submarine sand bars covered alternately by dunes during storms and by ripples during intervening periods.

The sequences may be capped by coarse-grained sandstones in tabular crossbeds that fill channels cut into bar surfaces. Dips of foresets possess a bimodal orientation, and the presence of numerous reactivation surfaces and cut-and-fill structures suggests a complex pattern of flow events in subtidal channels cut in bar surfaces.

Bars were probably most active during storm periods when tidal currents enhanced by storm-generated currents induced traction transport on bar surfaces when dunes migrated across the bars. Bar surfaces were reworked during fair-weather periods, and fines were winnowed out and carried into deeper water where they settled out of suspension in an oscillatory wave regime, forming the basal beds of the sequence.

The evolution of sand bodies formed on regressive shelves is less well known. The few examples that have been described from the rock record generally lack critical data concerning their complete geometry and coeval relationships (for example, Swift and others, 1987) or the details of their lithofacies and internal architecture. Phillips (1987), however, used extensive data from geophysical logs and cores to describe the geometry, lithofacies, and relationships to coeval shoreline deposits of an Upper Cretaceous delta plume sand body in East Texas.

Of the three lithofacies associated with the sand body, sandstones in wavy beds up to 3 cm thick are predominant. The beds are generally separated by millimeter thick shale laminae, but some intervals show a gradation to flaser bedding by an increase in sand bed thickness and a decrease in the thickness of interlaminated shales. Little evidence of bioturbation is present, possibly as a result of high sedimentation rates associated with regression. Interlaminated sandstones and shales, consisting of lenses and laminae of light gray sandstone in dark gray shale, underlie the wavy-bedded sandstones. They grade downward into laminated shales consisting of thinly bedded mudstones with thin, wavy-bedded to lenticular beds of very fine grained sandstone.

Regional stratigraphic correlations suggest a progressive stacking and amalgamation of sand beds toward the delta shoreface, which implies a proximal to distal evolutionary transport path in which sand is swept off the shoreface and transported onto the shelf during storms (Fig. 11.32).

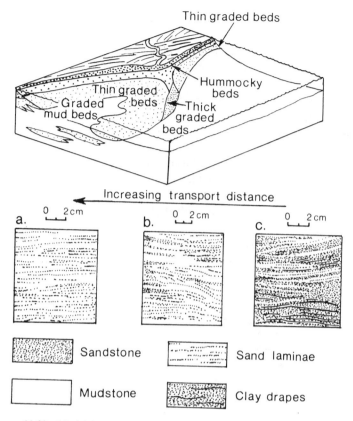

Thin graded beds

Thin graded beds

Graded mud beds

Graded beds

Hummocky beds

Thick graded beds

Increasing transport distance

a. 0 2cm

b. 0 2cm

c. 0 2cm

Sandstone

Sand laminae

Mudstone

Clay drapes

Figure 11.32 Model for development of a delta plume regressive sand body and the typical lithofacies that can develop in one. Lithofacies, including (a) laminated mudstone, (b) interbedded sandstone and mudstone, and (c) wavy-bedded sandstone, are from the Upper Cretaceous Woodbine Formation of east Texas (after Swift and others, 1987; Phillips, 1987).

Transgressive Shelf Sequences

Transgressive deposits on modern shelves are thin because the Holocene sea-level rise was of a speed and magnitude that exceeded the capacity of sediment sources to keep pace with it. In the last 7000 years, however, the rate of sea-level rise has slowed dramatically; and even though many shorelines continue a slow retreat, the surfaces of some shelves are presently being aggraded by sediments bypassing the shore zone. This section will examine the sequences left when sediment is continuously provided to a shelf experiencing a more moderate sea-level rise.

Shelf deposits of the Upper Cretaceous Cape Sebastian Sandstone of southwestern Oregon can be divided into four facies (Bourgeois, 1980). Basal deposits consist of shelly boulder conglomerate overlain by sandstones in plane beds and trough crossbeds (Fig. 11.33). These sediments were deposited in a shore zone fronting steep beach cliffs not unlike parts of the present-day Oregon coast (Fig. 11.34).

The overlying facies consists of medium- to fine-grained sandstone in hummocky crossbeds. Hummocky crossbeds in the lower part of the Cape Sebastian Sandstone are commonly associated with parallel-laminated sands in nonbioturbated sequences. Transport episodes must have been frequent, suggesting that deposition took place just below fair-weather wave base, where the bottom was mobilized by most passing storms.

About 60 m above the basal conglomerate, the first burrowed zones appear. These zones increase in thickness and abundance upward, indicating a progressive increase in the ability of infauna to rework sediment. Water depth was probably increasing and bringing the substrate into a shelf zone that was scoured only by major storms (Fig. 11.33).

The uppermost part of the unit consists of parallel laminated sandstone and bioturbated sandy siltstone. These beds were deposited in an outer shelf zone be-

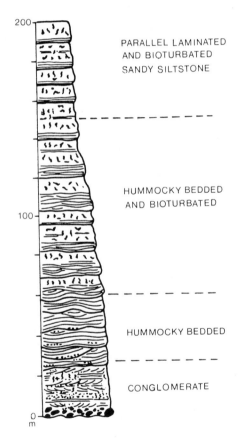

Figure 11.33 Composite section of the Cape Sebastian Sandstone showing change in bedding structures and increase in bioturbation upward (after Bourgeois, 1980).

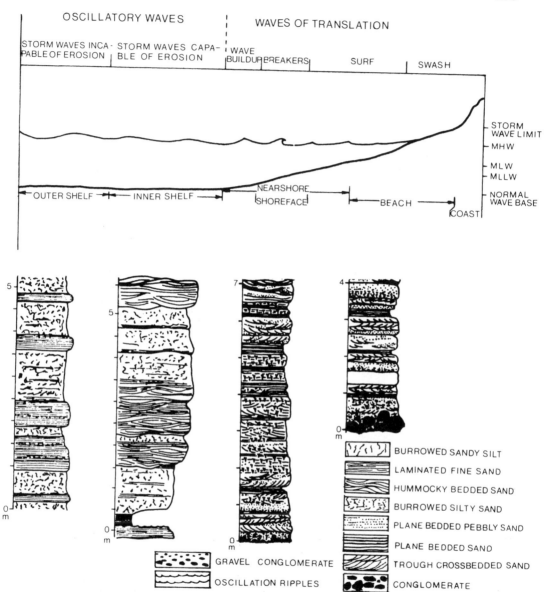

Figure 11.34   Relationship of the facies of Cape Sebastian Sandstone to processes on a storm-graded shelf (after Bourgeois, 1980).

yond the influence of nearly all storm waves. Clayey organic silts were the normal deposits in the zone, and they accumulated at a rate slow enough for infauna to mix the sediment. Sand was eroded from farther up the shelf during intense storms and was transported into the area by density currents (Fig. 11.33).

Decrease in grain size, increase in burrowing activity, and decrease in evidence of active sediment transport upward in the sequence all suggest that the Cape Sebastian Sandstone was deposited in progressively deepening water. The thickness of a sequence deposited during transgression, however, suggests that this was not a shelf starved of extrabasinal sediments. The amount of sediment delivered to the shelf was apparently large enough to overcome the tendency of sediments to be trapped in the shore zone during periods of rising water level.

The Oregon shelf during the Upper Cretaceous was smooth, but other modern shelves contain large-scale bathymetric features. Although their formation, for the most part, appears to be linked to processes active during the initial stages of transgression, evidence from ancient deposits suggested that their formation may be a continuous process on some shelves.

Lower Paleozoic quartz arenites of the Peninsula Formation of South Africa represent a 750-m-thick sequence of transgressive coastal and shallow shelf sediments (Hobday and Tankard, 1978) (Fig. 11.35). Basal beds of the formation were deposited during the initial stages of transgression. Sandstones deposited on wash-

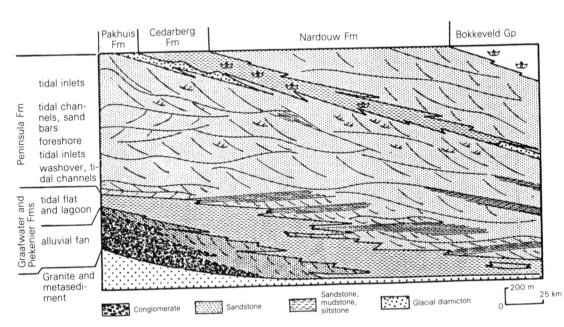

Figure 11.35 Lateral and vertical facies relationships within the Cape Supergroup (after Hobday and Tankard, 1978).

over fans and in tidal inlets overlie and interfinger with lagoonal siltstones and mudstones. They, in turn, are locally overlain by sandstones in plane beds with multiple low-angle discordances. These are interpreted as foreshore sands, and the whole sequence represents a transgressive barrier complex.

The bulk of the Peninsula Formation is dominated by tabular or lenticular sandstone bodies as much as 10 m thick consisting of a complex arrangement of smaller-scale structures superimposed on mega-crosssets. The smaller-scale structures include plane beds, trough crossbeds, and some herringbone crossbeds, and bedding planes of the master sets are marked by channels, scour surfaces, and lag gravels.

The sand bodies are interpreted to be deposits of tidal sand ridges and sand waves of the North Sea type. Transport was mainly in response to tidal currents, but periodic storm-induced currents, especially those produced by ebb surge, cut channels in the ridges, eroded flanks, and left lag gravels.

The great thickness of the shelf deposits, lacking any change in grain size and bedding style, suggests that the rate of deposition was nearly in balance with subsidence so that water depths over the shelf remained within critical limits of tidal current dominance. Eventually, however, shelf sedimentation was terminated by regression when coastal deposits of the upper Peninsula Formation and the Cedarburg Formation were deposited.

Regressive Shelf Sequences

A sequence of Jurassic sediments in southern Alberta may represent a transition from a transgressive to a regressive shelf system (Hamblin and Walker, 1979). The basal unit consists of glauconitic sandstones with abundant macerated wood fragments, bands of ironstone concretions, and scattered belemnites. These beds share many of the characteristics associated with slow deposition, and although not interpreted as such by Hamblin and Walker (1979), they may represent a sand sheet formed during transgression when the shelf was starved of extrabasinal sediments.

The basal unit is overlain by interbedded thin sandstones and somewhat thicker siltstones. Sandstones show sharp, flat bases with abundant oriented sole marks. Upward within each bed, parallel laminae change to ripple crossbeds, suggesting deposition by geostrophic storm currents (Fig. 11.36). The absence of a massive or graded interval at the base of each bed, however, indicates that the currents were slow moving, probably as a result of low gradients. Sandstones were introduced into a shelf zone normally dominated by deposition of silt from suspension. The degree of bioturbation in the siltstone beds suggests that the zone was below storm wave base.

These beds are succeeded upward by a unit consisting of thick beds of hummocky cross-stratified sandstones interbedded with thin bioturbated siltstones. The unit was deposited in a zone between storm and fair-weather wave base where finer

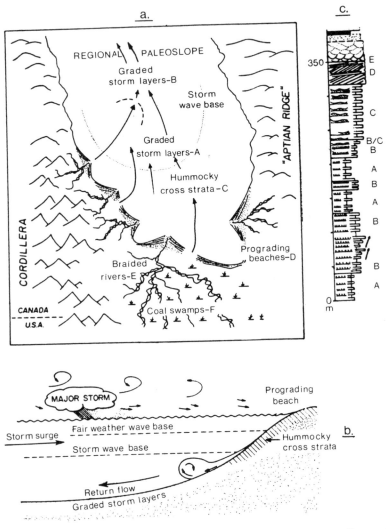

Figure 11.36 Sediments and inferred depositional environments of the Kootenai Formation of Alberta showing (a) paleogeography and lateral distribution of facies, (b) position of facies relative to processes on a storm-graded shelf, and (c) vertical facies relationships (after Hamblin and Walker, 1979; Walker, 1979).

sediment resided only temporarily before being moved farther offshore and where hummocky crossbeds were not modified by fair-weather wave processes.

The uppermost unit represents the culmination of the regression. The basal beds consist of plane beds deposited on the shoreface and foreshore of a strand-plain. They are overlain by trough crossbedded sandstones, siltstones, and coal deposited on an alluvial plain.

The sequence was deposited as a northward-prograding clastic wedge consisting of an aggrading shelf sequence overlain by prograding beach and coastal plain deposits. The shelf was storm dominated, and much of the sediment provided to the aggrading shelf was derived through shoreface bypassing during ebb surge. Large structures do not occur in the sequence. The shelf was probably smooth except for the low-amplitude undulations that formed the hummocky crossbeds.

Mudstones were not an important feature of any of the shelf examples discussed thus far. Either the water depth on these shelves was relatively shallow or the wave climate was intense enough to winnow clays and fine silts from the substrate. In addition, the shoreface was close enough that rip currents and downwelling currents were able to move sand from the shoreface and inner shelf into deeper water.

The Cretaceous Cardium Formation at Seebe, Alberta, however, is an example of a shelf where significant fair-weather currents did not exist and where aggradation took place far from any possible shoreline (Wright and Walker, 1981). The formation consists of five coarsening-upward sequences. Each sequence begins with a dark shale probably deposited well below even storm-weather wave base. The shales grade upward into bioturbated silty mudstones, and the sequences are capped by beds of hummocky crossbedded sands separated by bioturbated mudstones (Fig. 11.37). Foraminifera suggest that deposition occurred at depths shallower than about 50 to 60 m, but below fair-weather wave base, and the sequences represent construction of sandy shoals in an outer shelf environment. Sand and silt accumulated on shoals as graded layers during storms, and the shoals eventually aggraded to the point where storm waves could form hummocky crossbeds. At no time, however, did shoreface deposits ever prograde into the area, and the cyclic deposition was halted by a significant transgression during which marine shales of the Wapiabi Formation were deposited.

Another formation deposited on an aggrading shelf where sand was almost totally absent is the Shublik Formation of the Alaskan North Slope (Fraser and Clarke, 1976; Fraser, 1982). Deposition began with a sharp-based transgression over coastal alluvial plain sediments of the Sadlerochit Group. Bioturbated, glauconitic sandstones were deposited during initial phases of the transgression, which culminated in the formation of a lag deposit of chert and phosphatized limestone pebbles (Fig. 11.38).

Initial deposits of the aggradational phase consist of organic mudstones with abundant diatoms that probably accumulated at the shelf edge in an upwelling zone. Whole shells of very delicate pelagic flat clams occur in these mudstones, indicating that deposition occurred in an environment unaffected by strong currents.

Upward in the sequence, however, the proportion of mudstones to shells decreases, suggesting that muds were being held in turbulent suspension as water depth decreased. Eventually, storm wave base was intersected by the aggrading shelf surface, and an interbedded sequence of thick beds of bioclastic limestones and thin mudstone beds accumulated.

This aggradational phase was terminated by another sharp-based transgression. In some places a hard ground developed on the limestones, whereas in other

GRAPHIC
LOG   SEQUENCE

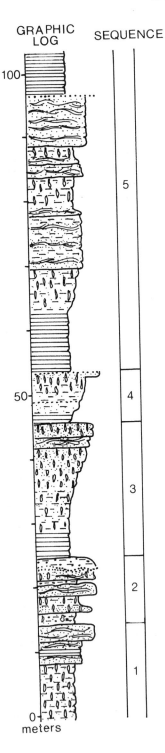

Figure 11.37 Stratigraphic sequence of the Cardium Formation of Alberta showing five aggradational sequences, each grading upward from mud to hummocky crossbedded sand (after Wright and Walker, 1981).

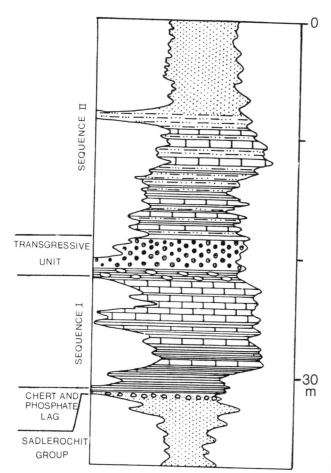

Figure 11.38 Vertical sequence in the Shublik Formation (Triassic) of north Alaska showing two packages of sediments deposited on an aggrading shelf. Each sequence rests on a transgressive unit consisting almost entirely of chemical sediments.

places mudstones containing nests of glauconitized and phosphatized fecal pellets were deposited. The ensuing aggradational phase was similar to the first one, except that lower shoreface deposits of the overlying Sag River Sandstone were deposited at the top of the sequence in place of the bioclastic limestones.

12

Epicontinental Seas

Introduction

The models developed thus far may be adequate for explaining the depositional dynamics of ancient marginal seas such as the Tertiary shelves of the Atlantic and Gulf coasts, but shallow marine environments also occurred in very broad, interior basins called epicontinental or epeiric seas where a significant portion of the sedimentary rock cover of the continents was deposited. The Baltic Sea and Hudson's Bay are often cited as modern counterparts of epeiric seas, but they are smaller than the broad, shallow marine basins that once occupied major portions of the cratons. Somewhat larger in extent are the Yellow Sea and Bering Sea, which have also been advanced as possible modern counterparts of ancient epeiric seas (Klein, 1982; Klein and others, 1982; Nio and Nelson, 1982).

Sedimentary Processes and Products

Epeiric seas differed from their modern shelf counterparts chiefly in their broad lateral extent and correspondingly low average gradient and, in many cases, their partial or complete isolation from deep ocean basins. These factors, in turn, probably imposed major constraints on the way most shallow marine processes operated in epeiric seas.

Because they were at least partially isolated from deep ocean basins, some of these basins may not have experienced significant astronomical tides. Howarth (1982) believes that tides could not be generated within these basins because of their shallow water depths, and because barriers to open ocean circulation probably prevented co-oscillation of basin waters with the tidal bulge generated in open oceans. Klein (1982) argues, however, that significant tides did exist and that they were especially important during the latter stages of transgression when shallow marine environments were at their broadest extent. He bases this interpretation on the work of Silvester (1964) and Cram (1979), who have shown that tidal range increases with shelf width. During low sea level stands, cratonic widths of submergence must have been narrow, and the tidal range would consequently have been small. Klein (1982) believes that weather-dominated shelves would exist at these times, but as transgressions occurred and progressively larger areas were submerged, tidal influences would increase.

When a tide is propagated onto a shelf, three factors begin to work on it. The kinetic energy of the tidal bulge is concentrated into a steadily decreasing cross-sectional area, causing the amplitude to increase (Swift and Niederoda, 1985). The amplitude also increases when shelf widths are such that resonant amplification occurs. Such widths occur at odd multiples of quarter-wavelengths ($\frac{1}{4}$, $\frac{3}{4}$, $\frac{5}{4}$, and so on) and local minima occur at even intervals ($\frac{1}{2}$, $\frac{3}{2}$, $\frac{5}{2}$, and so on) (Howarth, 1982).

Frictional forces, however, tend to reduce tidal amplitudes, and these increase with increasing shelf widths, setting up a competition between increasing energy density and increasing frictional loss as shelf width widens. For very wide shelves, the difference between maximum and minimum amplitudes is reduced, and maximum current velocities tend to be concentrated at the shelf edge (Howarth, 1982) (Fig. 12.1).

Present-day shelves are poor counterparts for epeiric seas when used to model tidal action in shallow marine environments because they are marginal to deep ocean basins, and they are mostly narrower than the quarter-wavelength necessary for maximum resonant amplification. Thus, on today's shelves any increase in shelf

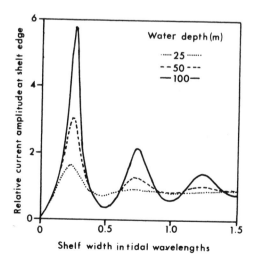

Figure 12.1 Variations in tidal current amplitude with increasing shelf width (after Howarth, 1982).

width will increase the tidal range at the shore (Howarth, 1982). This, however, does not necessarily apply to most ancient epeiric seas, which were at least partially isolated from ocean basins and were much wider and shallower than modern shelves. If tides cannot be propagated in at least some epeiric seas, however, an explanation must be sought for the fact that both carbonate and clastic shoreline sediments deposited under apparent tidal influence are common in cratonic basins (Klein, 1971, 1977; Klein and Ryer, 1978).

Tides may have been propagated into those epeiric seas having substantial access to open ocean basins. Slingerland (1986), for example, used numerical modeling to hindcast the hydrology of co-oscillated paleotides that might have been generated in the upper Devonian Catskill Sea of the eastern United States. These calculations suggested that a mesotidal range was possible, given the probable basin geometry and boundary tidal conditions, but they also indicated that the potential tidal range was highly sensitive to water depth.

Slingerland (1986) noted that the shape and orientation of the Catskill Sea lent itself to tidal wave augmentation through both convergence and resonance. Tides would have been substantially lower if the shoreline was less concave or if the shelf was wider.

Notwithstanding these calculations, evidence for the actual occurrence of tides in the Catskill Sea is ambiguous. Slingerland (1986) lists several features in the rock record indicative of a tidal regime, but Craft and Bridge (1987) interpret the same evidence differently and present other evidence to suggest that the Catskill Sea was dominated by intense tropical storms.

Slater (1986) also used numerical modeling to hindcast tides in the Cretaceous Seaway of western North America. His calculations incorporated various bathymetric parameters and boundary conditions, and they indicated that under most conditions only a small portion of the potential tides would be accounted for by co-oscillation with ocean basins to the north or south. Instead, the maximum tidal amplitude of 86 cm was achieved when an independent tide was generated within the seaway itself. Maximum current velocities sustained under this tidal regime were 10 cm/s, and it is unlikely that such velocities could have a significant effect on the seaway. Sedimentation, instead, appears to have been dominated by a storm-generated current regime.

Rhythmic, although nonperiodic raising and lowering of sea level in epicontinental seas may have been produced by other mechanisms. Storm winds, for instance, may push seawaters to temporary high levels along coasts. Depending on wind direction, wind-forced currents can add water to the coast faster than bottom flows can return it offshore until the barotropic gradient balances frictional forces at the bottom and equilibrium is established (Swift and others, 1985). Conversely, winds blowing in the opposite sense may force waters offshore, causing water levels at the coast to be lowered.

Storms presently cross the Atlantic coast of North America with a recurrence interval of 2 to 10 days, and there is no reason to suspect that storm periodicity was very greatly different in the past. In fact, many ancient shallow marine sequences consist mainly of storm deposits, suggesting that they were dominated by storm-induced processes (Dott and Bourgeois, 1982). Ginsberg (cited in Dott and Byers, 1981) also suggests that evidence of exposure in dominantly subtidal deposits of the early Paleozoic in the mid-continent may have been produced when storm winds forced water offshore and lowered sea level.

Distinction should be made, therefore, between epeiric seas and marginal seas. Marginal seas were probably much like modern shelves in their response to tide-generating forces, whereas tidal forces were restricted in epeiric seas by barriers between shallow and deep marine environments, or they were reduced by frictional forces.

Wind-forced currents may also have operated differently in epeiric seas. Swift and others (1985) divided the shelf into a friction-dominated zone, where onshore transport of water is balanced by offshore return flows, and a geostrophic zone, where water is sufficiently deep that the Coriolis force can effect offshore-directed return flows into shore-parallel core flows.

Shallow depths and low gradients in epeiric seas, however, might extend the friction-dominated zone much farther offshore than today's marginal seas. The dominant transport direction, therefore, may be perpendicular to shore except in littoral circulation cells of the surf zone. Because of the dominance of frictional forces in epeiric seas, it is also likely that the energy of most waves would be dissipated over broad areas of the basin and would have little effect on the substrate.

There is enough evidence of tractive transport of bedload material in epeiric seas, however, to assume that currents at times were sufficient to mobilize the substrate. These episodes must have occurred during rare severe storms (Dott and Bourgeois, 1982), and during intervening periods the normal background processes of suspension settling and bioturbation probably dominated. The resulting deposits consist of varying proportions of bioturbated layers, and beds with primary structures more or less intact. The exact proportion between the two bedding types would be dependent on the degree of exposure and the periodicity of severe storms. As on modern shelves, it is to be expected that episodes of tractive transport were more active in shallow water, but the occurrence of bioturbated layers within strandline deposits suggests that tractive transport was episodic even in very shallow water (Fraser, 1976).

Storm-generated downwelling currents are particularly important on modern shelves in transporting sediment from inner to outer shelf fields. The resultant deposits are much like those deposited by turbidity currents and are also the result of deposition from progressively waning currents. Graded storm layers are characteristic deposits on many modern shelves, and they have been identified on the Bering Sea (Nelson, 1982), which has been proposed as a modern counterpart for epeiric seas. It is probable, therefore, that such currents were an important process in epeiric seas. Graded beds must also have occurred where incoming waters were characteristically turbid. Such areas would commonly occur in front of deltas, and the Mississippian Borden delta of the east-central United States is an example in which turbidite sands and silts occur (Chaplin, 1982).

It should not be assumed from the foregoing discussion that epeiric seas were stagnant except during storm periods. Light-colored rocks with abundant and diverse faunal remains tend to dominate stratigraphic successions, although restricted circulation in some basins produced thick sequences of evaporitic rocks or organic-rich rocks deposited under anoxic conditions. Minor differences in elevation caused by broad-based tectonism may have been sufficient to isolate portions of epeiric seas, and widespread climatic changes may have been sufficient to form a long-lasting temperature stratification that would produce stagnant bottom waters. The Devonian black shales of the eastern United States, for instance, extended from the Appalachian to the Illinois basins and must have formed in response to changes in climate and tectonically induced changes in circulation patterns (Cluff, 1980).

Slow-moving currents must have produced a continuous exchange of waters, and they probably were caused by the same processes that operate on modern shelves. Prevailing winds of long duration were sufficient to induce slow movement of water masses despite frictional effects. They were probably able to keep the finest sediment in suspension, but they may not have been able to induce traction transport. Density-forced currents induced either by horizontal salinity gradients or freshwater overflow might also be effective in exchanging water masses, but not in moving sediments.

A possible modern counterpart for epeiric seas may be found on the shelf margin of the Gulf of Mexico. The shelf is essentially a coastal plain undergoing mioclinal downwarping along a curved hingeline. Shelf gradients are considerably in excess of those found in epeiric seas, and the shelf is fronted by a deep basin. Tidal range is negligible, however, and except during severe storms, the wave climate along the coast is mild, making the hydraulic regime similar to that which probably characterized epeiric seas.

The Mississippi Delta dominates the shelf along the Louisiana coast. It is an extreme case of river domination because of the very low wave energy expended at the delta front. Coastwise circulation carries mud downdrift to the west, where a chenier plain dominates the coast. The shelf along this portion of the gulf is covered by muds delivered by surface sediment plumes that can travel long distances as baroclinic currents because of the low level of mixing that occurs between the plume and shelf waters.

The remainder of the coast to the south and southwest is characterized by very long segments of barrier islands broken only in a few places by inlets. Sand is concentrated within a few kilometers of shore, and the remainder of the shelf, except for those areas of relict sand, is covered by muds with sand layers that thin and become less abundant down-gradient (Berryhill and others, 1977) (Fig. 12.2). Originally thought to have been emplaced as density flows (Hayes, 1967) (Fig. 12.3), much of this sand is probably delivered to the shelf by downwelling currents (Morton, 1981). The degree of bioturbation increases in an offshore direction as the frequency of transport episodes declines (Fig. 12.2) to the outer shelf where bottom sediments are mobilized only once every five years. Even though tidal range is negli-

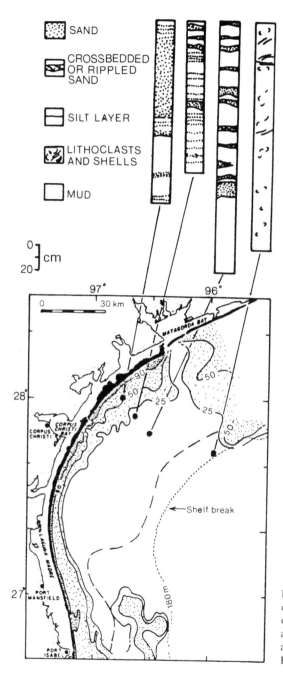

Figure 12.2 Distribution of sediments on the Texas Gulf Coast showing limit of discrete sand layers at the shelf edge and distribution of sand and mud layers at different locations of the shelf (after Berryhill and others, 1976).

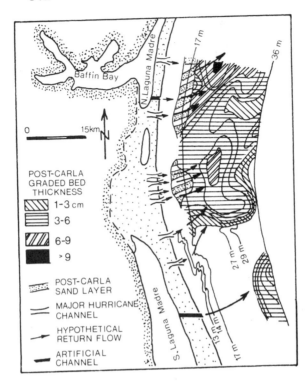

Figure 12.3 Thickness of graded beds deposited by currents flowing on the Texas Gulf Coast following Hurricane Carla (after Hayes, 1967).

gible, tidal flats do exist. Wind tidal flats in Laguna Madre, for instance, bear many similarities to mudflats produced by normal diurnal tides (Miller, 1975).

The Gulf Coast, however, cannot be used to model the effect that the low gradients of epeiric seas have on the evolution of stratigraphic sequences. Previous discussions have demonstrated that various types of prograding coasts have an intrinsic threshold built into them. As a coast progrades into progressively deeper water, the volume of sediment needed to maintain progradation increases. Given a constant sediment influx, a delta, for instance, may evolve from fluvial domination to wave domination, even to the point where an equilibrium between sediment influx and redistribution processes is attained. The same threshold also exists in prograding strandlines as the rate of progradation slows and the relative effectiveness of basin processes increases.

The nearly flat gradients in epeiric seas, however, mean that initial rates of progradation are faster than those occurring in marginal seas, and the volumes of sediment needed to maintain a constant rate of progradation need not increase geometrically. The result of low gradients, therefore, is that depositional systems can cover significantly wider areas than those on modern shelves. This effect is enhanced by the reduced ability of marine processes in epeiric seas to redistribute incoming sediment.

Sediments of the Ordovician Starved Rock Barrier and Glenwood Lagoonal

Systems, for instance, cover portions of four states today (Fraser, 1976) (Fig. 12.4), and the Mississippian Borden delta covered what is now parts of Kentucky, Ohio, and Illinois (Fig. 12.5). Correlative strata of Pennsylvanian age can be traced entirely across mid-continent basins, suggesting the occurrence of deltas of vast dimensions relative to today's standards (Fig. 12.6).

Special variants of epeiric seas were elongate bodies of water called epicontinental seaways. Like other epeiric seas, they were shallow and were characterized by low gradients, but because they were open at one or both ends, they share some features with modern shelves because of their open access to oceanic processes. For example, the western margin of the seaway that existed in the western interior of North America during the Cretaceous possessed its own shelf and slope system (Asquith, 1974), although the deep basin was only locally in excess of a few hundred meters deep (Asquith, 1970).

One such series of seaways occupied the western part of North America from the Triassic to Late Cretaceous (Bouma and others, 1982). At times the seaways were open only to the north, but at other times they were open at both ends (Brenner, 1980) (Fig. 12.7). The shorelines, consisting of deltas, barrier islands, and lagoons, and strandplain systems have already been described in previous sections.

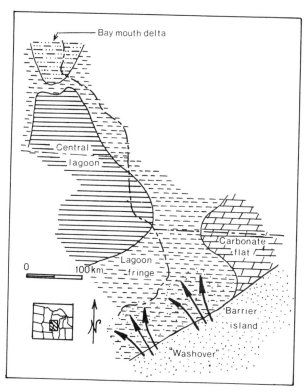

Figure 12.4 Middle Ordovician barrier island–lagoonal system in the Upper Mississippi Valley (after Fraser, 1976).

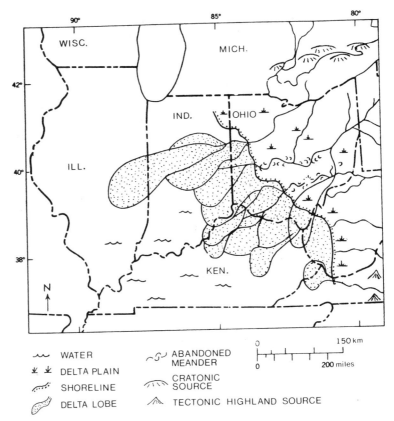

Figure 12.5 Extent of the Borden Delta (Mississippian) in Kentucky, Indiana, and Illinois (after Kepferle, 1977).

These shorelines prograded onto broad muddy shelves on which sand ridges and sand waves were interspersed. The sand ridges migrated in response to storms, tides, and oceanic circulation, and they were supplied with sand by shoreface bypassing (Brenner, 1980). These sand bodies characteristically are enclosed by muddy sediments, and some may have been localized on topographic highs standing above the surface of muddy shelves (Slatt, 1984).

These sand ridges deposited sediments in sequences characterized by upward-coarsening textures and a change from ripple bedding to larger-scale bedding structures indicative of deposition under shoaling conditions produced by aggradation of the bedform (Fig. 12.8). Sand waves migrated across ridge crests during storm periods, depositing a variety of cross strata with polymodel transport directions. Flow separation occurred at the crest of the waves, and turbulence produced at the point of reattachment produced winnowed lags (Brenner and others, 1985). Migration of the scour zones resulted in deposition of sheetlike lags that originally were

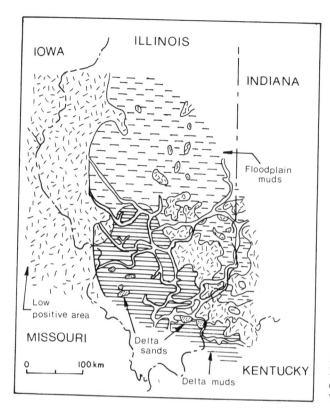

Figure 12.6 Extent of some Pennsylvanian age deltas in the mid-continent of the United States (after Wanless and others, 1970).

interpreted to have formed in channels cut through sand ridges (see Brenner, 1980, for instance). Migration may also have ceased at times because of a decline in sand supply or a prolonged fair-weather period during which bioturbated mudstones were deposited on the bars (Cotter, 1975).

These sand bodies apparently formed at major or minor transgressive horizons (Swift, written communication) and were subsequently buried during shelf regression. Nio (1976) described a process of sandwave construction that is dependent on transgression, and there is a causal link between sand-wave and sand ridge construction and a sand yield provided by retrograding shorelines. Several mechanisms, however, can be invoked to provide sand to aggrading muddy shelves, and there is no reason to suspect that large-scale sand bodies cannot be built and maintained during the regressive phase of shelf evolution.

Those built during regression phases, however, are different from those formed during transgressions. Swift and others (1987), for example, interpret beds of fine-grained sandstone enclosed by marine mudstones in the Upper Cretaceous sequence at Hatch Mesa, Utah as delta plume deposits. These beds are as much as 10 km from coeval shoreline deposits, and they display evidence of deposition by downwelling storm currents.

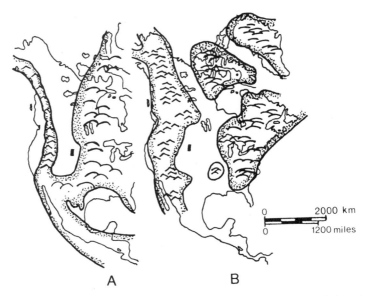

A   B

Figure 12.7 Seaway configurations during the Mesozoic in western North America. The seaway was closed to the south during the Oxfordian (A), but open during the Late Cretaceous (B) (after Bouma and others, 1982, reprinted with permission of the American Association of Petroleum Geologists).

Sandstone beds alternate with beds of laminated siltstone and shale (Fig. 12.9). They are as little as 5 cm thick near the base of the sequence, but thicken to as much as 1.5 m at the top. Thinner beds are planar laminated and cross-laminated, and thicker beds display a consistent sequence of massive or hummocky crossbedded basal sands overlain by planar-laminated and cross-laminated sands. Tool marks occur on the base of the beds, and upper surfaces of the thicker beds are covered by linguoid ripples or straight-crested symmetrical wave ripples. Upper surfaces of the thicker beds also show low-amplitude undulations with a spacing of about 100 m that probably represent incipient bedforms that did not develop avalanche faces.

Discussion: Interaction of Eustasy and Sediment Influx on Evolving Shelf Sequences

The gross lateral and vertical distribution of shelf facies can be explained in terms of the balance between sediment yield to a shelf and the rate at which accommodation space is created. The details of a given succession are also dependent on factors such as shelf width and gradient, the magnitude of sea-level rise, position with respect to migrating facies tracts, and variations in the rate of sea-level rise, subsidence, and sediment influx.

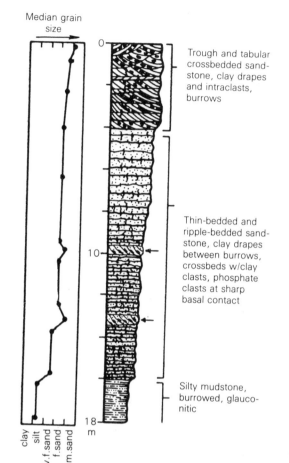

Median grain size

Trough and tabular crossbedded sandstone, clay drapes and intraclasts, burrows

Thin-bedded and ripple-bedded sandstone, clay drapes between burrows, crossbeds w/clay clasts, phosphate clasts at sharp basal contact

Silty mudstone, burrowed, glauconitic

clay   silt   v.f.sand   f.sand   m.sand

0

10

18 m

Figure 12.8 Vertical change in sand size and bedding structures in a sand body in the Shannon Sandstone (after Harms and others, 1975).

Rapid transgressions occur when sea level rises much faster than sedimentation can fill the space made available. Hiatal surfaces or condensed horizons, usually consisting of the finest sediments in a given succession, occur at the base of such sequences. These are succeeded upward by regressive deposits that accumulate as the rate of sediment influx first equilibrates with sea-level rise and then exceeds it (Fig. 12.10).

The Kootenai Formation of southern Alberta represents shelf sedimentation under conditions of initial rapid sea-level rise and slow rate of sedimentation. A basal condensed sequence is overlain by middle to outer shelf deposits that grade upward into inner shelf, shoreface, and coastal plain sediments.

Other units may also have been formed under conditions of sea-level rise and sediment accumulation similar to those of the Kootenai, but the resultant deposits

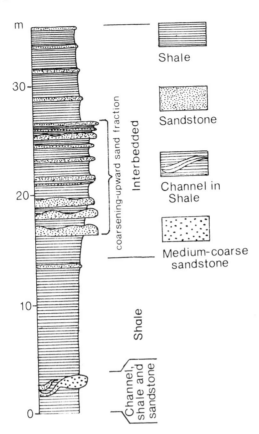

Figure 12.9 Columnar section in the upper Mancos Shale at Hatch Mesa, Utah, showing the vertical sequence that might develop in a delta plume regressive sand body (after Swift and others, 1987).

differ in response to variations in the other factors that affect evolving shelf sequences. Both the Shublik and Cardium formations, for example, were deposited under conditions of rapid sea-level rise and slow sediment influx. Condensed horizons at the base of aggradational sequences occur in both formations, but in neither case are the aggradational sequences punctuated by coastal or nonmarine sediments. Instead, aggradation proceeded only to the point when inner shelf deposition was initiated before relative sea-level rise recurred (Fig. 12.10).

The Cape Sebastian Sandstone, on the other hand, was probably deposited on a shelf where sea level rose only slightly faster than accumulation took place (Bourgeois, 1980). Coastal sediments are overlain by progressively deeper water facies, with no condensed horizon at the base. Each facies is somewhat thicker than the water depth in which they were deposited, suggesting that the rate of sediment accumulation was only slightly less than the rate at which accommodation space was being created.

It is also possible that accumulation rates may balance the rate at which sea level rises. The Narrow Cape Formation of Kodiak Island, Alaska, for instance,

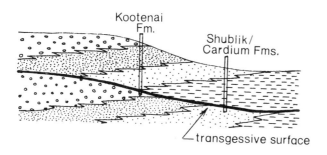

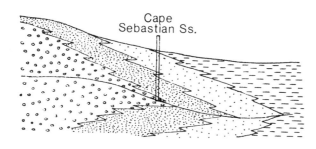

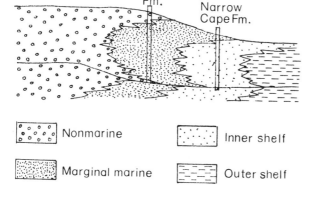

Figure 12.10 Diagrammatic cross sections relating the types of shelf sequences that can develop at various locations on a shelf undergoing varying rates of sediment influx and sea-level rise.

may be an example of shelf deposition when such a balance was maintained. It consists of approximately 700 m of bioturbated clayey siltstones that show little or no vertical change in texture or sedimentary structures (Nilsen and Moore, 1979). The sequence is much thicker than the potential water depth of the shelf on which it was deposited, and the uniformity of the vertical sequence indicates that conditions did not vary significantly during deposition. Facies boundaries apparently migrated only within narrow zones, and neither very pronounced transgressions or regressions occurred, even as accommodation space was being created.

Like the Narrow Cape Formation, the Peninsula Formation of South Africa was deposited on a rapidly aggrading shelf, but in a position relatively closer inshore. Shoreface deposits are overlain by an overthickened sequence of inner shelf sediments showing no pronounced vertical facies variations. The shoreface must periodically have prograded onto the shelf during deposition in order to generate the transgressive sand bodies characteristic of the unit, but vertical aggradation must otherwise have predominated over lateral migration of facies tracts.

Part **V**

CONTINENTAL SLOPE TRANSITION ZONE

Introduction

The continental slope is the leading edge of the continental margin, and it forms a physiographic transition between the shelf and deep water environments of the ocean basins. Typically, it is also coincident with a structural transition between continental and oceanic crusts (Bouma, 1979). In addition, it represents a transition zone between the sedimentary processes most active on shelves, such as wave-, storm-, and tide-induced currents, and those of deep ocean basins, where gravity-induced currents and ocean-scale thermohaline circulation predominate.

Unlike the coastal transition zone, which traps a considerable amount of sediment that is provided to it and which significantly modifies that sediment passing through, the slope is almost an open conduit for sediment. Sediment provided directly from the shelf may spill over the shelf break and collect on the upper slope, but this material is unstable and is carried into deep basins by mass transport processes. Submarine canyons that are present on slopes also act as open conduits for sediment transport from shelf to deep basinal environments. Long-term storage, however, can occur in intraslope basins created by tectonism, diapirism, or sediment mass movements, and most slopes are progradational during at least part of their history.

Sediments may accumulate on constructional slopes so slowly that they are as much as ten times thinner than coeval deposits in adjacent environments. This effect is exacerbated on most slopes, where periods of net deposition alternate with periods of erosion when sediments are removed from the slope and redeposited in deeper water.

Studies of modern slopes have emphasized three lines of investigation. Surficial deposits have been sampled with grab samplers or gravity corers, the morphology has been studied using sonar, and the internal structure has been determined through shallow and deep seismic surveys. Sediment sequences, however, are poorly known, because only a few deep cores have been recovered from modern slopes. At present, our primary source of information regarding the architecture of continental slope deposits must come from inferred slope deposits in the rock record.

13

Morphology and Sedimentology of Continental Slopes

Morphology

The continental slope ranges from 10 to 100 km in width and exceeds 300,000 km in total length (Vanney and Stanley, 1983). It is separated from the shelf on its landward margin by the shelf break, where a pronounced change in gradient occurs, and it is bounded on its basinward side by a variety of features, including submarine fans, rises, and deep trenches (Fig. 13.1).

Slope gradients may be as low as 1.3° off major deltas, but off fault-bounded coasts the gradient averages 5.6°, and as a whole, slope gradients average about 4° (Shepard, 1973). This gradient is considerably steeper than most shelves, making the shelf break easy to define. The break occurs at an average distance of about 75 km from shore and most often consists of a convex-up surface where the degree of curvature may be gentle, gradual, sharp, or abrupt (Southard and Stanley, 1976). The gradient change at the slope base is usually less well-defined, especially where the slope grades into a rise or deep-sea fan.

Slope trends tend to be straight or gently curving, even where they front irregular coastlines, probably reflecting the prograding nature of the slope or the structural control of its configuration (Bouma, 1979). The shelf break, however, is rarely straight for any distance and is especially irregular where it is cut by submarine canyons.

Three morphological elements appear to be common to all slopes, although their origin may differ. Channels of all sizes cut into slope surfaces. Some breach the shelf break and head on the shelf, whereas others head on the slope itself. Major canyons possess numerous tributary canyons and gullies that enter the trunk canyon

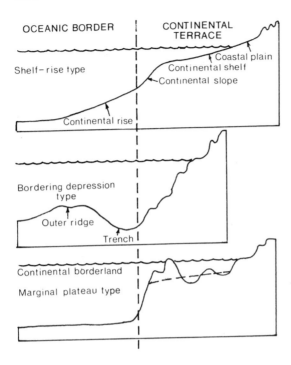

Figure 13.1 Diagrammatic cross sections showing the relationship of slope morphology to tectonic setting (after Curray, 1969).

from both sides (Shepard and Dill, 1966) (Fig. 13.2). Canyons heading on the shelf tend to show sinuous courses, whereas those heading below the shelf break display straighter courses (Twichell and Roberts, 1982), and the possibility exists that this difference in morphology reflects a difference in stage of development that parallels subaerial channels (McGregor and others, 1982). Many small gullies form parts of the tributary systems of main canyons, but others exist by themselves as small, low-relief features that head on the slope and normally do not extend to the base.

The origins of slope canyons are varied (Shepard and Dill, 1966). Some are the result of erosion during low sea-level stands when large amounts of sediments were being delivered to them. Some are probably fault-controlled, and some probably originate in slump scars that work their way upslope by headward erosion. In all cases, however, the canyons are maintained by the erosive effects of currents produced by tides, internal waves, and turbidity flows (Southard and Stanley, 1976). Erosion of the canyon floor steepens the margins and promotes slumping and avalanching on canyon walls. These mass movements contribute coarse sediment to the canyon floor, trigger additional down-canyon currents, and act to initiate the formation of gullies that eventually evolve into tributaries (LePichon and Renard, 1982).

Canyons are important conduits of sediment derived not only from their own lateral margins, but also from the shelf where sediment moving in coast-parallel fashion is commonly intercepted by canyons cutting across the shelf (see, for exam-

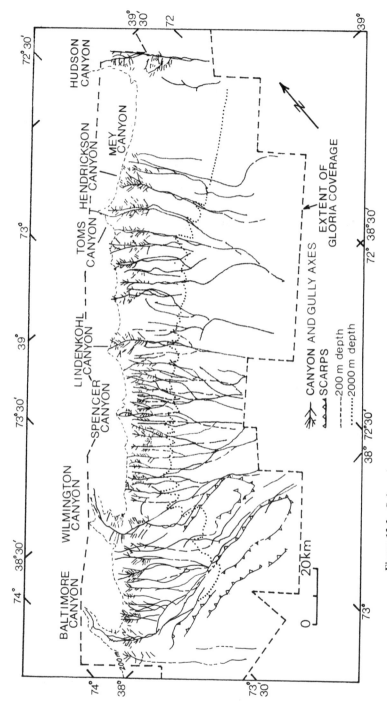

Figure 13.2 Submarine canyons on the continental slope off the Middle Atlantic coast of North America showing their lateral distribution along the slope and the density of lateral canyons in their upper reaches (after Twichell and Roberts, 1982).

325

ple, Kulm and others, 1975). Canyons heading in inshore areas may even trap sediment moving in littoral circulation cells (Dill, 1964; Moore, 1969). These sediments experience only short residence times in the canyons. Instead, the canyons are used to bypass the slope as the sediments are transported into deep marine basins.

Some sediments moving down the slope are trapped, however, in small basins that are also a feature of continental slopes (Fig. 13.3). They are particularly common on slopes facing subduction zones where narrow, elongate basins form as an integral part of the evolution of accretion complexes (Moore and Karig, 1976). They also form where block faulting or diapirism occur, and small basins form behind dams consisting of slump debris that clogs canyons (LePichon and Renard, 1982).

The third major morphological feature on slopes is the intercanyon areas. The nature of these areas has been a matter of controversy, principally because of the varying degrees of resolution provided by different survey methods. The broad areas between major canyons were originally thought to be relatively smooth, but new survey methods, including side-scan sonar, have shown them to be cut by previously unrecorded secondary canyons and gullies (Fig. 13.2). The slope surface may also be marked by a fourth morphological element consisting of the products of sediment mass movement, including slump and slide scars and the hummocky accumulation of debris.

Classification

Most classifications of continental slopes are based on morphology, and because of the diversity of slope shapes, they tended to be somewhat complex in their effort to be all-inclusive (see Vanney and Stanley, 1983, for instance). Classifications that are

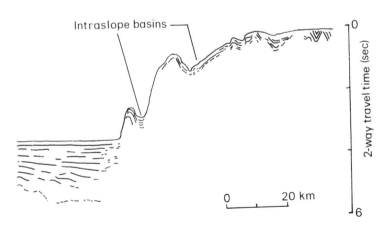

Figure 13.3 Cross-sectional view of a portion of the Oregon continental margin showing position of intraslope basins on the accretionary complex (after Kulm and Fowler, 1974).

most useful in categorizing ancient slope deposits, however, are those based primarily (Inman and Nordstrom, 1971) or secondarily (Emery, 1977) on tectonic setting. Not only do such classifications give some indication of the internal structure (and, therefore, the external shape), but the tectonic setting controls to a large degree the type of sediment being delivered to the slope.

Slopes on passive, or Atlantic-type, margins are characterized by relatively smooth relief, reflecting a thick accumulation of sediment (Fig. 13.4). Adjacent shelves are most often wide, and shelf breaks tend to be shallow. Slope gradients tend to be low, and they merge gradually onto submarine rises with very low gradients.

The most common internal structure consists of a series of blocks that were downfaulted basinward during plate separation (Fig. 13.4). Continued subsidence of the margin has produced a basinward-thickening wedge of sediments that drape

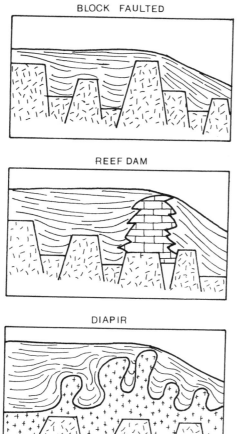

Figure 13.4 Relationship of slope morphology to the internal structure of passive continental margins.

this underlying structure. On some passive margins, the shelf break is commonly controlled by a reef dam, and in others thick salt deposits that accumulated during initial phases of rifting are now involved in diapirism that controls slope morphology (Emery, 1977) (Fig. 13.4).

On older margins, the sediment cover is thick and relatively undeformed, and the slope is smooth. On younger margins, such as the Gulf of California, however, the sediment cover is thin and the slope may be broken by low, parallel ridges and basins resulting from initial uplift followed by subsidence. The structural control of these types of slopes is still very evident because depositional smoothing of the topography has not yet occurred.

Adjacent hinterlands tend to be low-lying (except along juvenile pull-apart margins), and drainage networks are well-developed. Many of the world's largest rivers, such as the Mississippi and Amazon, empty onto passive margins. The sediment yield from such sources tends to be fine-grained, but abundant. Progradational margins are common.

Active, or Pacific-type margins, can be subdivided into those along convergent plate boundaries and those involved in transform faulting. At convergent margins, one plate boundary is subducted beneath another, producing an arc–trench system. As subduction proceeds, increments of sediment consisting of a mixture of oceanic and trench sediment are added in successive slices, forming a subduction complex (Seely and Dickinson, 1977) (Fig. 13.5). As it grows upward, the complex forms a slope characterized by an irregular surface with few large canyons, but many smaller ones that commonly terminate in intraslope basins (Fig. 13.5). The subduction complex is eventually uplifted above sea level and itself becomes a source of sediment for more recently formed parts of the slope.

The shelf break is relatively deep, and the slope descends to a deep trench at a steep angle. Horsts, grabens, dikes, and terraces combine to form a complex tectonic terrain that is reflected in an irregular slope topography. In addition, slopes

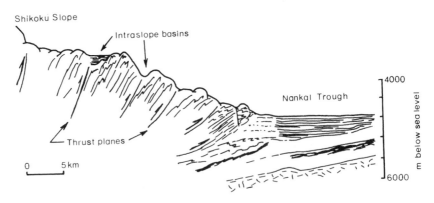

Figure 13.5   Internal structure of a subduction wedge complex on the Japan continental margin (after Ingle and others, 1973; Leeder, 1982).

are cut by numerous gullies and canyons that transport sediments into the deep trench or into the intraslope basins (Underwood and Karig, 1980). Narrow shelves are backed by youthful coastal mountain ranges that shed coarse debris. Rivers tend to be small and drainage networks poorly developed, but high regional gradients may still cause large amounts of sediment to be delivered to adjacent slopes (Klein, 1984). In addition, large rivers, such as the Columbia River or the Copper River of Alaska, that rise farther inland may drain large basins and be capable of delivering large quantities of sediment to adjacent slopes.

Transform margins occur where there is relative lateral motion between oceanic and continental lithosphere. The motion produces a wide zone of anastamosing fault segments that break the margin up into small basins and uplifted blocks (Howell and von Huene, 1981) (Fig. 13.6). Basins can be separated by uplifted blocks, but most are connected by submarine canyons that act as conduits for sediment derived from uplifted source areas. The blocks themselves may serve as minor sources, but most sediment is derived from the adjacent coast. Basins closest to the shelf fill first with relatively coarse sediment, and those farthest away fill last with sediment that finds its way to them by complex pathways (Gorsline, 1981).

Sedimentary Processes

Mass Gravity Movements

Mass gravity movements are probably the sedimentary process most commonly associated with slopes, but currents caused by surface and internal waves, tides, and oceanic circulation systems are important transport processes as well (Southard and Stanley, 1976). Gravity mass transport is the downslope movement of sediment either as individual grains or as coherent sediment bodies under the influence of gravity (Field, 1981). The important distinction is between sediment slides that move as discrete, internally cohesive blocks, and sediment flows, where substantial reorganization of the internal fabric of the moving mass takes place.

Two types of slides exist. Glides fail along planar slip surfaces that correspond to bedding planes, and movement consists of simple downslope translation. Internal structure tends to be undeformed except near the base and margins, where folding and extensional deformation is evident. Slumps, on the other hand, fail along curved shear surfaces, and considerable rotational movement occurs in the sediment mass. Slumps remain coherent during movement, but the internal fabric of the mass may show some degree of deformation, especially near the toe, where thrusting and folding can occur.

Both types occur when the downslope vector of the gravitational force exceeds the shear strength of the material. This is especially common on active margins where cyclic loading by earthquakes can disturb the sediment, but it may also happen where excess pore pressures are created by rapid sedimentation, artesian flow of gas or water, or interstitial generation of gas in organic sediments (Fig. 13.7).

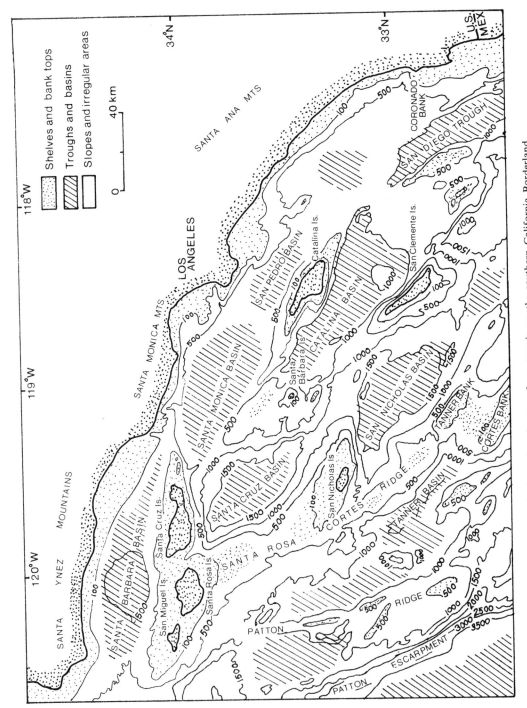

Figure 13.6 Physiography of a transform margin on the southern California Borderland showing the complex distribution of basins, banks, and islands (after Field and Edwards, 1980).

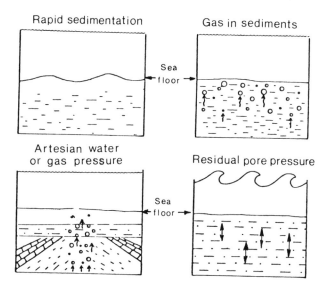

Figure 13.7 Triggering mechanisms for mass sediment movements (after Sangrey, 1977; Field, 1981).

Large coherent slide blocks may measure several square kilometers and be resolvable by current surveying techniques on modern slopes (Cook and others, 1982). Smaller blocks may not be presently detectable, but evidence from the rock record suggests that small mass movements are also common on slopes.

Sediment gravity flows retain little or no evidence of their original internal organization. They may be subdivided into plastic flows, such as debris flows, that begin to deform only after a threshold strength has been exceeded, and fluid flows that deform continuously after a shear has been applied. Fluid flows, in turn, may be divided into grain flows, fluidized flows, and turbidity currents that show a progressive decrease in the degree of mechanical grain support during movement (see Chapter 1 for a more complete description of fluid gravity flows and their resultant deposits).

Basin-generated Currents

Continental slopes are also affected by currents induced by oceanic processes. Waves, for instance, are capable of influencing sediments at the upper slope, even though the shelf break occurs at depths below the influence of most waves. Komar and others (1972) reported the occurrence of oscillation ripples at depths of as much as 200 m, and Ewing (1973) suggests that refraction of very large surface waves around irregularities may reinforce oscillatory bottom currents at the shelf break. Slope surfaces can also be affected by waves generated along density interfaces within the water column. Sediment motion can be induced by shoreward-moving internal waves where upslope amplification occurs. Where waves break on the slope, even larger near-bottom velocities can be generated (Cacchione and Southard, 1974) (Fig. 13.8). Experiments with breaking internal waves suggest that an upslope cur-

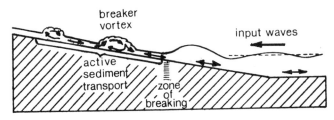

Figure 13.8 Effects of internal breaking waves on slope substrates (after Southard and Cacchione, 1972).

rent can be induced by the passage of a decaying wave after it breaks and that a downslope return flow is set up by the passage of the breaker (Southard and Stanley, 1976). Both types of currents may be capable of transporting sediment in sand waves and ripples.

Wave activity on the inner shelf and in the littoral zone can also affect slope sedimentation where canyons cut across shelves and intersect inshore circulation cells. Sandy sediments can be provided to canyons in this manner, and rhythmic, down-canyon currents can be induced where seaward-directed jet flows, produced by the return flow of onshore-directed waves, enter heads of canyons (Reimnitz and Gutierrez-Estrada, 1970; Reimnitz, 1971).

The effect on shelf sedimentation of large oceanic circulation cells has previously been discussed, but these currents are particularly effective in their impact on the slope. They are generated by thermohaline forcing mechanisms, and they flow along isopycnic surfaces that are parallel or subparallel to bathymetric contours. The currents flow at velocities sufficient only to move fine sediment at the bottom or to prevent it from settling out of suspension, but in areas of flow acceleration, such as through straits or other narrow passages, they can erode slopes and remold sandy substrate into migrating bedforms (Kelling and Stanley, 1976).

Tidal currents are an integral part of the spillover process whereby sediment is carried over the shelf break and onto the upper slope (Stanley and others, 1972a), and they may also be effective transporting agents in slope canyons. Shepard and others (1974), for instance, measured currents flowing both upcanyon and down-canyon with a diurnal period. Shepard and others (1977) also showed that the normal background processes operating in submarine canyons produced a net up-canyon flow that was interrupted periodically by turbidity currents produced by stream flooding or storms.

Suspension Settling

Suspension settling of pelagic (produced by the ocean) or hemipelagic (derived from both land and ocean) fine-grained sediments has long been thought to be the major process by which sediment is added to the slope. The water column is not homogeneous, however, and instead is composed of layers of differing densities that can form effective barriers to sediments settling from suspension. In fact, shear-induced turbulence may be produced at the interface between two water masses, and this can be especially effective in preventing the bulk of fine sediments from settling from suspension (Gorsline, 1981). Fine-grained sediments can settle from suspension, however, when they collect into aggregates by the pelleting action of pelagic microfauna, when they adhere to organic filaments, or through flocculation.

Nepheloid Layers

The fine laminations commonly encountered in slope muds are not easily accounted for by simple suspension settling, but a more active process of fine sediment deposition may be provided by slow-moving, dilute turbidity currents or nepheloid layers (Moore, 1969; Rupke and Stanley, 1974). Some of these greatly diluted suspensions of sediment occur at relatively shallow depths and tend to be ephemeral features that are only provided episodically with sediment. Those that occur in deeper water, however, are semipermanent features that are continuously being replenished (Eittreim and Ewing, 1972).

Plumes of sediment can originate on the shelf during storms when large runoff from source areas occurs or when sediment on the shelf substrate is resuspended by wave activity. There tends to be a component of movement down the shelf gradient, but the net movement is dependent on local topography and the circulation patterns of both surface and bottom currents. Some plumes may simply spill over the shelf break and deposit sediment on the upper slope, and because the positions of the plumes may be relatively constant (Karl, 1976), sediments may accumulate rapidly along some portions of the slope, making them more susceptible to mass sediment movement (Gorsline, 1981). Other sediment plumes move down the slope in canyons, and these may be important sources of sediment for the deep nepheloid layers (Eittreim and Ewing, 1972).

Deposits of dilute turbidity flows or nepheloid layers may be finely laminated and graded and contain no sand-sized particles. Suspensate deposits, on the other hand, are structureless or bioturbated, poorly sorted, and contain abundant sand-sized particles, most of which are biogenic (Rupke, 1975).

Slope Sediment Facies

Cook and others (1983) described three main sediment types on modern slopes, including (1) undisturbed pelagic and hemipelagic sediment, (2) chaotically deformed sediment, and (3) sediment gravity flows. These sediment types occur on both active and passive margins, and they can occur in both canyon and intercanyon areas. Undisturbed sediment, however, is more characteristic of intercanyon areas, whereas gravity flows and deformed sediment are more characteristic of canyons.

Undisturbed Sediments

Sediments on open slopes consist of clays and fine silts with very subordinate amounts of coarse silt and sand (Stanley and others, 1972b). In some cases, these sediments are finely laminated, but in others they have been homogenized by burrowing organisms. A pronounced increase in clay content occurs near the shelf break at the so-called mudline. Above the mudline, the sand content decreases steadily from shore; but at the mudline, the proportion of clays increases substantially and remains relatively constant downslope (Stanley and others, 1983).

The origin of the mudline and its position relative to the shelf break are a function of a number of complexly interrelated factors. The line may occur well below the shelf break, where sediment delivered by shelf spillover is small and the winnowing capability of ocean currents is great. On the other hand, the mudline may occur at the shelf break or even on the shelf in front of large deltas, where the sediment influx is large and basin processes are incapable of moving all of it off the shelf.

Undisturbed sediments may collect in submarine canyons, where they tend to contain a larger proportion of sand than adjacent intercanyon areas. Sediments in middle and lower reaches of canyons may consist of poorly graded or ripple-bedded sand interbedded with bioturbated silty mud (Kelling and Stanley, 1976), and some canyons may even contain thick beds of well-graded sand (Field and Pilkey, 1971). Active canyons heading on narrow shelves may contain coarser sand, or even gravel, in relatively thick, graded beds, but canyons that became inactive are floored by burrow-mottled silty clays with rare thin layers of sand.

Some canyons show a pronounced axial facies distribution. In the Oceanographer Canyon off the east coast of the United States, for instance, sediment facies are arranged more or less symmetrically with respect to the canyon axis (Valentine and others, 1984) (Fig. 13.9). Shelf sand is transported by currents into the canyon. At the canyon rim it mixes with immobile gravels to form deposits of gravelly sand. Lower on the walls it passes over muds, where it is mixed by burrowing organisms to form sandy muds. The sand collects on the canyon floor, where it is molded into small- and medium-scale bedforms that reflect both upcanyon and downcanyon currents. The sands are relatively well sorted because silt and clay are winnowed by active axial currents, and bioturbation is not pervasive because of the mobility of the substrate.

Organic matter is a common constituent of slope sediments. It is particularly abundant in anoxic intraslope basins, but it can also accumulate in areas on the slope that intersect the oxygen minimum layer (Fig. 13.10). Oxygen is being depleted constantly in this zone as organic matter is oxidized, and because it is below the level where photosynthesis occurs, little oxygen is added to the zone, and levels of dissolved oxygen can fall to as little as 0.1 ml/l.

Sediments deposited within such zones may be finely laminated because of the absence of infauna, and they tend toward darker colors because of the presence of organic matter and reduced iron. Where the rate of organic productivity is large, such as where upwelling occurs, the proportion of terrigenous sediment can be small, and the sediment may consist primarily of pelagic constituents, including diatoms and coccoliths. Diatoms are composed of a form of silica that is easily dissolved, and where they are abundant the sediments may be siliceous. Organic slope sediments may also contain abundant phosphate nodules and pellets apparently formed during very early stages of diagenesis.

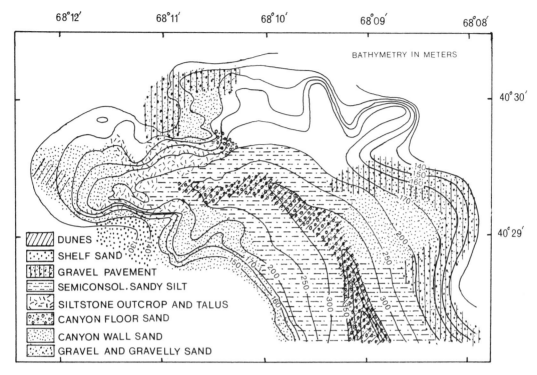

DUNES
SHELF SAND
GRAVEL PAVEMENT
SEMICONSOL. SANDY SILT
SILTSTONE OUTCROP AND TALUS
CANYON FLOOR SAND
CANYON WALL SAND
GRAVEL AND GRAVELLY SAND

Figure 13.9   Facies distribution in part of the Oceanographer Canyon on the continental slope off the Middle Atlantic coast of the United States (after Valentine and others, 1984).

Deformed Sediments

Deformed bedding is a common constituent of slope sediments, and Field (1981) has divided the deposits that result from coherent mass gravity movements into three types (Fig. 13.11). A slide block is a discrete mass of sediment that has undergone transportation without disaggregation. Glide blocks are a type of slide that slip

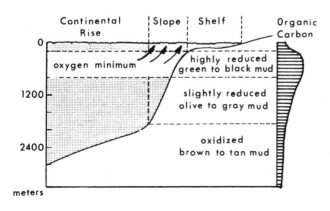

Figure 13.10   Diagrammatic representation of the water column off a continental margin showing the position of the oxygen minimum layer and its effects on sediments (after Fischer and Arthur, 1977; von Stackelburg, 1972).

Processes   Deposits

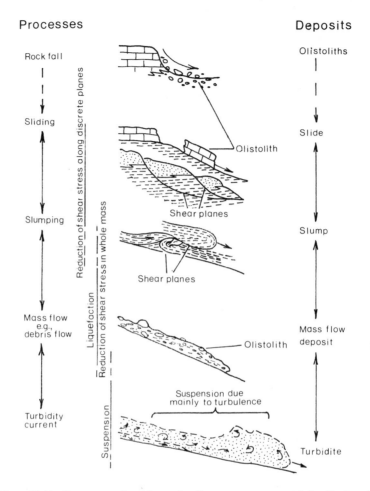

Figure 13.11   Processes and products of sediment mass transport (after Kruit and others, 1975; Field, 1981).

along horizontal or near-horizontal slide planes, whereas slump blocks fail along curved surfaces. Detailed bottom profiling of slopes shows numerous areas of hummocky topography where the mass has come to rest, as well as various types of scars and slide planes where the movement has occurred. Internal organization of these masses, revealed by seismic reflection profiling, ranges from virtually undisturbed in glide blocks to greatly deformed at the toe and along the margins of both glide and slump blocks.

Slides may be grouped together into zones that can be either organized or disorganized. Organized zones consist of blocks that retain their individual identity. The zones show a hummocky surface, and internal reflectors are often chaotically arranged except for multiple parallel or subparallel failure planes. Organized zones occur in an upslope retrogradational pattern produced when the movement of one block triggers the failure of a sediment mass farther up the slope (Fig. 13.12).

Disorganized slide zones are recognized by their hummocky topography and apparent lack of internal structures. Seismic records may show a chaotic, disrupted pattern of reflectors at best, but the zones are normally seismically transparent. Although cores taken in submarine slide deposits normally show no evidence of internal deformation, except where the basal shear plane was penetrated, cores in disorganized slide zones show all stages of plastic deformation, including initial stages of breakup into individual clasts.

Disturbed sediments are particularly common in submarine canyons where the gradient of the canyon walls may exceed 10°. Lateral movement of material takes place from the walls into channel axes, leaving talus aprons and slump blocks along margins and transporting coarse clastic debris into the thalweg. Lateral mass movements in turn trigger longitudinal movement of sediment down the canyon as grain flows, debris flows, and turbidity currents (LePichon and Renard, 1982).

Chaotic Sediments

Deposits of rockfalls and avalanches consist of a grain-supported framework and a variable amount of a matrix (Cook and others, 1982). The surface expression of the deposits is strongly hummocky, and the internal structure may be massive or chaotic. Acoustic records show a weak or chaotic internal return indicative of the structureless nature of the deposit.

Except for debris flows, sediment flows are not well represented on continental slopes. The more fluidized mass movements can travel over very gentle gradients, and slope gradients normally are far in excess of that; so flows initiated on the slope normally can transport sediment well into the adjacent basin. Even debris flows can move over slopes of as little as 1°, so the deposits of only a few debris flows have been confirmed in modern slope deposits (Cook and others, 1982).

Sediment flows do occur in canyons, however, and their deposits may even be abundant in intraslope basins. Even sand turbidites may be found in these settings,

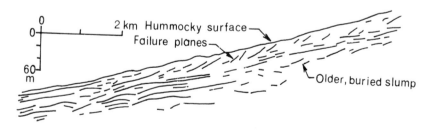

Figure 13.12   Internal structure of a retrogradational slump zone (after Cook and others, 1982, Fig. 23, reprinted with permission of the American Association of Petroleum Geologists).

although they do not consist of complete or even nearly complete Bouma sequences. Turbidites more commonly consist only of the c, d, and e intervals, especially at the present time when sources of coarse sediment have retreated from the shelf break following Holocene sea-level rise (Kulm and Scheidegger, 1979; Bouma and others, 1978).

14

Evolution of
Slope Sequences

Intrinsic and Extrinsic Controls on the Evolution of Slope Sequences

Introduction

Slope sequences form at various scales in response to variations in extrinsic controls, such as eustatic changes in sea level, changes in position or strength of oceanic currents, and tectonism, as well as variations in intrinsic controls that mainly affect the ability of the slope to bypass or accumulate sediment.

Eustatic changes in sea level alter the way sediment is delivered to the slope, which in turn can cause the development of first-order sequences (Vail and others, 1977a). During low sea-level stands, rivers are rejuvenated and deliver large amounts of stored sediment to the shelves. Shelves become narrower and water depths over them become shallower, so sediment is transported across shelves more easily and delivered to the slope, resulting in progradation. During exceptionally low stands, the shelf may be eroded and act as a sediment source.

During sea-level rise, sediment is trapped in estuaries or aggrading river valleys, and the sediment that is provided to the shelf moves across it more slowly because the shelf is wider and water depths are greater. Thus, progradation ceases during high sea-level periods, and episodes of erosional retrogradation may occur.

Retrogradation can also be caused by large-scale shifts in oceanic boundary currents. Since the Miocene at least, for instance, the Gulf Stream has shifted laterally in response to global sea-level changes (Pinet and Popenoe, 1982). The Gulf Stream shifted westward against the Florida–Hatteras Slope during high sea-level stands, resulting in erosion of the slope (Fig. 14.1). During low stands, however,

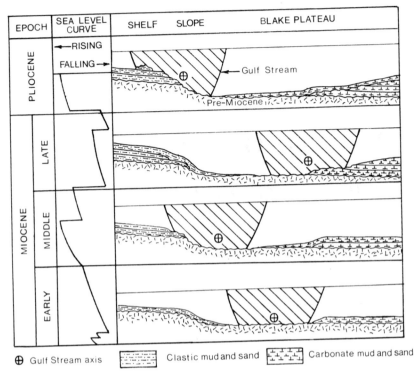

Figure 14.1 Diagrammatic representation of the sequence of events occurring during the Late Tertiary that resulted in periodic reworking of the slope off the southeast coast of the United States (after Pinet and Popenoe, 1982).

the current shifted eastward over the Blake Plateau, allowing deposition of terrigenous clastic sediments inshore of the axis and causing progradation of the Florida-Hatteras Slope.

Most knowledge of how continental slopes adjust to these changes comes from interpretation of seismic reflection data supplemented by only a few long cores. Although very many short gravity cores and piston cores have been taken of slope sediments, they sample only the surface layers, and these may be misleading because of effects of Holocene sea-level rise on slope sedimentation.

The seismic pattern from prograding passive margins consists most often of reflectors that diverge in a downslope direction. Superimposed on this pattern are intraformational truncation surfaces that separate bundles of parallel to subparallel strata (Fig. 14.2). Truncation surfaces represent periods of erosion that occur during periods of relatively low sea level, reduction in rates of sediment influx, or change in position or strength of oceanic currents.

Oblique progradational patterns form when the shelf is a zone of sediment bypassing or erosion, and deposition occurs seaward of the shelf break. Such conditions may exist during periods of tectonism when the shelf is uplifted or during low

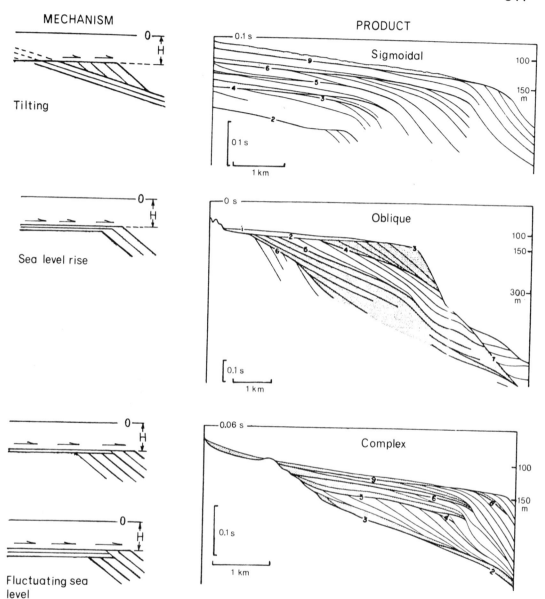

Figure 14.2 Examples of the internal structures of prograding shelf breaks showing modes of origin and arrangement of strata of sigmoidal, oblique, and complex types (after Mougenot and others, 1983).

sea-level stands (Mougenot and others, 1983). The truncation surfaces form during uplift or eustatic lowering of sea level, and deposition occurs during periods of tectonic quiescence or high sea-level stands (Fig. 14.2). Sigmoidal patterns, on the other hand, are produced when contemporaneous deposition occurs on both shelf and slope, and this occurs most often when the whole margin is subsiding.

Sequences on Passive Margins

The internal structure of most prograding passive margins is usually a combination of sigmoidal and oblique patterns because the rates of subsidence, eustatic sea-level changes, and deposition normally differ enough that the shelf is alternately a zone of deposition and erosion or bypassing.

The few deep cores that have been taken on passive continental margins suggest that rates of sedimentation are small and that the sediment permanently stored on the slope is dominantly fine grained. One 550-m-thick sequence of sediments cored off the Spanish Sahara, for instance, represents the time period from the Quaternary to the Lower Cretaceous, with only two relatively short hiatuses (Lancelot and others, 1977). The sequence consists predominantly of pelagic sediments with abundant planktonic floral and faunal debris and only rare coarse detritus (Fig. 14.3).

The volume of the terrigenous component is surprisingly small with approximately two-thirds of the sediment of biogenic origin. The paucity of terrigenous sediment coupled with small rates of accumulation suggests that this slope is dominated by pelagic rather than hemipelagic sedimentation. The lack of terrigenous sediments on this slope may be attributed to reduced sediment yield from an arid hinterland or to trapping by a cross-shelf canyon, but these factors do not explain the similar lack of terrigenous sediments in a core drilled on the North American continental slope (Hollister and others, 1972).

The relative importance of the terrigenous component in slope sediments may be a function mainly of the source relative to shelf break. Pleistocene-age sediments, for instance, tend to contain a larger terrigenous content (Stanley and Unrug, 1972), presumably because sea level was lower and spillover was a more important process.

Proximity of source has influenced sedimentation on slopes near major deltas. The Mississippi Delta, for example, has prograded almost to the shelf break, and sediments on the slope west of the delta consist dominantly of silty mud of terrigenous origin. In some areas of the slope, the sediments are characterized by continuous sparker reflectors that formed in response to slow, continuous sedimentation. Sand is absent in these sediments except for occasional tests of foraminifera, but silt may comprise up to 40% of the total (Woodbury and others, 1978). Sand content is greater, however, in areas of disturbed sediment where reflectors are chaotic and the bottom topography is hummocky. These sediments were deposited by mass transport mechanisms, and the presence of neritic foraminifera suggest that they were displaced from shallower water.

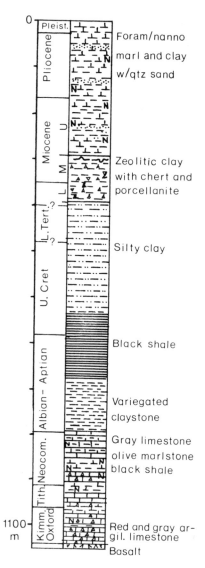

Figure 14.3 Stratigraphy of Mesozoic and Cenozoic sediments underlying the continental slope off the Spanish Sahara illustrating the dominance of pelagic sediments in the sequence (after Lancelot and others, 1977).

Tectonism along passive margins normally takes the form of thermally forced subsidence (Watts, 1981). Simple downstepping along normal faults may result in steepening of slope gradients, leading to an incidence of slope failure and erosional retrogradation, especially at the slope break, until a new equilibrium profile is attained. On the other hand, complex faulting may occur and result in the formation of intraslope basins. Such basins also result from diapiric movement of salt commonly deposited during initial stages of pull-apart and subsidence along passive margins.

Sediments in intraslope basins tend to be somewhat coarser than surrounding slope surfaces, and they also tend to show a large terrigenous content, probably because they trap sediment that might ordinarily bypass the slope. Sediments in the Gyre Basin of the northwest Gulf of Mexico, for instance, consist dominantly of detrital muds with lesser amounts of nannoplankton (Bouma and others, 1978). Layers of detrital quartz silt and sand are common in the deeper parts of the basin, where they probably were deposited by sediment gravity flows. Strata show an on-lapping relationship in intraslope basins. Reflectors are nearly flat and parallel in the deeper parts of the basin, but become progressively more curved as they onlap onto the margins with a slight angular uncomformity (Sangree and others, 1978).

Sequences on Active Margins

Sediment sequences on slopes along active margins differ from those on passive margin slopes. Shelves tend to be narrow and source areas are higher. Rates of sedimentation are greater, spillover processes are more important, and the sediment provided to the slope is coarser. Because of the narrowness of the adjacent shelves, however, and the greater depth of the shelf break, slopes on active margins may not respond as dramatically to eustatic changes in sea level (Jacka and others, 1968). The tectonic climate is more severe, and the response of the slope to tectonic epi-sodes is commonly more complex. Uplift in source areas may increase sediment yield to the slope (Klein, 1984), but tectonic steepening of the slope may result in a greater degree of sediment bypassing (Underwood and Karig, 1980). Complex fault-ing along margins, however, may result in the formation of intraslope basins that can trap some of this sediment (Moore and Karig, 1976).

Slope sequences in the Japan fore-arc, for example, rest on sedimentary and volcanic rocks that were subaerially exposed during the early Tertiary and began to subside only during the Oligocene (Arthur and others, 1980). Initial deposits consisted of shallow-water sandstones and siltstones, but rapid subsidence quickly established open slope conditions where vitric and diatomaceous mudstones were deposited (Fig. 14.4). Coarse sediment tended to bypass the slope until the Plio-Pleistocene, when renewed tectonism formed intraslope basins that trapped this ma-terial. On open slope surfaces, however, fine-grained, mostly terrigenous sediments continued to accumulate.

The nature of the fill in intraslope basins of active margins, however, is source dependent. Basins filled by canyons heading on shelves may contain considerable coarse sediment, but those filled by canyons heading on the slope, or those fed only by spillover over the margins, may contain predominantly fine-grained sediments (Underwood and Karig, 1980).

Holocene sediments of the Oregon slope, for instance, consist of diatomaceous clayey silt or silty clay (Kulm and Scheidegger, 1979). Sediments of the upper slope may contain considerable relict sand from Pleistocene low sea-level stands, but the bulk of the muds of the lower slope contains less than 2% sand-sized material, some of which is biogenic (Fig. 14.5).

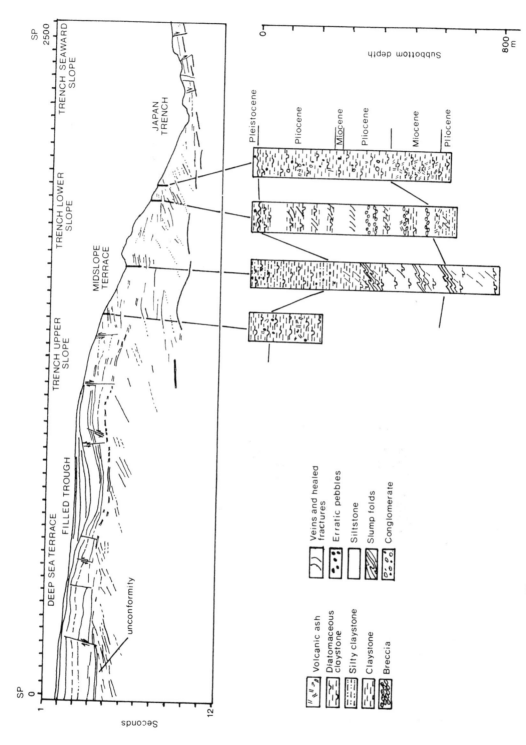

Figure 14.4 Stratigraphy of sediments underlying various parts of the Japan continental margin (after Arthur and others, 1980).

345

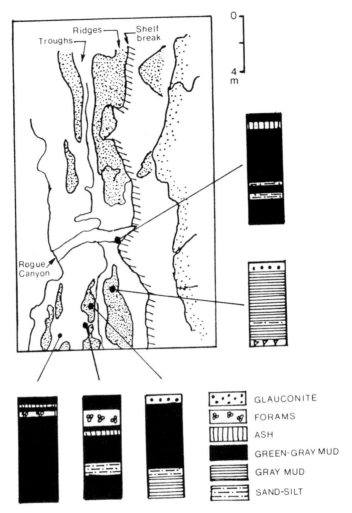

Figure 14.5 Shallow cores from the Oregon continental margin showing relationship of surficial sediments to water depth (after Kulm and Scheidegger, 1979).

A core from an accretionary basin on the lower part of the Oregon slope contains sediments spanning the latter part of the Pleistocene (Fig. 14.6). They consist of dark greenish-gray muds with only rare fine sand or coarse silt layers, suggesting that even during the Wisconsinan low sea-level stand, the basin lacked access to coarse debris. Lower in the core, however, silt turbidites become increasingly abundant, but these antedate the formation of the intraslope basin and were probably deposited on the adjacent deep basin plain prior to uplift.

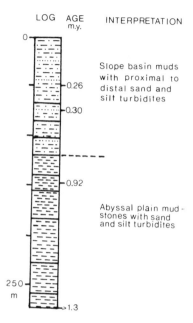

LOG   AGE   INTERPRETATION
       m.y.

Slope basin muds
with proximal to
distal sand and
silt turbidites

Abyssal plain mud-
stones with sand
and silt turbidites

Figure 14.6 Stratigraphy of sediments filling an intraslope basin on the accretionary complex of the Oregon continental margin (after Kulm and Scheidegger, 1979).

Sequences in Canyons

There is surprisingly little known about the stratigraphy of the sediments in canyons on slopes of active or passive margins, although the nature of the surface sediments is fairly well known. Field relations suggest that canyons must have transported considerable amounts of relatively coarse sediment to submarine fans and rises, and yet the surface sediments of many canyons, especially of those on passive margins, consist mainly of hemipelagic sediments and probably do not reflect the nature of the deeper fill.

These canyons have responded to the Holocene sea-level rise by transporting less sediment. During the transgression, sediment sources retreated from the shelf edge, where they were actively providing sediment to canyons, and the sediment yield from continents became trapped in river valleys and estuaries. It has been estimated that present-day canyons transport only about 10% of the sediment that passed through them during low sea-level stands (Heezen and Hollister, 1964), and the bulk of that is derived from shelf spillover, pelagic settling, and lateral erosion of canyon walls (May and others, 1983). Surficial sediments of the Gully Canyon of the Nova Scotian shelf, for instance, consist mainly of bioturbated mud with only occasional layers of graded and ripple-bedded sand. Mud clasts and gravel are present, suggesting active mass sediment movement, possibly down the flanks of the canyon (Stanley, 1967). Active sediment transport was episodic, but the dominance of bioturbated hemipelagic muds suggests that the episodes are widely spaced and that accumulation rates are normally quite small.

Channel fills in front of the Mississippi Delta also tend to be fine-grained. Seismic patterns consist of gently curved reflectors that lap onto margins (Fig. 14.7), and the sediments consist largely of clay with a few scattered sand laminae (Sangree and others, 1978). Some seismic patterns, however, display nearly chaotic patterns, suggesting the occurrence of mass sediment movements.

Canyons on active margins, like those on passive margins, may experience periods of active sediment transport alternating with periods of relative inactivity caused by tectonism or eustatic changes in sea level (Howell and von Huene, 1981). The Rogue and Astoria submarine canyons, for instance, head on the outer shelf of Oregon, and are presently veneered with the fine-grained sediments currently being transported along the shelf by storm-induced currents. The bulk of the sediments, therefore, consists of green or gray muds with occasional silty turbidite layers (Fig. 14.8) (Kulm and Scheidegger, 1979). The fine texture of these surface sediments is in apparent contradiction to the relative coarseness of those in the fan-channel systems fed by the canyons. Apparently, the canyons have not been particularly active since the beginning of Holocene sea-level rise.

Other canyons, however, can remain active if they are in a position to maintain access to abundant coarse sediment (Kelling and Stanley, 1976). The Hueneme canyon, for instance, heads within 200 m of the southern California coast and intercepts sediments being transported in littoral circulation cells (Scott and Birdsall, 1978). Sediments in the canyon axis consist of fine sands and silts in planar beds, asymmetric ripples, and graded beds that are indicative of a relatively constant process of

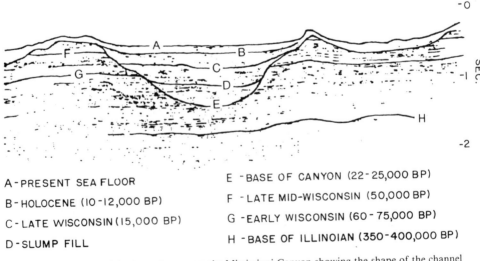

A - PRESENT SEA FLOOR

B - HOLOCENE (10 -12,000 BP)

C - LATE WISCONSIN (15,000 BP)

D - SLUMP FILL

E - BASE OF CANYON (22 - 25,000 BP)

F - LATE MID-WISCONSIN (50,000 BP)

G - EARLY WISCONSIN (60 - 75,000 BP)

H - BASE OF ILLINOIAN (350 - 400,000 BP)

Figure 14.7 Seismic section across the Mississippi Canyon showing the shape of the channel and the architecture of the infilling sediments (after Coleman and others, 1983).

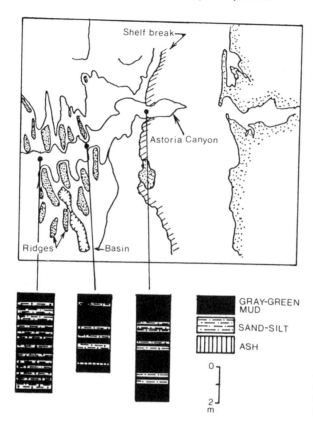

Figure 14.8 Shallow cores from the Astoria submarine canyon showing distribution of surficial sediments down the axis of the canyon (after Kulm and Scheidegger, 1979).

downcanyon traction transport. On the floor of the canyon away from the axis, however, the sediments are muddy and extensively bioturbated, suggesting that active transport is more or less confined to the axis of the canyon and that sediments flanking the axis are the result of infrequent extreme flows.

Thus, two types of submarine canyons can be defined on the basis of their relationship to sediment sources (Dill, 1981). Proximal canyons are those that head close to sources. They characteristically accumulate large masses of poorly sorted sediment until a threshold of instability is reached and failure occurs that generates slumps and gravity flows that travel downcanyon. These events are primarily erosive (Shepard and Dill, 1966), and the sediments that are in transit do not necessarily represent the sediment that would accumulate during canyon filling. Instead, this material bypasses the slope and is deposited at canyon mouths (Underwood and Karig, 1980). Proximal canyons occur on both passive and active margins during low sea-level stands, but during high stands they preferentially occur on active margins where shelves are narrow and steep.

Distal canyons are those that are not in contact with an active sediment source (Dill, 1981). They receive an appreciable proportion of their sediment from shelf spillover that works its way down canyon walls into canyon axes, where it is re-worked by indigenous processes (Valentine and others, 1984). Lateral input from slumping canyon walls (LePichon and Renard, 1982) and from tributary gullies (Twichell and Roberts, 1982) is also an important source, as is longitudinal supply from headward erosion as the canyon works its way landward (Farre and others, 1983). Hemipelagic sedimentation also provides sediment to distal canyons (Kelling and Stanley, 1976), where it probably acts as a background process in between epi-sodes of more active transport by gravity flows and clear-water currents. Distal can-yons characteristically occur on passive margins during high sea-level stands, but they may also occur on those parts of active margins where shelves are wide or where canyon piracy occurs (Herzer and Lewis, 1979).

Although the preceding discussion suggests that canyon filling is unlikely dur-ing periods when canyons are active transporters of large amounts of sediment by gravity flows, it will be shown later that some ancient canyons are filled, in part at least, with coarse-grained sediments deposited by a variety of mass movements (Stanley, 1975; Clifton, 1981b; May and others, 1983). Filling may not even occur during periods of quiescence. Most distal canyons continue to act as conduits for turbidity flows, although at a rate much reduced from more active periods when they were closer to sources (Shepard, 1981). Hemipelagic sediments tend to be flushed from canyon axes during such flows, but they continue to collect on mar-gins, where they actually serve to accentuate canyon topography (Slater, 1981).

Even though the similarities between rivers and submarine canyons have long been recognized (see McGregor and others, 1982, for instance), it is not altogether certain that the processes of valley aggradation in the two systems are similar. Sub-aerial canyons aggrade, for instance, when the rivers they contain are no longer capable of carrying the imposed sediment load. One way this may occur is when sediment yield is increased, but this mechanism is unlikely to cause aggradation in subaqueous canyons because of the positive feedback relationship between increased sediment influx and increased frequency and intensity of erosive gravity flows.

Subaerial canyons also aggrade, however, when a reduction in gradient occurs, and this mechanism is more likely to cause aggradation in slope canyons because it would act to slow downcanyon flows and reduce their ability to transport the im-posed load. The gradient in submarine canyons may be reduced in three ways, in-cluding (1) tectonism, (2) headward migration of the canyon, and (3) deposition in the basin, which is similar to raising the base level of subaerial streams (see Schumm, 1977, p. 260, for instance).

Tectonism might work to reduce the gradient on active margins where uplift occurs, but subsidence is the normal tectonic process on passive margins; and except where diapirism can affect base level locally by creating intraslope basins, tectonism most often would work to increase the gradient.

Headward migration of canyons would decrease the gradient by increasing the distance over which the descent to the basin floor occurs. Simple geometric consider-ations suggest that the process would accelerate rapidly once the canyon had breached the shelf break. It is noteworthy that canyons on the Atlantic continental slope begin to assume a sinuous course once they begin their headward erosion across the shelf (Twichell and Roberts, 1982) and that the gradients of straight canyons are significantly steeper than those of sinuous ones.

Deposition on the basin floor also reduces canyon gradients by reducing the difference in elevation between the heads and mouths of canyons. Thus, both headward extension and basin filling act to reduce canyon gradients, and the two processes act synergistically as canyons evolve. Youthful canyons contained within the slope provide only small amounts of sediment to the basin because they are not in a position to intercept shelf sediment streams (Underwood and Karig, 1980) and because their tributary canyon system is poorly developed (Farre and others, 1983). As canyons evolve, however, the amount of self-generated sediment increases, along with the increasing size of the tributary system; but the amount of externally derived sediment increases dramatically once the canyon extends itself onto the shelf and is able to intercept sediment in transit. Once the canyon breaches the shelf break, the rate of longitudinal extension increases, as does the amount of sediment provided to the basin, and the process of gradient reduction accelerates.

Sedimentation in subaerial canyons that occurs because of rising base level or reduction in gradient tends to start at its mouth and work its way headward. This process of backfilling produces a fining-upward sequence, and, presumably, subaqueous canyon fills may also fine upward when sedimentation occurs because of a reduction in gradient.

Canyons might also fill if they no longer gain access to sediment coarse enough to generate turbid, erosive sediment flows, but continue to be conduits for lesser-density turbidity currents and nepheloid layers. In the Hudson Canyon, for instance, storm- and tide-generated currents intensify toward the head of the canyon where they resuspend fine sediments (Drake and others, 1978). These sediments are transported downcanyon in dilute density flows, but deceleration occurs below the pycnocline at about 500-m water depth, and the sediments settle rapidly from suspension, resulting in rapid sedimentation in the upper canyon. Clear-water currents, however, predominate in the lower canyon and rates of accumulation are very small (Cacchione and others, 1978).

These studies of the Hudson Canyon suggest that canyon filling can occur entirely with fine-grained sediments. The specific mechanism of deposition in this case, however, is triggered by a change in fluid density, rather than a change in downcanyon gradient. The canyon would fill with a progradational sequence that, except for the possible occurrence of coarse lags, slumps, and slide blocks at the base, would coarsen upward within a narrow range of grain sizes.

Slope Sequences in the Rock Record

One of the most complete descriptions of an assemblage of continental margin sediments is that of a Late Paleozoic and Mesozoic arc-trench sequence in New Zealand (R. M. Carter and others, 1978) in which a complete sequence of slope sediments occurs (Fig. 14.9). Upper trench slope deposits consist largely of fine-grained sediments, particularly siltstone. A general absence of internal structures is evident, and many massive siltstones are thoroughly bioturbated. Nonbioturbated siltstones, however, contain abundant silt and fine sandstone layers, and graded mud turbidites suggesting that these sediments were deposited by dilute turbidity currents and other smaller-velocity flows, including contour currents. Channel sediments consist of rhythmically interbedded sandstone turbidites and mudstone beds up to 20 cm thick. Channel deposits pass upward into thinly bedded turbidites that represent overbank deposits.

Mid-slope basins are represented by very thick accumulations of mudstone interbedded with thin, graded sandstone layers and ripple-bedded siltstone and sandstones. Mudstones are red, green, or black and are parallel-laminated on a millimeter scale. Sandstone beds are graded and may contain abundant intraformational detritus. They locally are convoluted, and other suggestions of mass sediment transport include the presence of resedimented sandstone and conglomerate in lensoid masses within the mudstone.

The background sediments of the lower trench slope deposits consist of hemipelagic mudstones, siltstones, and pelagic chert. Associated sediments, however, consist of traction deposits, including mud turbidites and contourites, channelized sandy turbidites, and conglomerates deposited by sediment gravity flows.

The slope sediments are overlain by a heterogeneous assemblage of shelf sediments, including bioturbated siltstones, sandy traction deposits, and thick conglomerates deposited in channels cut in other shelf sediments. They are underlain by thick sequences of sandy turbidites probably deposited as submarine fans and longitudinal trench fills at the base of the slope.

Another assemblage of continental margin sediments has been described from the Neogene Rio Dell Formation of California by Piper and others (1976), who were able to differentiate three slope facies. Upper slope sediments consist of bioturbated muddy silt and silty mud (Fig. 14.10). One unique feature of these sediments is the occurrence of lensoid masses of laminated mudstones with occasional thin-graded sand layers that appear to fill slump scars (Fig. 14.10).

Sediments of the middle slope contain assemblages of in situ middle and upper bathyal forminifera, but few displaced shallow-water fauna, suggesting that this zone was primarily one of bypassing. Sediments consist of dark, silty mudstones that may be structureless and bioturbated or that may occur in beds with irregular horizontal laminae.

Lower-slope sediments overlie deep-sea fan deposits and consist of a heterogeneous assemblage of sediments. Mudstones and muddy siltstones are the dominant lithology, but there are some graded siltstone turbidites with T_{bcd}, T_{bc}, or T_b sequences and some allocthonous mudstone blocks that were displaced by slumping from the middle and upper shelf.

A number of slope sequences rich in organic material were deposited along margins of the Pacific where upwelling occurred during the Miocene (Ingle, 1981), and the most intensely studied of these is the Monterey Formation of California. In

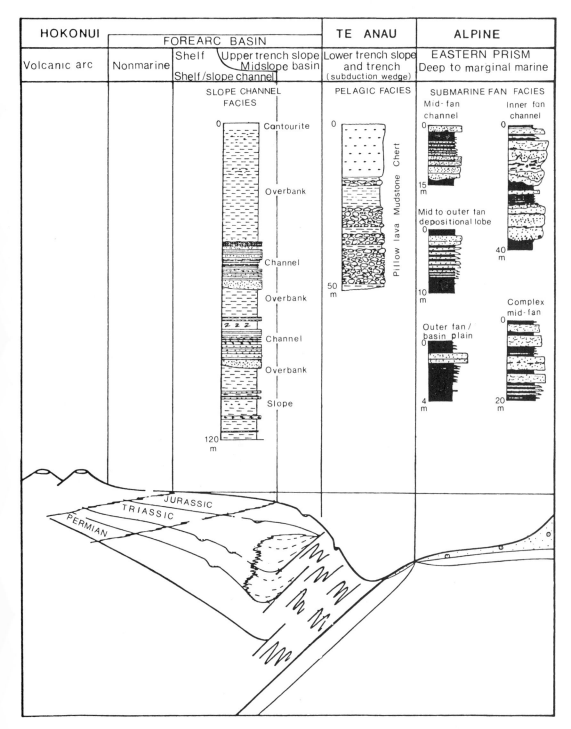

Figure 14.9 Diagrammatic cross section of the Rangitata Geosyncline of New Zealand showing the arrangement of the facies within the geosyncline and their typical characteristics at the surface (after Carter and others, 1978).

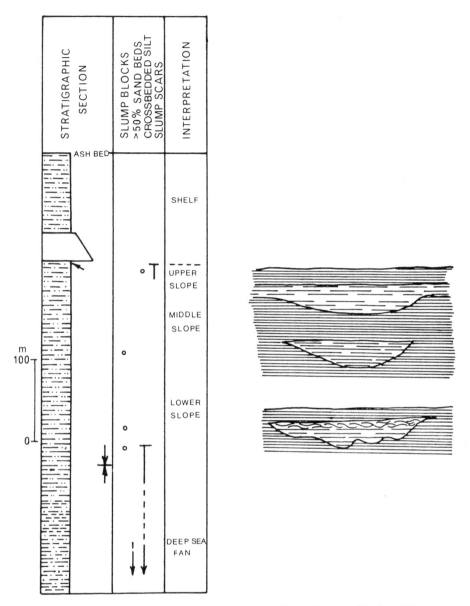

Figure 14.10 Summary stratigraphic column and depositional model of the Rio Dell Formation of California (after Piper and others, 1976).

most places the formation can be divided into three units (Pisciotto and Garrison, 1981). The basal unit consists of a calcareous mudstone in which coccoliths are prominent. Mudstones may be massive or platy, and they may contain thin-graded siltstone layers with displaced fauna or graded and crossbedded arenites with erosional bases (Fig. 14.11). This unit was deposited in newly subsided basins beneath warm, moderately fertile water masses. Suspensate settling may have been the background sedimentary process, but there was evidently some transport of sediments as gravity flows, and some of the muds may even be deposits of fecal pellets transported by turbidity currents.

Phosphate is characteristic of the middle unit, which was deposited during a period of climatic deterioration when climatic zones were compressed and strengthened oceanic circulation caused an increase in upwelling. The unit consists predominantly of phosphate that may be crudely cyclic with a basal laminated zone, a middle zone where laminae are contorted, and an upper zone where laminae are destroyed by intense deformation. Deposition occurred in an oxygen minimum layer by suspensate settling, but occasional episodes of mass sediment flow deposited conglomerate layers and graded, crossbedded sand layers with load structures at the base.

The upper zone consists of diatomites and diatomaceous mudstones. Accessory constituents included sandstones, claystones, shales, breccias, and pelletal and nodular phosphorites. This zone is especially known for its rhythmic bedding, which is present in several scales and may involve different constituents (Fig. 14.12). The most common consist of alternating beds of clastic and biogenic sediment. First-order cycles consist of millimeter-thick alternations of clay mudstone and biogenic silica caused by seasonal variations in runoff and upwelling. Second-order cycles are composed of a clastic unit consisting of graded and crossbedded sand and silt layers and a unit of siliceous mudstone. Clastic units were deposited by turbidity

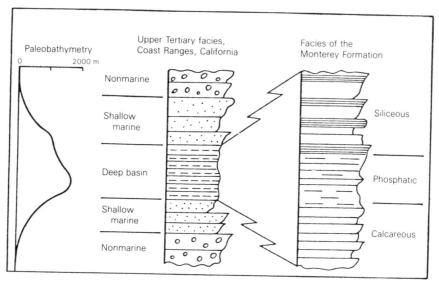

Figure 14.11 Generalized stratigraphy of the upper Tertiary rocks of the California Coast Ranges, including an expanded column of the deep-marine facies of the Monterey Formation (after Pisciotto and Garrison, 1981).

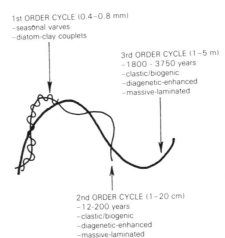

1st ORDER CYCLE (0.4–0.8 mm)
–seasonal varves
–diatom-clay couplets

3rd ORDER CYCLE (1–5 m)
–1800 - 3750 years
–clastic/biogenic
–diagenetic-enhanced
–massive-laminated

2nd ORDER CYCLE (1–20 cm)
–12-200 years
–clastic/biogenic
–diagenetic-enhanced
–massive-laminated

Figure 14.12   Causes of the various
orders of cyclicity in the siliceous facies
of the Monterey Formation (after
Pisciotto and Garrison, 1981).

currents during periods of increased rainfall and runoff, and siliceous members were
deposited during periods of reduced terrigenous input. Third-order cycles are also
composed of alternating siliceous and clastic members, but the cycles are thicker
and presumably represent climatic cycles of longer duration.

Alternating zones of massive and laminated beds occur in the Monterey Formation. Both second- and third-order cycles consist of alternating beds of laminated
and massive (bioturbated) siliceous mudstones that record temporal fluctuations of
the oxygen minimum zone in response to changes in climate and circulation patterns
and related eustatic fluctuations in sea level (Pisciotto and Garrison, 1981).

Although relatively few descriptions of sediment sequences exist from modern
canyons, numerous examples from the rock record may provide some insight into
the mechanisms involved in canyon aggradation. One such ancient channel fill occurs in the Pigeon Point Formation of California (Lowe, 1972; Howell and Joyce,
1981). The formation can be divided into two sequences, a basal one consisting of
alternating mudstone and sandstone members and an upper one that fines upward
from conglomerates at the base to mudstones at the top (Fig. 14.13).

The basal sequence was deposited on a slope and upper rise. Sandstone members in the unit are as much as 100 m thick and consist of thick beds of coarse-grained pebbly sandstone. Individual beds may be graded, and some contain dish
structures indicative of transport as grain flows. Mudstone members are as much as
500 m thick and consist of mudstone and thin layers of sandstone with sole marks
and partial Bouma sequences. Minor debris flows and slumped units that occur
within the mudstone suggest that mass movement of sediment occurred in a setting
dominated by hemipelagic sedimentation.

The overlying unit consists of three members deposited as a fining-upward
channel-fill sequence capped by a fourth member deposited on a marine shelf. The
basal member consists of well-bedded, closed-framework conglomerate that may
occur in reverse graded beds or cut-and-fill structures (Fig. 14.13). These are over-

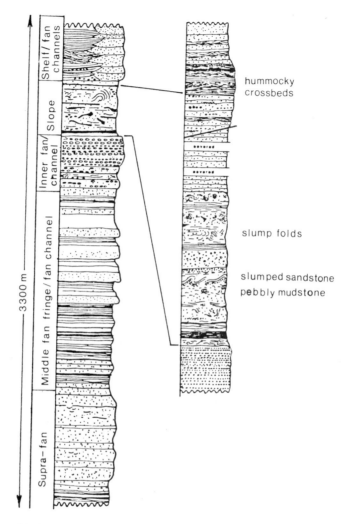

Figure 14.13   Generalized stratigraphy of the Pigeon Point Formation of California showing an expanded lithologic column of slope facies (after Howell and Joyce, 1981).

lain by interbedded and interfingering mudstones, conglomerates, pebbly mudstones, and sandstones. Most beds show some evidence of synsedimentary deformation and probably were rapidly deposited as grain flows, debris flows, and turbidites alternating with periods of hemipelagic sedimentation. Rapid deposition led to instability, which resulted in mass sediment movement.

The uppermost unit of the channel fill consists of dark gray mudstones with pebbly mudstones and contorted mudstones at the base passing upward into evenly bedded mudstone with occasional graded sand and silt layers. Hemipelagic sedimentation began to predominate as this unit was deposited, and episodes of sediment mass transport became less frequent as the canyon filled.

The aggradational sequence is capped by shelf sandstones. They are horizontally bedded and contain mudstone units and rare conglomerates near the base, but mudstones are absent in the upper part of the sequence, and the sandstones occur in crossbeds and hummocky beds.

The Eocene Torrey Canyon system is also filled with a fining-upward sequence (May and others, 1983). The basal part of the fill consists of structureless to faintly laminated sandstones probably deposited as grain flows (Fig. 14.14). They occur in amalgamated cut-and-fill structures and contain floating mudstone clasts as much as 5 m long.

The intraclastic sandstones pass upward into planar or convolute-laminated sandstones that display various types of dewatering structures. These sandstones were probably deposited by fluidized flows and may have been postdepositionally altered by downslope creep.

The upper unit consists of mudstones in channels hundreds of meters wide. The channels display complex crosscutting relationships evidently caused by lateral migration and by alternating periods of hemipelagic sedimentation, episodic erosion by turbidity flows, and reworking by indigenous currents (Fig. 14.14). Some channels are filled with interbedded hemipelagic mudstones and sandstone turbidites, and occasional fining-upward sequences may have been produced by meandering thalwegs. The diversity in channel shape and sediment fill in the upper unit was caused by relatively short term changes in type and quantity of the sediment supply, rate of eustacy relative to tectonism, and shelf hydrodynamics. The overall fining-upward sequence, on the other hand, was the result of a long-term process of eustatic sea-level rise that deprived the canyon of a source of coarse sediment.

Another fining-upward canyon-fill sequence occurs in the Paleocene Carmello Formation of California, but the facies relationships in these rocks suggest that aggradation of the canyon floor was accompanied by lateral migration of the channel (Clifton, 1981b). Both organized and disorganized conglomerates were deposited by sediment gravity flows in the channel axis. Movement of coarse sediment was episodic, however, and laminated or highly bioturbated mudstones were deposited during intervening periods (Fig. 14.15).

Sandstones were deposited on the flanks of the channel. Massive or laminated sandstones with rippled upper surfaces may have been deposited as fluidized flows, but some sandstones display partial Bouma sequences that were probably deposited by turbidity flows.

Transitional beds, consisting of thin-bedded sandstones in well-defined T_{bc} sequences were deposited on levees by tractive currents that overflowed the channel, and interbedded sandstone and mudstones were deposited in interchannel areas affected only by spillover.

During aggradation of the canyon floor, the channel migrated laterally, leaving relatively thick channel and interchannel sequences, but only thin levee deposits (Fig. 14.15). The primary factor responsible for initiating aggradation in the canyon may have been a gradual decline in the availability of coarse sediment.

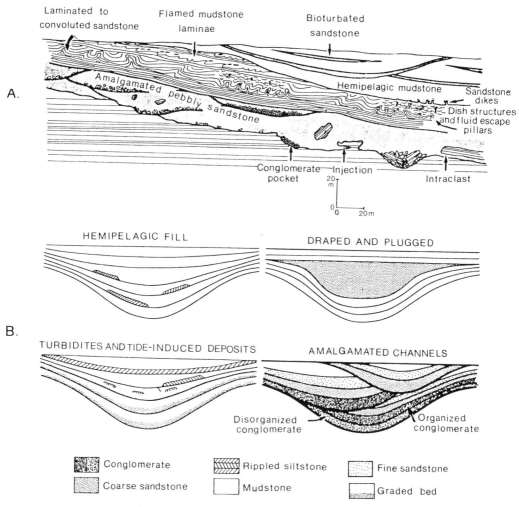

Figure 14.14 Sediments of the Eocene Torrey Canyon in southern California showing the internal organization of the fill in a tributary of the canyon (A), and the types of channel-fill sequences deposited at the head of the canyon in response to various transport processes (B) (after May and others, 1983).

Discussion: Controls on Cyclic Sedimentation in Evolving Slope Sequences

The three canyon-fill deposits described in the last chapter are all variations on a similar theme. The fining-upward sequences are the result of a change from depositional processes involving debris flows and grain flows to those involving turbidity currents and hemipelagic sedimentation. The long-term process involved in the evolution of these fining-upward sequences clearly is one that causes a diminution of

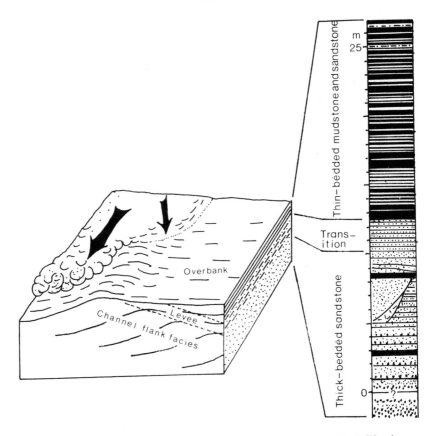

Figure 14.15 Stratigraphy of the Carmello Formation at Pebble Beach, California, with a depositional model of the three main units present (after Clifton, 1981b).

coarse sediment influx to the canyon, but other shorter-term processes may superpose cycles of sedimentation on the main sequence evolving in the canyon.

The processes by which these cycles are produced mirror similar processes that act on open slope surfaces. The cycles are the product of a complex interaction of climate, tectonism, and geomorphology that combine to affect the amount and type of sediment provided to the slope and the manner and timing of its delivery.

Climate can impose cyclic sedimentation of several scales on the slope. Yearly rainfall cycles, for instance, can cause alternating periods of large and small sediment influx, and passage of storms or alternating periods of frequent and infrequent storms can produce alternating periods of transport and deposition at the shelf break. Longer-term climatic cycles may involve alternating periods of strong and weak oceanic circulation that may affect the intensity of upwelling currents or the ability of thermohaline currents to erode or remold slope sediments. More extreme climatic fluctuations may result in eustatic raising and lowering of sea levels, as well as variations in the strength of oceanic circulation patterns. Together these two may affect patterns of sedimentation on the slope, and extreme variations may even expose the shelf to erosion or cause it to be primarily a zone of bypassing.

The evidence of these climatic variations in slope sequences is the alternation of tractive and hemipelagic sediments or terrigenous and biogenic sediments. Short-term variations may produce cycles on a millimeter scale, whereas longer-term and more extreme variations can produce cycles hundreds of meters thick, with shorter-term cycles superposed on them.

Geomorphic processes may also affect the amount of sediment provided to slopes and canyons. On active margins where shelves tend to be narrow and canyons may head far inshore, changes in coastline configuration can change the shape of the littoral circulation cells that provide coarse sediment to the canyons. As canyons extend themselves on the slope, they may become involved in channel capture and abandonment (Graham and Bachman, 1983), much like subaerial stream valleys, and when they breach the shelf break, they can intercept sediment flow on the shelf, increasing their own sediment load and causing progressive abandonment of canyons farther down the shelf.

Subtle topographic features on the shelf may influence the movement of turbid layers and cause them to flow over the shelf break at a particular place for a long period of time. The region of the slope down gradient from this place may experience large rates of sedimentation relative to adjacent regions, but changes in shelf topography may cause the flows to shift position and deliver their sediment to a different slope area, causing alternations of hemipelagic and biogenic sediments.

Tectonism can change the sediment yield in the source area, and it may change the way sediment is delivered to the slope if it changes the position of the source relative to the shelf break. It may also cause the shelf to be alternately a sediment sink and a zone of erosion or bypassing, so the slope may be cyclically provided with and starved of terrigenous sediment.

Tectonism may also cause changes in the way sediments accumulate on active margins. Slides and flows may dominate sedimentation at the slope base in trenches, but as this zone is uplifted into the accretionary wedge, it may pass into an area of the slope dominated by hemipelagic processes, resulting in a broad fining-upward sequence. Similarly, when a midslope basin is initially formed, it is a sediment trap for turbidity currents that normally bypass the area. Coarse sediment carried by these currents is deposited in the basin, but as it fills and gradients are reduced, turbidity flows may become less effective agents of sedimentation. After the basin fills, hemipelagic processes may come to dominate, and, again, a fining-upward sequence results.

These are the simplest cases relating the effect of one or perhaps two variables on slope sedimentation. The reality, however, is undoubtedly more complex; and because integrative studies of slope stratigraphy are only now being made, much of what has been discussed is based on limited data from modern slopes and by drawing comparisons with ancient slope deposits. A more complete understanding of slope stratigraphy must await the acquisition of detailed knowledge of slope sequences that presently are in the process of formation.

DEEP MARINE ENVIRONMENT

Introduction

The deep marine environment is synonymous with the open ocean. The landward limit of the environment occurs at the base of the slope and extends outward from continental margins into ocean basins that lie at depths in excess of 3000 m. Average oceanic depth is about 4000 m, but numerous positive bathymetric features exist, including mid-ocean ridges, rises, and plateaus and isolated seamounts. At active continental margins, the sea floor may descend into trenches in excess of 10,000 m deep.

In central ocean basins, much of the deep marine environment receives relatively little terrigenous sediment, except airborne clays, volcanogenic particles, and very fine clays carried in suspension by ocean currents. Shallower regions of the sea floor in central ocean basins may be covered by pelagic clays and carbonate and siliceous fossils. At greater depths, however, seawater is undersaturated in calcium carbonate, and carbonate fossils are dissolved before they can accumulate.

The depth at which the rate of dissolution of calcium carbonate exceeds the rate of accumulation is called the calcite compensation depth (CCD). It normally occurs at about 4000-m depth, but the depth can change with respect to latitude, position relative to the continents, water temperature, and the rate of organic production in surface waters. Above the CCD on plateaus and flanks of ridges and seamounts, relatively large amounts of calcite may accumulate as tests of nannofossils and foraminifers, but in deeper parts of ocean basins below the CCD, sediments accumulate much more slowly and may consist entirely of oxidized clays and siliceous fossils, especially diatoms.

Although the deep marine environment is the most extensive depositional regime, the largest part of it is beyond the influence of terrigenous sources of sediment, and accumulation rates may be as low as a few millimeters per 1000 years (Kulm and others, 1973). Rates of accumulation in basins receiving sediment from nearby landmasses, however, may be several orders of magnitude greater, and this section will concentrate on those basins, including trench systems, continental rise and abyssal plains, and borderland basins.

Two components commonly occur in these deep marine settings. Their deepest parts are generally floored by flat, nearly featureless abyssal plains, and submarine fans form at points of large sediment influx. Although their shapes may differ locally in response to variations in sediment delivery and basin shape, the processes in these two components are common to all deep marine systems.

Part VI

15

Components

Submarine Fans

Submarine fans occur along both active and passive continental margins where they form in front of submarine canyons at the base of slopes. Present understanding of sedimentary processes operating on submarine fans is derived mainly from modern fan morphology and sediment distribution, but processes involved in the evolution of fan sequences have been inferred mainly from the study of ancient fan deposits. Some inconsistencies were apparent between the two lines of study, especially where details of mid-fan morphology could not be reconciled with mid-fan depositional sequences; but as more detailed knowledge became available from studies of various fan systems, remarkably similar conclusions concerning the geomorphic dynamics of fan evolution emerged.

Sedimentary Processes and Products

A variety of fan shapes and sizes exists because of diversity in the controls exerted on their growth. Sediment yield and basin shape are the primary controlling factors, but tectonic setting and climate also influence fan morphology. Despite wide variations in fan shape, two basic types are defined (Nelson and Nilsen, 1984). Small fans, with diameters less than 100 km, tend to grow toward a radial shape where unhindered by basin obstructions (Fig. 15.1), whereas larger fans, ranging in diameter from 100 to 2500 km, tend to be elongate (Fig. 15.2).

Both fan types can be divided into three relatively distinct geomorphic terrains (Normark, 1978; Howell and Normark, 1982). The inner part of both radial and

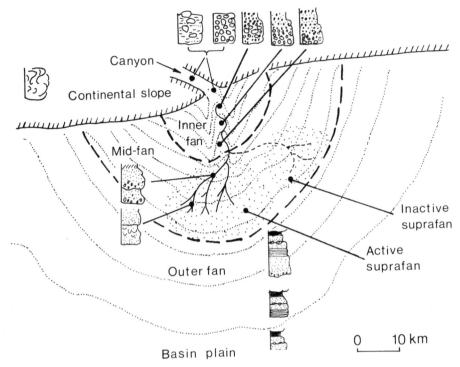

Figure 15.1 Map of a sand-rich submarine fan showing its shape, morphologic zonation, and facies distribution (after Normark, 1978; Walker, 1976).

elongate fans is characterized by the presence of one or more fan valleys with steep channel walls, well-developed levees, and steep longitudinal profiles (Nelson and Nilsen, 1984). The character of the mid-fan, however, varies with respect to fan type. Valleys on sand-rich radial fans lose their definition in the mid-fan (Nelson, 1983). Levees are poorly developed and channels are braided (Belderson and others, 1984). Channels in the middle part of mud-rich, elongate fans, on the other hand, retain their identity, with one or more channels occurring within well-developed levees. Channels tend to meander, and apparently only one is active at any one time, except perhaps during extreme flow events (Nelson and Nilsen, 1984). This channel may extend to the lower part of large fans (Damuth and others, 1983; Garrison and others, 1982), although numerous smaller channels are present that may be part of distributary or braid channel systems. The lower part of sand-rich fans, however, is characterized by a smooth, concave-upward profile that usually grades imperceptibly into the basin plain (Fig. 15.1).

The inner fan begins at the mouth of the submarine feeder canyon, but much of the sediment brought to the fan by the canyon bypasses the inner part because it is confined during transport within a deep, wide valley bounded by prominent levees. Valleys may be tens of kilometers wide on large fans (Normark, 1978), and

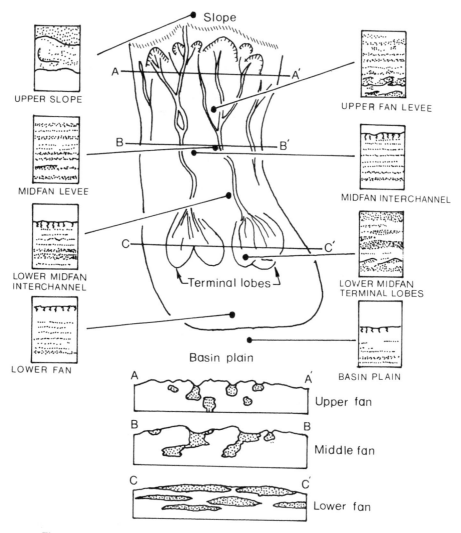

Figure 15.2 Map of a mud-rich submarine fan showing shape, morphological zonation, and facies distribution (after Stow, 1981).

valley floors may be cut by one or more smaller channels up to hundreds of meters wide. Immature sands in thick beds accumulate on the valley floor, raising it up above the level of the fan. Many of these beds are massive layers of sand or pebbly sand that may or may not be bounded by mud layers, and most are evidently the products of various types of mass sediment movement (Fig. 15.1). Undercutting of valley margins during flow events leads to slope failure, resulting in the presence of slump and slide deposits along valley margins (Nelson and Nilsen, 1984).

The upper fan channel is flanked by prominent levees. The levees are built by overbank flow from out of the channels, and they consist mainly of rhythmically laminated silt and sand turbidite layers interbedded with mud turbidites and hemipelagic deposits. Silt and sand turbidites exhibit T_{cde} sequences near the channels, but they become thinner-bedded and finer-grained away from the channel, where they grade to T_{de} sequences (Nelson and others, 1978) (Fig. 15.1).

Levees are more prominent and turbidite layers are better developed on the right side of channels in the northern hemisphere (left side in southern hemisphere) because overflows from the channel are deflected to the right by the Coriolis force. The bulk of inner fan sediments consists mainly of muds and fine sands deposited on levees and in interchannel areas, even though the focus of sediment transport is in the fan channel. Levees are so well-developed that mass sediment flows carrying coarser material are confined to the channel and thus form only a small proportion of inner fan sediments.

Channels dissipate into suprafan lobes in mid-fan areas of sand-rich radial fans (Normark, 1970). These lobes form at the terminus of inner fan valleys, and they occur as convex-upward bulges on mid-fan surfaces. The upper part of suprafan lobes is crossed by numerous low-relief distributaries filled with coarse, poorly sorted, and massive beds of sand and gravel similar to those of inner fan valleys (Fig. 15.1).

The upper part of distributaries of the middle fan may be filled with mass flow deposits, but much of the remaining area is a site of classic turbidite deposition. Nearly complete Bouma sequences are deposited farther down distributary channels, but as flow expands from the channels, the basal traction layers are progressively lost and the deposits become progressively finer grained. Channel patterns, however, are unstable because of rapid rates of aggradation on the middle fan, and the distributaries form a complex of shifting channels that may spread channel deposits of rhythmically bedded turbidites over much of the upper mid-fan area.

Channelized flow is less well developed on more distal areas of the suprafan lobe, and deposits there may consist mainly of interbedded sands and muds in T_{ede} or T_{de} sequences, and hemipelagic muds may accumulate in inactive areas of the mid-fan (Fig. 15.1).

Well-defined channels continue to confine flow in mid-fan areas of large, mud-rich fans (Nelson and Kulm, 1973). Mid-fan channels are shallower than those of the inner fan, and they show a pronounced tendency to meander. Deposits within the channels consist of sorted sands in graded beds that may display T_{ab} sequences. Overbank flows build levees of hemipelagic muds and turbidite sands with T_{cde} sequences, and interchannel areas are covered with muds and thin turbite sands showing T_{de} sequences (Nelson and others, 1978).

Nonchannelized deposits of finer-grained turbidites are the characteristic deposit of the lower part of radial, sand-rich fans. Traction units of the Bouma sequence may be deposited under the axial parts of turbidity flows, but lateral parts of these distal flows transport predominantly fine grained sediments in suspension, and outer fan sediments consist mainly of T_{de} sequences interbedded with hemipelagic muds.

Major active channels extend to the outer part of mud-rich elongate fans. Outer fan channels are sinuous and shallow, and they appear to branch out into distributaries and braided channels where they terminate in outer fan depositional lobes. In some cases, however, a fan valley may extend out onto the abyssal plain (Maldonado and others, 1985).

Channel, levee, and interchannel deposits may occur at the head of depositional lobes, but laterally continuous turbidites characterize the lower, nonchannelized parts of lobes (Nelson and Nilsen, 1984) (Fig. 15.2). Deposits consist of numerous turbidite layers with well-developed Bouma sequences (Nelson and Kulm, 1973; Nelson and Nilsen, 1974). Individual beds thin and fine downfan and grade laterally into basin plain or fan fringe deposits.

Submarine ramps are another base-of-slope deposit, but they differ significantly from submarine fans in the way sediment is delivered to them and the way it is transported across them. Submarine fans are fed by a major canyon and are characterized by a well-defined channel system. Submarine ramps, on the other hand, are fed by multiple sources, and they lack major persistent channels (Heller and Dickinson, 1985). Submarine ramps build where deltas prograde to the shelf break and deliver coarse sediment directly to slopes from multiple points along the delta front. Delta platforms are typically sandy, and they prograde onto a narrow marine shelf. They are fronted by prodelta slopes that also form the flanking slopes of adjacent basins.

The two major components of submarine ramps are proximal zones flanking the delta slopes and distal zones that grade outward into basin plains. High-density turbidity currents are the dominant process in proximal zones. These flows may locally be channelized, but they occur primarily as sheet flows. Distal zones are dominated by low-density, turbid sheet flows that are rarely channelized.

Submarine Fan Facies

Beginning with the work of Jacka and others (1968), and continuing with that of Mutti and Ricci-Lucchi (1972) and Walker and Mutti (1973), attempts have been made to understand the dynamics of fan growth through the study of ancient fan sequences with only partial knowledge of modern fan sequences. These studies have combined current knowledge of the morphology and sediment distribution of modern fans with Walther's principle, and they relied on comparisons with modern alluvial plain and alluvial fan sequences to predict the sequences that would develop during submarine fan growth and abandonment.

Walker and Mutti (1973) and Mutti (1979) recognize a number of lithofacies within ancient fan sequences that possess genetic implications, because they appear in associations coincident with the sediment assemblages that occur on modern fans (Fig. 15.3). Thick-bedded conglomerates, conglomeratic sandstones (facies A), and

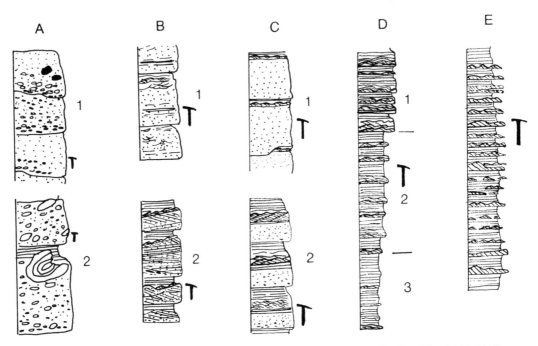

Figure 15.3 Characteristic sediment facies associated with submarine fans (after Mutti, 1979).

massive sandstones (facies B) commonly occur together in ancient fan deposits. Conglomerates may be normally or inversely graded and matrix or clast supported. Conglomeratic sandstones are well-bedded in horizontal or planar or trough cross-beds, and they may occur in laterally discontinuous, channeled beds. Shales are rare in both conglomerates and conglomeratic sandstones, and the deposits consist of amalgamated beds that were probably formed by various mass flow mechanisms.

Massive sandstones occur in thick amalgamated beds commonly separated by scour surfaces and layers of mud intraclasts. The beds tend to be parallel-sided and are more laterally continuous than those of the coarser-grained facies. The occurrence of dish structures within these beds suggests that they originated as fluidized flows.

The coarseness of these lithofacies, their probable mode of deposition, and their common association suggest that they were deposited in inner fan valleys or perhaps upper mid-fan channels. This interpretation also agrees with sediment distribution patterns of many modern fans, where the coarsest sediments are concentrated in these channels.

Another common lithofacies (facies C) consists of thick-bedded sandstones separated by thin mudstone beds. Where the sandstone beds are amalgamated, the upper pelitic interval of the Bouma sequence is missing, but thinner beds may consist of T_{a-e} sequences typical of classic turbidites. By comparison with modern fans and by their occurrence below inner fan deposits in progradational fan sequences, this lithofacies is normally interpreted as a deposit of mid-fan depositional lobes of sand-rich radial fans or distributary channels of mud-rich elongate fans (Nelson and Nilsen, 1984).

Interbedded sandstones and mudstones consisting of the upper units of Bouma sequences (facies D and E) are the most commonly occurring lithofacies within fan sequences. Where they occur below mid-fan deposits of prograding fan sequences, they are interpreted as the products of distal turbidites deposited on the outer fan. They may also be associated with thick-bedded conglomerates and sandstones, and these are interpreted as spillover deposits from channels in inner fan and upper mid-fan areas of radial fans and outer areas of elongate fans. Thinly interbedded sandstones and mudstones were probably deposited near channels or on the outer fan, but depositional sites more distal to channels are suggested by sequences where thin sandstone beds are interbedded with thicker mudstones. Where it is possible to differentiate them, the mudstones may consist of a terrigenous component (T_{et}) and a hemipelagic component (T_{ep}).

Intrinsic and Extrinsic Controls on the Evolution of Fan Sequences

It is apparent that the assignment of specific depositional sites for these various lithofacies is based not only on sediment distribution on modern fans, but also on the sequential arrangement of the facies in stratigraphic succession. Detailed analysis of these ancient successions led to the development of three fan models in which vertical trends in facies characteristics and associations were linked to processes of fan evolution (Mutti, 1979; Nelson and Nilsen, 1984). Variations in these models appear to be directly related to the type of sediment being supplied to a fan.

Outer fan or fan fringe areas are not well-developed on sand-rich fans. Deposition, instead, is concentrated on a mid-fan depositional lobe characterized by a complex of shifting braid or distributary channels. A progradational sequence deposited by such a fan would consist of a basal unit of basin plain hemipelagic mudstones interbedded with occasional sand turbidite layers (Fig. 15.4). The basal unit would coarsen and thicken upward with the addition of more numerous sandstone beds, but the transition into amalgamated sandstone beds of the suprafan would be abrupt. Suprafan deposits would consist mainly of channelized thinning and fining-upward cycles deposited in response to rapid establishment of channels, followed by gradual abandonment in an environment characterized by unstable channel configurations.

Mixed fans are composed of a mix of grain sizes. Mutti (1979) refers to these as efficient fans where mud-rich turbidity currents are capable of transporting sand well out into the basin. For this reason, fan fringe and outer fan deposits are important components of mixed fan sequences.

The mixed fan model, developed initially by Mutti and Ricci-Lucchi (1972), envisions a basic coarsening-upward progradational sequence in which sandstone bed thickness and grain size increase upward. Superimposed on this basic sequence,

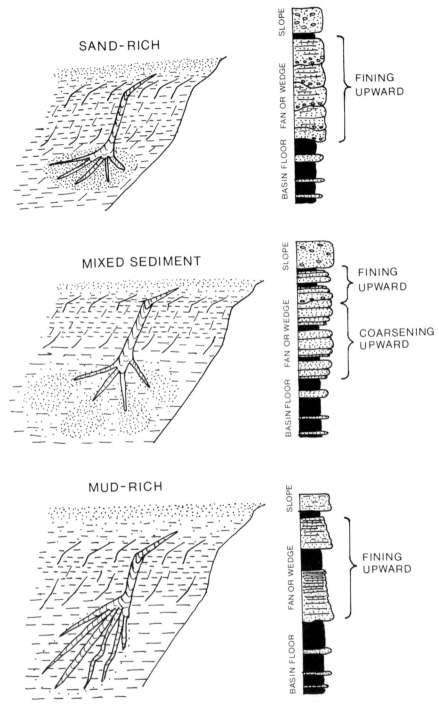

Figure 15.4 Models of stratigraphic sequences deposited by sand-rich, mixed-sediment, and mud-rich fans (after Nelson and Nilsen, 1984).

however, are thinner coarsening-, thickening-upward sequences and sequences where sandstone bed thickness and grain size decrease upward (Fig. 15.4).

Basin plain and fan fringe deposits, consisting of mudstones interbedded with thin sandstone beds, occur at the base of the sequence. Sandstones thicken and become slightly coarser grained upward in response to progradation of outer fan facies over basin plain sediments as the fan grows.

The outer fan deposits consist of a series of coarsening-upward, thickening-upward sequences deposited by individual prograding depositional lobes. Sand beds thicken and coarsen upward as more proximal parts of the lobe prograde over more distal ones or where a lobe shifts its direction of growth and axial parts of the lobe migrate over lateral ones (Fig. 15.4).

Sandstone beds of the first thickening- and coarsening-upward sequences that appear in the section normally are parallel-sided and are interpreted to be the deposits of the unchannelized portion of the lobes. With continued fan progradation, however, the channeled inner portion of the lobe migrates over the smoothed outer fan. Sandstone beds thicken considerably and begin to occur in fining-upward channel-fill sequences. Sandstone beds at the base of these sequences may consist of thick, amalgamated grain flow deposits; but as the channel aggrades and carries progressively smaller flows, grain size diminishes, and beds thin and become interbedded with mudstones. If channel abandonment was gradual, the top of the channel fill may consist of interbedded sandstones and mudstones deposited first on levees and then in interchannel areas. Avulsion, however, may cause an abrupt change in depositional processes so that channel-fill facies may change upward from thick, amalgamated sandstone beds to thinly bedded, fine-grained turbidites and hemipelagic muds.

The top of the coarsening-upward sequence is occupied by thinning- and fining-upward channel-fill sequences of the inner fan. Basal beds consist of coarse-grained debris flow and grain flow deposits that become thinner and finer-grained upward as the channel aggrades. Continued aggradation, however, produces instability, and when the channel is abandoned, the remainder of the sequence is completed with levee and interchannel deposits consisting of interbedded sandstones and mudstones (Fig. 15.4).

Mud-rich fans are not well-known from the rock record, but it is expected that they would consist of thinning- and fining-upward cycles deposited by channel–levee complexes that extend the length of the fan and out onto basin floors (Nelson and Nilsen, 1984) (Fig. 15.4). A prograding sequence deposited by such a fan might consist of basin plain hemipelagic mudstones and sandstone turbidites overlain by a series of channelized thinning- and fining-upward channel sandstones enclosed by interchannel deposits (Fig. 15.4).

The primary distinction between submarine ramps and submarine fans is the absence of a channelized inner and middle fan. Ramp sediments consist mainly of thick, massive beds of turbidite sandstone typical of facies B of Mutti and Ricci-Lucchi (1972, 1978) (Heller and Dickinson, 1985). Bed thicknesses are generally random, although rare asymmetric thickening- and thinning-upward cycles suggest local buildup and abandonment of depositional lobes near the base of slope. An overall coarsening- and thickening-upward sequence, however, is produced during ramp progradation.

The basic coarsening-upward sequence described above is the depositional product of a prograding fan. The pattern of sedimentation that results in such a sequence is dependent on the complex interaction of a number of variables, and any variation in the way these operate during fan growth will alter this simple sequence.

A retrogradational sequence can develop, for instance, where there is a gradual decline in sediment yield, perhaps caused by eustatic or tectonic rise in sea level, change in the location in the feeder channel, tectonism in the source area, or a change in climate (Howell and Normark, 1982). In such a case, the basic sequence would fine upward, with the coarse-grained canyon and inner fan sediments at the base overlain by progressively finer grained middle and outer fan deposits (Fig. 15.5).

The minor cycles, however, might respond differently. Channel abandonment would probably continue to occur on a retrograding fan, and thinning- and fining-upward sequences would continue to be deposited in inner and mid-fan channels,

(A) PROGRADING   (B) RETROGRADING

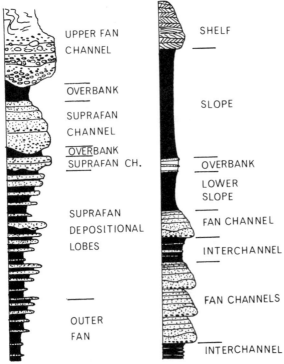

Figure 15.5 Models of stratigraphic sequences deposited by prograding and retrograding submarine fans (after Walker, 1976, reprinted with permission of the Geological Association of Canada; Ingersoll and Graham, 1983).

and thickening- and coarsening-upward sequences would be deposited by mid-fan or outer fan depositional lobes. In sequences deposited by retrograding fans, the channelized sands of the lobes would occur lower in the sequence than the parallel-sided turbidite sands of the outer, smooth portions of the lobes. Fan fringe deposits would thin and fine upward in response to fan retrogradation.

The Balleny Group of New Zealand was apparently deposited by a retrograding fan (Carter and Lindqvist, 1977). Conglomerates and conglomeratic sandstones at the base of the fan sequence overlie shallow-water sandstones and are, in turn, overlain by channelized sands in thinning-upward sequences (Fig. 15.6). The channelized sandstones are overlain gradationally by nonchannelized sandstones in a sequence that fines and thins upward, but within this sequence there are a number of smaller thickening- and coarsening-upward cycles. The fining-upward sequence is capped by thinly interbedded sandstones and argillaceous chalk marls (Fig. 15.6).

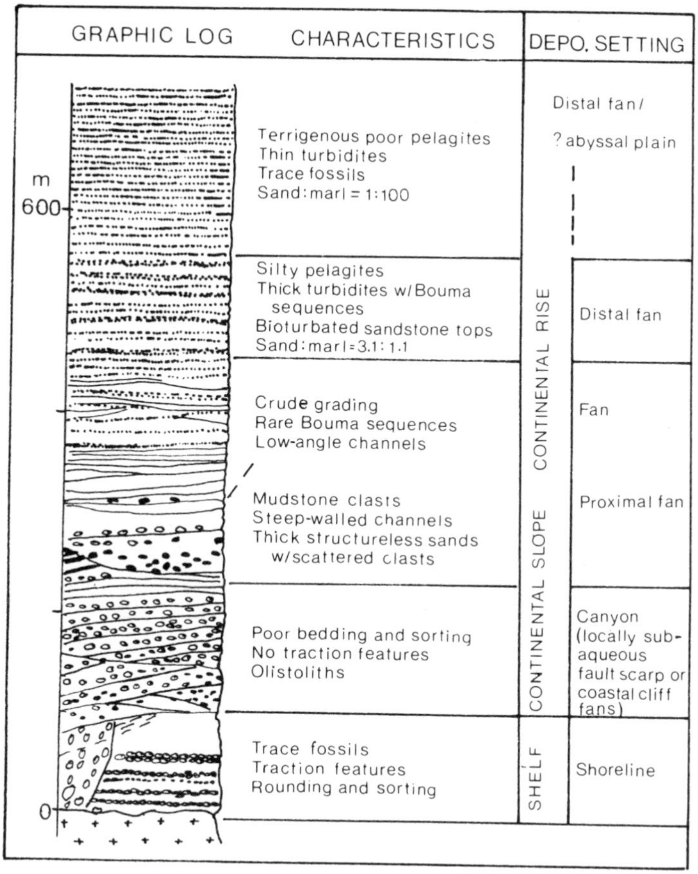

Figure 15.6 Retrogressive sequence of deep-water sediments in the Balleny Group of New Zealand (after Carter and Lindquist, 1977).

Fans may also be abandoned rapidly when the sediment yield is abruptly cut off. The Holocene sea-level rise has left many fans without sources of coarse sediment, and these fans are currently being covered by pelagic and hemipelagic muds (Kelling and Stanley, 1976). The Navy Fan of San Diego, for example, is fed by canyons that presently head on the outer shelf and upper slope. The entire fan was probably built during the last glacial interval when the shoreline was closer to the canyon heads, and since the postglacial sea-level rise, the fan has been provided mostly with terrigenous muds (Normark and Piper, 1972).

Size of turbidity flows may change in response to sea-level changes. On Navy Fan, turbidity currents were small but relatively frequent events during Pleistocene low stands, and suprafan deposition predominated (Piper and Normark, 1983). After Holocene sea-level rise, however, the fan received its sediment through a shelf-edge canyon that carried little sand. Large, muddy turbidity currents predominated during the Holocene, and because these are subject to flow stripping, turbidites are best developed away from the presently active channel.

Response of submarine fans to changing sea level is complicated by the control exerted on intrinsic fan processes by changes in sediment yield. Yield to fan and basin, especially of coarse grades, is at a maximum during falling sea level when source area streams are rejuvenated, slope canyons are reactivated, and shelves are eroded (Nardin, 1983). Deposition on suprafans is favored at this time, because turbidity flows tend to be coarser grained and relatively inefficient. Fan progradation may also be important at this time, however, because large flows, capable of long-distance transport onto lower fan and basin areas, recur more frequently during periods of low sea level. The potential, therefore, exists at this time for initiation of both suprafan and outer fan lobe deposition.

Sediment yield is smallest during rising sea level as coastal sediment traps form and shelves widen. Two conditions may prevail on fans during low sea levels. Fans may become inactive if they are cut off from sediment sources, but where feeder canyons continue to receive sediments, fans may remain active, although the locus of deposition may shift (Nardin, 1983).

Underlying basement structures may also control fan growth and place constraints on the way fan sequences evolve. The positions of the main feeder systems for the La Jolla fan system, for instance, are controlled by the presence of structural troughs or ridges that prevent lateral migration of channels (Graham and Bachman, 1983). Multiple upper fan channels also alter the pattern of simple radial growth and facies distribution normally associated with sand-rich fans. With further growth, these channels may produce interleaving sediment wedges, but fault-controlled channel patterns are presently producing multistory linear sand bodies encased in fine-grained sediment.

Intrinsic mechanisms, such as canyon piracy, can dramatically alter fan growth patterns by increasing the sediment delivery to one area of the fan and causing others to be abandoned. The bulk of the Monterey fan, for instance, was built with sediment provided by the Ascencion and Monterey Canyon systems. Sometime during the Quaternary, however, a tributary of the Monterey Canyon intercepted the Ascension Canyon by headward erosion, causing the lower part of the canyon to be abandoned. In addition, the tributary became the main valley of the Monterey system, forcing that part of the Monterey Canyon above the tributary to be abandoned (Normark, 1970; Normark and Hess, 1980). The change in major feeder systems has caused abandonment of much of the northeast part of the fan by redirecting sediment mainly to the south.

Avulsion may also shift fan depocenters, and when this occurs, the way in which turbidity currents transport and deposit their load may also be altered. When nodal avulsion of a fan channel occurs, an abrupt bend in the channel may result. Lower, coarse-grained parts of flows encountering such a bend may change directions with the channel, but upper, finer-grained parts may overflow the channel and continue downfan in a straight line (Piper and Normark, 1983). Flows that are stripped of their fine-grained component decelerate quickly, and the coarse load is deposited in active channels. The fine-grained component, however, eventually is deposited on the "inactive" part of the fan.

Flows of different sizes respond in different ways to this process. Small flows may remain confined to the channel during their downfan transit. Such flows, triggered mainly by processes such as slope failure and storm episodes, are the most common flows on fans (Piper and Normark, 1983), and they may be most responsible for lobe construction in front of newly established channels. Larger flows are triggered by seismic events. They occur less frequently, but because of their magnitude, they are primarily responsible for fan morphology. Large muddy flows are capable of channel spillover, and they actively aggrade upper mid-fan levees and abandoned fan segments. Large sand-rich flows, on the other hand, can erode levees and deposit much of their sediment in distal environments away from active lobes.

Fan abandonment may occur where gradient changes or increases in sediment yield cause incision of the main leveed valley even of sand-rich radial fans. When this occurs, sediment bypasses the middle and upper fan through the deepened canyon, and new depocenters are established farther basinward. Sediment currently bypasses the inner and middle parts of the La Jolla fan, for instance, because the main feeder valley was incised into the fan in response to Pleistocene sea-level changes (Normark and Piper, 1972).

Profound changes in sediment yield to fans, or at least major parts of fans, may cause abrupt rather than gradual changes in sedimentary processes. Mid-fan and upper fan channel sequences, for instance, might consist only of coarse channel deposits overlain by hemipelagic muds and fine-grained interchannel turbidites, and the transitional elements deposited during gradual channel abandonment would be missing. Similarly, sequences of outer fan facies would consist of incomplete coarsening- and thickening-upward sequences overlain by hemipelagic muds and distal turbidites.

Submarine Fan Sequences in the Rock Record

Link and Welton (1982) attribute the Eocene Matilija Sandstone of southern California to a major regressive event in which sand-rich submarine fans prograded over basin plain sediments and were subsequently covered by shallow marine and nonmarine deposits of a major delta complex. Fan fringe and outer fan depositional lobe facies are poorly developed, and the primary locus of deposition, instead, appears to have occurred on a braided suprafan.

The lower part of the formation consists of alternating sandstone and shale sequences deposited on depositional lobes and fan fringe settings and in interlobe areas. Lobe fringe and interlobe deposits consist of thin sandstone beds randomly interbedded with finely laminated or bioturbated mudstones. Depositional lobes are represented by apparent thickening- and coarsening-upward sequences as much as 7 m thick. Together, these deposits overlie basin plain sequences of the Juncal Formation consisting of thin-bedded, noncyclically arranged beds of mudstone and thin sandstone (Fig. 15.7).

Inner and midfan facies make up the bulk of the formation and consist of channel and interchannel deposits. Channel deposits consist mainly of facies A or B sandstones in laterally discontinuous beds with channelized bases. They occur in apparent thinning- and fining-upward megasequences separated by thinner sequences of interbedded mudstone, siltstone, and sandstone deposited on channel margins and in interchannel areas (Fig. 15.7).

The fan facies of the Matilija Sandstone are dominantly composed of sand. Although depositional lobe deposits apparently occur within outer fan sediments, they are minimally developed compared to the much thicker channelized deposits of the mid-fan and inner fan. These data suggest that the formation was deposited on a sand-rich, inefficient fan system characterized by extensive development of braided aggradational midfan channels.

It should be noted that Heller and Dickinson (1985) have substantive disagreements with the interpretation of Link and Welton (1982) of the Matilija Sandstone as a submarine fan deposit. They point out, for example, that a statistical analysis of the bedding characteristics of the unit shows no preferred packaging of sediments in the middle or lower parts of the succession. The lack of clearly defined bed thickness or grain size trends in the Matilija Sandstone and its association with a major delta complex suggests to Heller and Dickinson (1985) that the unit, instead, may represent deposits of a submarine ramp. They recognize, however, that the presence of channel-fill sandstones require some modification of the ramp model (Fig. 15.8).

In contrast to the sand-rich deposits of the Matilija Sandstone are the turbidite facies of the Eocene Hecho Group of northern Spain, which were deposited on mixed sediment fans fed through channels incised into a prodelta slope. Rather than possessing one dominant inner fan channel, the surface of the fans were incised by numerous smaller distributaries that linked the delta slope with the outer fan (Mutti, 1977; 1983–1984).

Inner fan channel deposits consist mainly of coarse-grained, thick-bedded sandstones in stacked channel fills marked by scoured contacts (Fig. 15.9). Amalgamated sandstones are overlain gradationally or abruptly by levee and interchannel deposits forming thinning- and fining-upward sequences deposited during phases of channel abandonment. Levee deposits consist of thin-bedded sandstones and mudstones, and interchannel sequences consist dominantly of mudstone with subordinate thin sandstone beds.

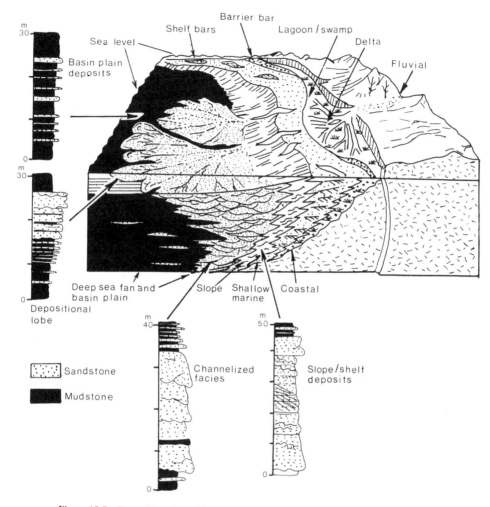

Figure 15.7 Depositional model for the Matilija Sandstone of southern California showing typical stratigraphic sequences and their three-dimensional arrangement in a deep-sea fan complex (after Link and Welton, 1982, Fig. 16, reprinted with permission of the American Association of Petroleum Geologists).

A unique feature of this fan system is the occurrence of sand bars at the mouths of channels. Sand beds in these bars do not display Bouma sequences, but instead show an upward progression from plane beds through large-scale cross-lamina to ripple cross-laminae. The beds apparently accumulated as traction deposits, but without substantial fallout, and they probably represent reworked turbidite sands.

Outer fan deposits are the best-developed facies of the Hecho Group. They form a very thick sequence of nonchannelized sandstone bodies and associated thin-bedded deposits arranged in thickening- and coarsening-upward cycles deposited as

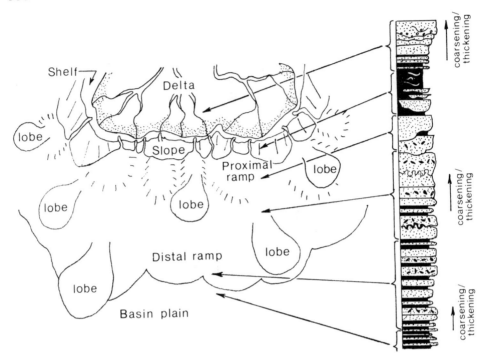

Figure 15.8 Model showing how the various segments of a typical depositional sequence of the Matilija Sandstone relate to the various facies tracts of a shelf-edge delta–submarine ramp component (after Heller and Dickinson, 1985, reprinted with permission of the American Association of Petroleum Geologists).

lobes prograded over fan fringe sediments. The fan fringe deposits consist of mudstones and thin sandstones formed from dilute turbidity currents. The sand beds of the outer part of the depositional lobes are parallel-sided, but thicker beds deposited in the inner part of the lobe have channeled bases.

The fan prograded into an east–west trending elongate basin filled with a uniform sequence of alternating thin sandstone, mudstone, and hemipelagic marl beds. The basin filled from the east, producing a consistent westward decrease in turbidite bed thickness, and since the basin axis occurred nearer the north side, hemipelagic deposits are substantially thicker along the southern margin.

Both the Matilija Sandstone and Hecho Group are progradational sequences, but the late Precambrian Konigsfjord Formation of northern Norway represents a broadly retrogradational sequence deposited by a mixed sediment fan. Inner fan deposits occur at the base of the formation. They consist mainly of poorly sorted pebbly sandstones in amalgamated beds separated by channeled bases. They were deposited by a variety of gravity-flow processes in inner fan channels, where they accumulated in fining- and thinning-upward sequences separated by thin-bedded

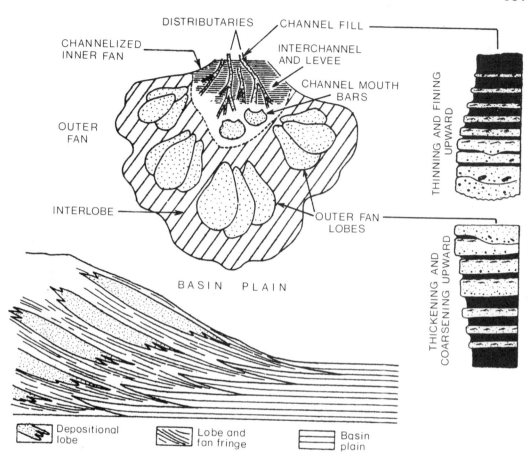

Figure 15.9 Depositional model for the Hecho Group of northern Spain showing distribution of major fan elements in plan view and cross section, and typical stratigraphic sequences deposited at proximal and distal locations on depositional lobes (after Mutti, 1977; 1983–1984).

mudstones and siltstones deposited as overflows in interchannel areas (Pickering, 1981a) (Fig. 15.10).

The middle portion of the formation consists of sandstones in medium- to thick-bedded turbidite layers interbedded with thin-bedded siltstones and mudstones. Individual turbidite layers are separated by channelized contacts commonly marked by intraclastic lags, and they form thinning- and fining-upward channel-fill sequences as much as 40 m thick. Interchannel deposits consist of thin-bedded mudstones and siltstones deposited as overbank flows in sequences as much as 25 m thick. Occasional thicker-bedded and coarser-grained sandstone beds are found within the interchannel facies, and they probably represent crevasse splay lobes or channels (Fig. 15.10).

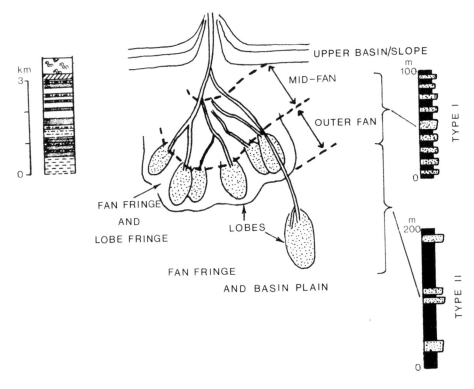

Figure 15.10 Diagrammatic plan view of the Konigsfjord Fan system showing the two types of outer fan lobes that developed during fan evolution, and the stratigraphic sequence deposited by the fan system (after Pickering, 1981a, 1981b).

Outer fan deposits constitute the bulk of the Konigsfjord Formation. Deposits consist of nonchannelized, sheetlike beds of turbidite sandstones in thickening- and coarsening-upward sequences formed by prograding lobes. Occasional symmetrical cycles that thicken and coarsen upward, then thin and fine upward probably record episodes of lobe progradation and retreat (Fig. 15.10).

Two types of depositional lobes occur within outer fan facies (Pickering, 1981b). Both types range from 2.5 to 15 m thick, but the thickness and character of the associated facies are quite variable (Fig. 15.10). Associated with one lobe type are fine-grained, thin-bedded fan fringe deposits in packets 3 to 5 m thick. Deposition of this lobe type was probably controlled by regularly recurring processes intrinsic to the mid-fan, such as channel migration and avulsion. In contrast, the associated deposits of the other lobe type are finer-grained and thicker and probably were deposited on more distal parts of the fan where lobes prograded only sporadically. Progradation of such lobes may have been triggered by fluctuations in sediment yield to the fan, as suggested by Nardin (1983).

The Konigsfjord Formation forms a broad retrogradational sequence, with basal inner fan facies being successively overlain by mid-fan and outer fan sediments. The process responsible for the retrogradation was a decrease in the rate of fan growth during a period of basin subsidence.

Basin Plains

Basin plains are the ultimate resting place for most terrigenous deep-sea sediments. They are found near continents, and the largest occur along passive margins in front of major drainage systems. Their chief distinguishing feature is a flat floor, which is a product of depositional smoothing of an underlying irregular topography. Small plains, such as those that occur in borderland basins may be only a few hundred square kilometers in extent, but large ones cover thousands of square kilometers, and a segmented basin plain occurs along almost the entire margin of the North and South Atlantic oceans (Heezen and Hollister, 1971).

Sedimentary Processes and Products

The sediments that form the plains consist of pelagic and hemipelagic muds in part, but deposition of these sediments is only a background process to the turbidity flows that are the primary cause of basin plains. In fact, the boundaries of a basin plain are coincident with the extent of the widespread distribution of turbidity flows (Horn and others, 1972).

Sand and silt beds are only a few millimeters thick, but mud beds are much thicker. Some muds are the result of pelagic and hemipelagic sedimentation, but a significant proportion of the muds is deposited from the dilute tails of turbidity flows.

Terrigenous sediment is usually delivered to the plains through submarine canyons that cut across fans and rises. Once turbidity currents reach the plain, they flow over gradients as low as 1:1000, but even subtle topographic variations may direct the currents to various depositional sites on the plain because basinwide correlation of turbidite layers has been found to be impossible (Horn and others, 1971; Rupke, 1975). Instead, the turbidite fill of the basin consists most likely of a series of overlapping flows interbedded with layers of pelagic and hemipelagic muds deposited from suspension or from slow-moving, dilute nepheloid layers.

In addition to topography, the distribution of turbidite sands in deep ocean basins is dependent on sediment grade, frequency of tectonic activity, size of source area, number and distribution of sediment point sources, and size and geometry of the basins (Pilkey and others, 1980) (Fig. 15.11).

Accumulation of abundant thin beds with limited lateral continuity is favored where tectonic activity is frequent because large bodies of unstable sediment are unable to form at basin entry points. Thick, laterally extensive beds are deposited in basins experiencing relatively little tectonic activity and where the size of the

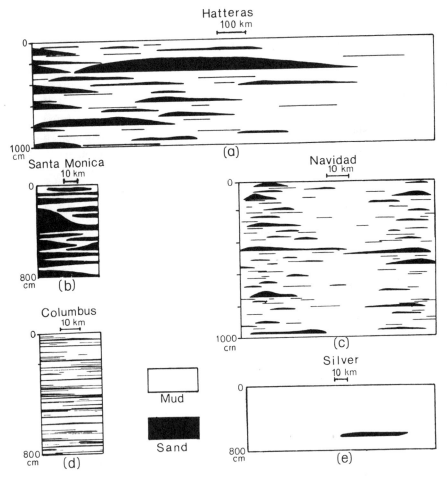

Figure 15.11 Shape of sand bodies on deep marine basin plains as a function of relative size of source area and basin plain, frequency of seismic events, and the number and distribution of sediment sources: (a) large basin (Hatteras) with single source; (b) small basin (Santa Monica) with numerous sources and frequent seismic activity; (c) small basin (Navidad) with large source area and frequent seismic activity; (d) small basin (Columbus) relative to size of source area with several point sources; and (e) large basin (Silver) supplied only with mud (after Pilkey and others, 1980, reprinted with permission of the American Association of Petroleum Geologists).

source area is large relative to that of the basin floor. If there are numerous points of entry, however, sand body continuity is reduced, but their abundance increases.

In some basins, turbidity flows are confined to channels. Sand is largely confined to a channel and its levees in the Labrador Sea, for instance (Chough and Hesse, 1980). The only sediment that regularly reaches the basin floor away from the channel is largely derived from body spillover. Head spillover from the channel, which could contribute sand to the basin, is a relatively infrequent event.

Dispersal patterns may be hard to discern in basin plains with multiple sediment sources, but where correlations can be made confidently, a number of features can be used to define proximal, distal, or lateral relationships. Turbidite beds are thickest and coarsest along the axial trends of the flows and closest to their point of origin. The beds may vary greatly in thickness and show only a slight tendency toward grading (Horn and others, 1972). Well-graded coarse silts are deposited farther along primary flow paths and thinly bedded, graded fine silts are deposited laterally to the main flow or in distal parts of the plain.

Proximal–distal relationships are enhanced where source areas are small relative to basin size, especially where sediment sources occur exclusively at one point on the basin. In such situations, sand bodies thin and become less numerous with increasing distance from sediment sources. Mud layers may also predominate over the entire basin where turbidity currents are largely mud rich.

Sediment thicknesses vary considerably within basins with respect to distance from a sediment source and because of irregular bottom topography. Thicknesses vary among basins because of variations in sediment yield and the age of the basin. Thicknesses of as much as 300 m are not uncommon, and thicknesses in excess of 1000 m have been recorded in some basins (Horn and others, 1972).

Basin plains may be primary, secondary, or relict (Rupke, 1978). Primary basins are the most numerous and include those basins that receive sediment directly from a terrigenous source. A secondary basin plain is one that is isolated from a primary sediment source and instead receives its sediment through an abyssal gap that connects it to another basin. Sediments in secondary basins tend to be very fine grained because primary basins trap most of the coarse sediment. The Silver Abyssal Plain, for instance, is covered by mud turbidites and hemipelagic clays because it receives sediment through the Vema Gap only after it has traversed the Hatteras Abyssal Plain (Fig. 15.11). The sand content, on the other hand, may be relatively large where transport paths are short. An appreciable amount of sand occurs in some of the outer basins of the California Borderland (Gorsline, 1981), because transport paths from sediment sources on the shelf are only tens of kilometers long even though the paths traverse several basins. Secondary basins may also begin to receive coarse sediment after the primary basin has been aggraded to the level of the sill separating the two.

Relict plains are those that were once active but have become isolated from any source of terrigenous sediment. Such plains are relatively rare because the circumstances necessary for their formation are complex. The Aleutian Abyssal Plain, for instance, is underlain for the most part by a thick turbidite fill; but it has been cut off from its sediment source by the formation of the Aleutian Trench, and almost 100 m of pelagic sediments cap the basin-fill sequence (Horn and others, 1972; Kulm and others, 1973).

Basin plains may also be classified according to their geomorphic expression, which in many cases is a reflection of their tectonic setting. Large basin plains in open oceans are most common along passive margins where they tend to occur in broad, elongate troughs bounded by the continental rise and the flanks of the mid-ocean ridge. Arc trenches are narrow, elongate plains associated with subducting margins. Although they are narrow, arc-trenches may extend for thousands of kilometers, and because uplifted source areas can provide large sediment yields, trenches can accumulate large volumes of sediment.

Enclosed basin plains vary greatly in shape and size, but they tend to share in common a complex sediment dispersal system. The Sigsbee Abyssal Plain, for instance, covers about 150,000 square kilometers in the central Gulf of Mexico. It receives a large contribution of terrigenous sediments from the Mississippi Delta and the Texas–Louisiana shelf, but it also receives abundant carbonate detritus from the Campeche Shelf. Underneath the plain, these sediments are interbedded (Fig. 15.12).

Even small basins may be provided with sediment from a number of different sources. The San Nicholas Basin is a secondary basin in the California Borderland. Turbidite flows enter the basin across sills at the western and northern ends of the basin, as well as from canyons cut into the slopes off San Nicholas, San Clemente, and Santa Barbara islands. In addition, hemipelagic sediments are provided by dilute plumes originating on the California shelf (Gorsline, 1981).

Terrigenous sediment yield can be variable to enclosed basins, and where it is small or where organic productivity in the surface waters is great, (hemi) pelagic sediments may comprise an appreciable proportion of the basin fill. The Balearic Abyssal Plain in the Western Mediterranean is somewhat larger than the Sigsbee Plain, but the amount of terrigenous sediment supplied to it is much smaller. In fact, nearly half of the late Pleistocene and Holocene sediments under the plain consist of pelagic sands and (hemi) pelagic muds (Rupke, 1975).

The (hemi) pelagic muds are poorly sorted and show no systematic vertical grain size trends. There is a large proportion of pelagic microskeletons within the muds, and they are ubiquitously burrowed. Most significant, however, is the fact that layers from widely varying core locations and from varying depths in cores share similar characteristics, suggesting that they form a background process only occasionally interrupted (less than 2 every 1000 years on average) by turbidity flows.

Like some other enclosed basin plains, the Balearic Abyssal Plain is relatively shallow and the floor is affected by bottom currents. They act to winnow mud from (hemi) pelagic layers and concentrate skeletal remains into millimeter-thick layers.

Extrinsic Controls on Basin Plain Sequences

Lithofacies generally show only subtle lateral variations within a basin depending on distribution of sediment sources, basin morphology, and composition of the incoming sediment (Rupke, 1978). More abrupt vertical variations can occur, however, in response to variations in sediment influx. Climate, for instance, may play a large role in determining the quantity and caliber of sediment delivered to a basin. During heavy rainfall periods, the yield of terrigenous sediment might increase and

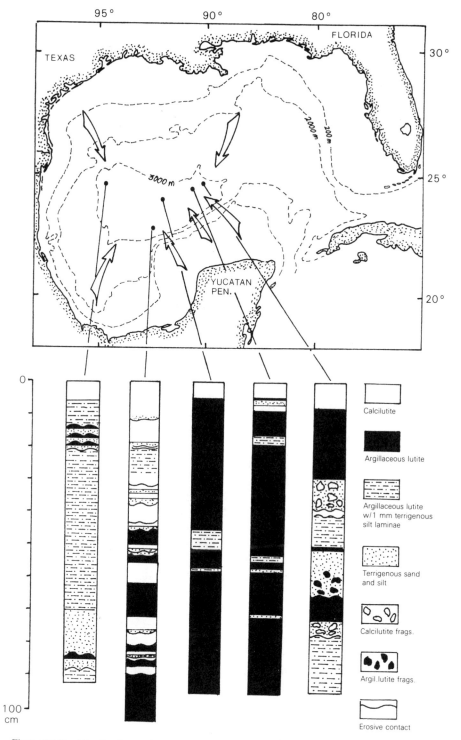

Figure 15.12 Distribution and provenance of sediments of the Sigsbee abyssal plain (after Davies, 1972, reprinted with permission of the American Association of Petroleum Geologists).

reduce the proportion of pelagic and hemipelagic sediments being deposited, and the resulting basinal deposits during that period might consist largely of sand and silt turbidites.

Tectonic or eustatic sea-level changes also might affect sediment yield. Many submarine canyons have been cut off from sources of coarse sediment by the post-glacial sea-level rise (Kelling and Stanley, 1976). Turbidites deposited in abyssal plains during the Holocene have been thinner and finer-grained than their Pleistocene counterparts, and the frequency of turbidite flows has decreased by an order of magnitude (Horn and others, 1972).

Tectonism may also cause abrupt vertical variations in abyssal plain sequences. Growth of the Aleutian Trench has isolated the Aleutian Abyssal Plain from sources of terrigenous sediment and the basal turbidite sequence is capped by sediments consisting primarily of pelagic material. Block faulting can also isolate basins from sediment sources by raising sediment dams or it can create new sediment sources for isolated basins by uplifting blocks. All these changes would be manifested in basin-fill sequences by variations in the relative proportion of terrigenous and pelagic sediment and by variations in composition and characteristics of the turbidite layers.

Basin Plain Sequences in the Rock Record

There are numerous examples of ancient basin plain sediments, but, for the most part, the study of these rocks has been ancillary to the study of the coarser-grained fan sequences with which they are associated. The characteristics usually cited for their recognition include the rhythmic interbedding of thin sandstones and thicker mudstones, lack of obvious cyclicity in addition to the coarse–fine couplets, lateral continuity, at least on an outcrop scale, parallel-sided beds, and lack of evidence of sediment mass movement.

These studies, with few exceptions, were also aimed at analyses of vertical sequences and not at basinwide analysis of sediments. One exception to this is the study of the Miocene Marnoso-arenacea Formation of the northern Apennines by Ricci-Lucchi and Valmori (1980), who defined the dispersal pattern of turbidites over an area of approximately 3600 km².

Basin-fill sediments consist primarily of thick- and thin-bedded turbidites, with less than 20% of the total sediment thickness represented by hemipelagic sediments. Thin-bedded turbidites possess a small sandstone-to-mudstone ratio and consist only of the upper units of Bouma sequences. Thick-bedded turbidites, on the other hand, include traction deposits of Bouma sequences with the b division dominating the thickest sandstones.

The main sediment source was at the northeast end of the elongate basin, and most of the thick-bedded turbidites occur there in submarine fan successions. Total sand content decreases away from the proximal end of the basin, and thin-bedded turbidites predominate throughout the remainder of the basin (Fig. 15.13). Thick-bedded turbidites occur only sporadically away from the basin margins, but those

SANTERNO V.

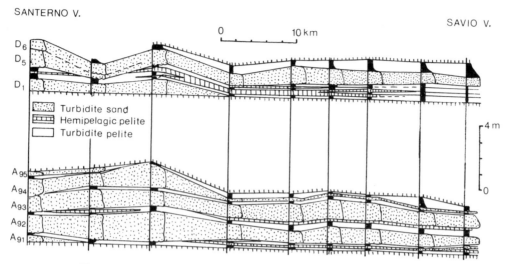

Figure 15.13 Example of detailed correlations of sandstone bodies in Marnoso-arenacea Formation of Italy (after Ricci-Lucchi and Valmori, 1980).

that do were traced over the whole basin (Ricci-Lucchi and Valmori, 1980). Even some of the thin-bedded turbidites display striking lateral continuity, suggesting that many of the flows were delimited only by basin margins and that local topography was insufficient to alter the paths of the flows.

16

Deep Marine Systems

Basins and fans are the basic components found in the deep marine environment. In those parts of the ocean dominated by terrigenous sedimentation, three basic types of fan–basin interactions can be recognized, including trench–fan systems, continental rise–abyssal plain systems, and continental borderland systems. The different tectonic regimes characteristic of these systems are the dominant control on geometry of the basin and the shape of the basin fill, the nature of the interaction of basin plain and fan components, and the resulting architecture of the basin-fill sequences.

Of these three types, borderland basins will not be treated here. They form along transform margins and occur as morphologically constricted basins with variable shapes, sediment supplies, and time of existence, and a strong morphotectonic control is imposed on the style of sediment accumulation in them (Stow and others, 1983–1984). A synthesis of sedimentation in these basins is probably not possible at this time because each appears to have its own characteristic facies distribution, morphologic evolution, and depositional sequence.

Trench–Fan Systems

Sedimentary Processes and Products

Trenches are elongate basins whose sediments undergo almost simultaneous deposition and deformation. They form at convergent margins and result from the subduction of an oceanic plate under a continental plate (Schweller and Kulm, 1978). The landward margin of the trench is a slope consisting of faulted and folded remnants of the accretionary wedge and is called the trench inner slope (Fig. 13.1). The outer margin is defined sometimes by a slight topographic high, called the outer rise, that is formed at the point of flexure of the oceanic plate. The dip of the oceanic plate is near-horizontal seaward of the rise, but landward of the rise the plate dips into the trench and forms the outer slope.

The trench floor is a narrow basin plain that is underlain by a variable thickness of pelagic and terrigenous sediment. Sediment thickness varies with proximity to sediment sources, and the trench itself may be broken up into segments that can control sediment distribution by starving parts of the trench and creating ponds in others.

Most sediment is transported laterally into the trench by mass movements across the inner slope and downslope canyons. Sediment slides are particularly important at base-of-slope settings, and turbidite fans can form at canyon mouths. Sediment transported down the inner slope meets the reverse gradient of the outer slope, and flow is directed one way or the other longitudinally along axes. Most trenches possess a longitudinal dip, and sediment is preferentially directed down the axial gradient. Variations through time in the rates of subduction along the trench, however, may reverse this dip, forcing flow in the opposite direction (Schweller and Kulm, 1978).

Sediments thin and become finer grained away from sources (Fig. 16.1), but because most flows on modern trench floors are confined to axial channels that occupy only small portions of the basin floor, a lateral thinning and fining away from the axial channel occurs as well. Reversals in flow direction may also add complexity to patterns of sediment distribution.

Axial channels are small and bounding levees are not well-developed, perhaps because of their relatively young age. Their present pattern of deposition was only established after glacial low stands when larger flows occupied the entire width of the trench and deposited sediments in broad sheets. Present-day channels carry a much diminished load and in this sense are analogous to underfit subaerial Holocene streams.

Schweller and Kulm (1978) recognize four main sediment assemblages within trench sequences, including pelagic and terrigenous plate facies, trench-wedge facies, and trench-fan facies. The pelagic plate facies consists of ocean-derived sediments deposited outside the trench, but brought into it during subduction of the oceanic plate it rests on (Fig. 16.1). The sediments vary according to the nature of the water column and ocean floor and may consist of oxidized clays and calcareous or diatomaceous oozes. Accumulation rates are slow, especially where the ocean floor is below the calcite compensation depth; but even where it is not, rates of accumulation of biogenic sediment are slow because productivity of most open ocean waters is small.

Terrigenous plate sediments are deposited on the outer slope and rise, and they may extend oceanward for several hundred kilometers as they grade laterally into the pelagic facies. They consist of clays and silts deposited by nepheloid layers and

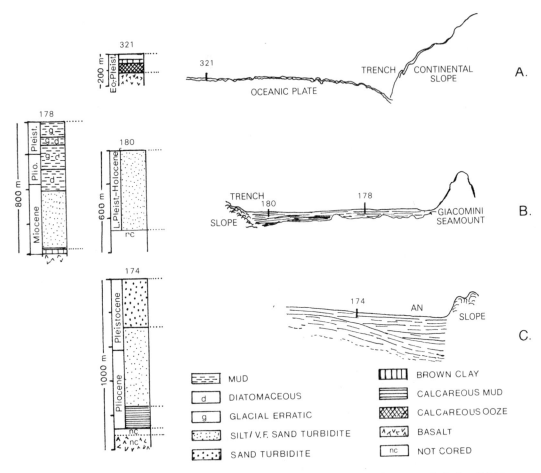

Figure 16.1 Variations in sediment sequences found in arc-trench systems, including (A) pelagic and hemipelagic sediments in the Peru-Chile Trench, (B) terrigenous and oceanic plate sediments overlain by trench-wedge sediments in the Aleutian trench, and (C) submarine fan sediments in the trench fill of the Washington–Oregon Trench (after Schweller and Kulm, 1978).

dilute turbidity currents that overflow the outer rise. Accumulation rates are 10 to 100 times that of the pelagic facies (Kulm and others, 1973), and they tend to be greatest in areas near the trench.

Sediments of the wedge assemblage are deposited in basins, along the axial channel, and against the inner trench slope (Thornburg and Kulm, 1987). Sheeted basins occur along those parts of the trench receiving sediment from submarine gullies and chutes that are so closely spaced they approximate line sources. An axial gradient persists in these basins, even though individual sheet turbidites are conformable and continuous across the entire width of the basins. However, lithofacies variations indicate the occurrence of core flows, depositing massive sand beds, and flank flows, depositing T_{abcde} to T_{bde} turbidite sequences, within individual events.

Ponded basins exist where reversals in axial gradient occur in areas isolated from slope trenches. These basins are also filled with sheet flow turbidites, but ponded basins have no axial gradient and their sediments are deposited on a flat surface. High-velocity flows capable of depositing complete turbidite sequences occur in proximal parts of basins, and lower-velocity flows, depositing T_{cde} to T_{de} sequences, occur in distal areas of the basin, as well as on bathymetric highs and along trench walls.

Axial channels occur at the base of the inner trench slope except where they are displaced basinward by fan lobes. Channels are filled with sand in laminated beds that are graded or ungraded or in beds that are massive or structureless. Sand-to-mud ratios may be as much as 4:1 in channels, and amalgamated sand bodies are common. Some beds consist of truncated turbidite sequences (T_{ae}) indicative of high-energy turbidity currents or grain flows.

Axial sediment lobes can accumulate where flow expansion or rapid deceleration occurs at discontinuities in the axial channel. They are morphologically similar to trench fans that occur at the base of the inner slope, but axial lobes form parallel to trench margins and are cut by an axial channel that lacks distributary channels.

Channels are bounded by levees deposited from concentrated sediment suspensions when turbidity flows spill over channel margins. Levee deposits are comprised of rhythmically bedded units ranging in thickness from 5 to 15 cm that consist of a basal graded sand layer overlain abruptly or gradationally by a graded pelite.

Contour currents are thermohaline-forced flows that follow bathymetric contours. Deposits of contour currents are most often associated with sediments on continental rises, and they will be described in more detail later in this chapter. However, they also occur on walls and floors of incised sediment-starved trenches and low on inner slopes next to filled trenches. Contourites in trenches consist of silt laminae displaying no obvious grading and no hierarchical sequence. Laminae are clustered within hemipelagic muds, and they appear to be the winnowed products of these muds.

Trench-wedge assemblages are not gradational with pelagic or terrigenous plate facies. Rather, they are confined to the trench axis; and because plate facies are carried into the trench by the subducting plate, wedge facies overlie them with a slight angular unconformity (Fig. 16.1).

Trench fans accumulate at mouths of submarine canyons where sediment delivery is large. Large fans tend to develop at point sources fed by major rivers. The Columbia River, for instance, is the primary source for the Astoria Fan, which completely fills the Washington–Oregon Trench (Kulm and others, 1973). However, only small fans occur in the Aleutian Trench, even though rates of sedimentation are large, because the trench is fed by numerous minor canyons (Piper and others, 1973).

Trench fans may show a variety of relationships with other trench facies. They may rest directly on plate facies with a discordant contact, they can grade laterally into trench-wedge sediments, especially where the two are genetically related, and they can overlie trench-wedge sediments with an erosive contact (Fig. 16.1).

Because of the presence of an axial gradient, fans in the Chile Trench develop asymmetrically, with smooth depositional lobes facing upgradient and lobes with pronounced evidence of erosion on the downgradient side (Thornburg and Kulm, 1987). The erosion may have occurred in response to an increase in the down-axial gradient, or it may have occurred during periods of lowered sea level or heightened tectonic activity when sediment yield to the trench increased. The increased yield may have raised velocities of turbidity flows to the point where they were sufficient to mobilize and erode fan sediments.

Controls on the Evolution of Trench–Fan Systems

Stratigraphic successions are produced in trenches either by migration of sediment facies or by migration of the depositional substrate due to plate migration (Schweller and Kulm, 1978; Thornburg and Kulm, 1987). Convergence produces large-scale coarsening-upward sequences, consisting of abyssal clays and hemipelagic muds overlain by distal and then proximal turbidites, that record the growing proximity of the plate to its primary sediment source.

At DSDP Site 174 in the Oregon–Washington Trench, for example, 284 m of sand turbidites are the distal facies of the Astoria Fan. They overlie thin-bedded silty clay and silt beds of the terrigenous plate facies with a short transitional contact, and an uncored 35-m section between the base of the sequence and acoustic basement may represent the pelagic plate facies (Kulm and others, 1973) (Fig. 16.1).

However, numerous complicating factors exist that may result in the deposition of incomplete sequences. Trench wedges, for example, may not develop in parts of segmented trenches that are isolated from longitudinal flows, and they are poorly developed along parts of trenches below intraslope basins that trap lateral flows from the trench inner slope (Schweller and Kulm, 1978). A wedge may not develop at all where rates of convergence and plate consumption far exceed the sediment yield or where axial flows become erosive at points of constriction in the trench or at places of increased axial gradient. In such cases, the sequence may consist solely of pelagic and terrigenous plate facies (Fig. 16.2).

At DSDP Site 178 in the Aleutian Trench, for instance, the Miocene to Holocene sequence consists of pelagic brown clay and chalk overlain by a varied section of hemipelagic muds, fine sand and silt turbidites, and ice-rafted erratics (Fig. 16.1). At nearby Site 180, however, the late Pleistocene sequence consists of 470 m of horizontally layered turbidites of the trench-wedge facies (Fig. 16.1).

Depositional sequences representing facies migrations form in response to changes in relative positions of trench facies tracts, and they record the evolution of proximal-to-distal facies relationships. In the trench fan component, one such relationship develops parallel to the continental margin from the canyon mouth to intercanyon areas; trench fans accumulate near canyons and basins develop away

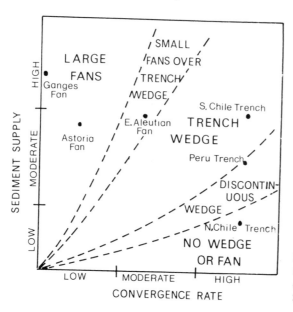

Figure 16.2 Model for relating rates of sediment influx and convergence to development of trench sequences (after Schweller and Kulm, 1978).

from them (Thornburg and Kulm, 1987). Another such relationship, however, can develop transverse to the trench in association with the axial channel.

The longitudinal dispersal system shifts downgradient in response to an increase in the axial gradient or an increase in sediment yield to the trench (Thornburg and Kulm, 1987). The axial channel will encroach on sheeted basins in intercanyon areas or ponded basins in areas of reduced sediment influx, and ponded basins can evolve into sheeted ones (Fig. 16.3). During periods of reduced sediment yield, the relief in the axial facies tract is enhanced by tectonism. Structural basins are created that change the axial gradient and produce local basins where axial channels formerly existed.

Axial channels normally carry the coarsest sediments in the trench, and the sediments of the trench-wedge facies normally fine laterally from them. Where the channel is fixed against the accretionary slope, as in the Aleutian Trench, it cannot migrate, so coarsest sediments accumulate preferentially against the slope (Fig. 16.4a). As the oceanic plate descends into the trench, a simple coarsening-upward sequence is superimposed on the plate sediments as the coarser channel deposits override the finer-grained trench-wedge sediments initially deposited nearer the trench outer slope (Piper and others, 1973).

This simple evolution, however, might be complicated by migration of axial channels (Schweller and Kulm, 1978; Thornburg and Kulm, 1987) (Fig. 16.4b). Axial channels near the canyon mouth display well-developed levee systems, so channel migration will produce a sequence consisting of basal channel sands overlain by levee deposits and capped by basinal T_{cde} or T_{de} turbidite sequences. Channels, however, are transient features with low relief and broad flat floors in intercan-

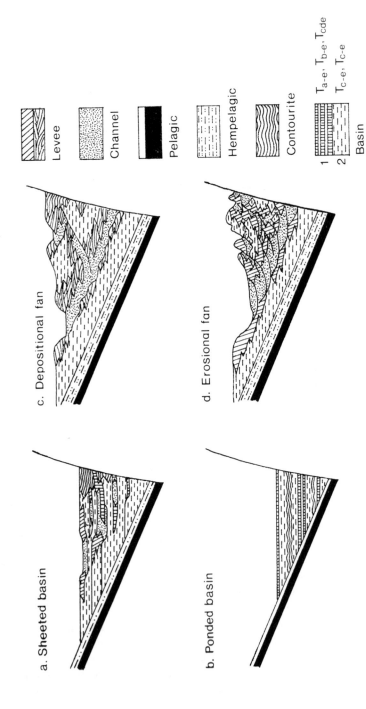

Figure 16.3 Diagrammatic cross sections showing sequences that might develop by longitudinal facies progradation down the trench axis (a and b), and by transverse progradation of trench fan facies (c and b) (after Thornburg and Kulm, 1987).

396

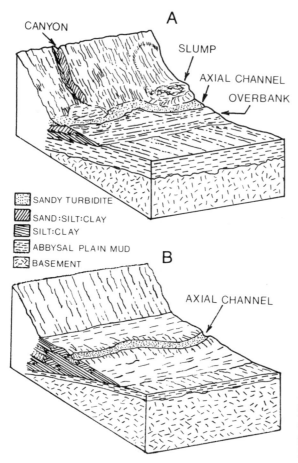

SANDY TURBIDITE
SAND:SILT:CLAY
SILT:CLAY
ABBYSAL PLAIN MUD
BASEMENT

Figure 16.4 Depositional models for trench fill sequences deposited where (A) the axial channel remains in a fixed position at the toe of the slope (after Piper and others, 1973); and (B) where the axial channel migrates across the trench floor (after Schweller and Kulm, 1978).

yon areas. Levees are poorly developed and facies changes are gradual transverse to flow (Thornburg and Kulm, 1987). Migration of more distal channels, therefore, will produce sequences consisting of basal channel deposits overlain by T_{abcde} turbidites and capped by T_{cde} or T_{de} turbidites deposited by lower-energy flows.

Some trenches do not contain axial channels (Underwood and others, 1980). In these cases, turbidite flows occur as broad sheets confined only by the landward and seaward slopes of the trench. Sediment facies vary longitudinally along the trench in response to distance from point sources, and no consistent trend develops across the trench floor. The trench-wedge sequence that would develop under such conditions would show no consistent vertical grain size variations.

Fans drive axial channels basinward into trench wedges as they prograde away from the inner trench slope, producing a time-transgressive erosive unconformity across the top of the coarsening-upward trench-wedge sequence. A fining-upward sequence representing axial channel abandonment and fan progradation will ideally develop above the unconformity (Thornburg and Kulm, 1987). Such a sequence will be produced by prograding channelized fans with basal channel sands overlain by levee deposits and capped by interchannel sediments (Fig. 16.3). Nonchannelized fans, on the other hand, are deposited by sheet flows, and their prograding depositional lobes will produce coarsening-upward sequences representing migration of proximal over distal fan depocenters.

Channels scour truncation surfaces across sediments deposited during depositional phases on erosional fan lobes. The channels initially follow courses established during depositional phases, but eventually they transgress interchannel areas, and at full maturity relief is eliminated and the fan surface is covered by a lag pavement (Thornburg and Kulm, 1987) (Fig. 16.3).

The degree to which the trench-wedge sequence develops is also dependent on relative rates of convergence and sediment influx (Schweller and Kulm, 1978; Thornburg and Kulm, 1987). Where sediment yield is large or rates of convergence are slow, large fans may develop and be included in the sequence as the uppermost part of a coarsening-upward succession (Schweller and Kulm, 1978). Only small fans and trench wedges can develop, however, with moderate rates of convergence, and where sediment yield is small or rates of convergence are fast, only plate facies may occur in the trench (Fig. 16.2).

Sediment is also continuously removed from the system as the oceanic plate converges on the trench, but a steady-state equilibrium can be attained when the rate of sediment yield to the trench balances the rate of sediment loss (Fig. 16.5). A trench is a dynamic environment, however, and steady-state conditions are short-lived. Disequilibrium is forced on the system by varying the rate of sediment accu-

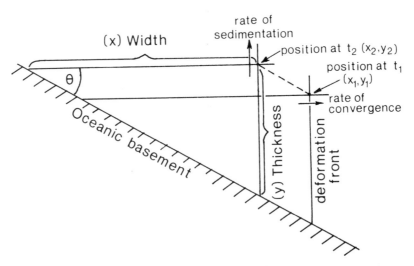

Figure 16.5 Diagrammatic representation of a trench wedge showing the relationship between sediment accumulation rate (dy/dt) and plate convergence rate (dx/dt) (after Thornburg and Kulm, 1987).

mulation, the convergence rate, or the dip angle of the subducting plate, and the trench wedge either grows or shrinks in response until equilibrium is restored (Thornburg and Kulm, 1987).

Trench-Fan Sequences in the Rock Record

Cretaceous sediments in the Flysch zone of the East Alps display several features that have been used to identify ancient trench sediments. Flysch beds in the Early Cretaceous Gault Formation, for instance, were traced for up to 115 km along strike (Hesse, 1974). The beds range in thickness from a few centimeters to beds of amalgamated turbidites up to 4 m thick. Intercalated mudstones consist of alternating thin gray-green and black layers. The gray layers normally overlie the sandstone beds and are interpreted as suspension deposits of the turbidites, and the green and black layers are interpreted as hemipelagic muds.

Most beds were traced the entire length of the outcrop belt, although a few pinch out downdip (Fig. 16.6). Beds tend to thin downdip and amalgamated beds are split by mudstones. The lateral continuity of the beds suggests a flat, nearly level substrate, and the lack of carbonates in an area where correlative units contain abundant carbonates suggests deposition below the calcite compensation depth. Consistent trends in sand bed thickness and sandstone–mudstone ratios along strike indicate a dominantly longitudinal transport direction, and reversal of transport directions twice later during the Cretaceous suggests that differential tilting of the axis reversed the gradient periodically.

The Shumagin Formation of the Alaska Peninsula consists of an interbedded sequence of sandstones and mudstones that have been interpreted as an arc-trench deposit (Moore, 1973). The sandstones occur in both thick and thin beds. Thin-graded beds and beds of intermediate thickness tend to consist of T_{ce} or T_{de} turbidite sequences. They show consistent transport directions along strike, suggesting that they were deposited longitudinally along the axis of the trench. Thick graded beds are up to 5 m thick and consist of amalgamated, structureless, nongraded layers of sand. Beds show sharp erosional bases and may contain deformed mudstone intraclasts up to 3 m long. These beds are interpreted as proximal turbidites and grain flows that flowed laterally downslope into the trench. Points of sediment influx into the trench occurred in those areas where concentrations of thick beds are enclosed by thinly bedded sandstones and mudstones.

Like the Cretaceous sediments of the Flysch Zone of the Alps, deposition of the Shumagin Formation took place in a narrow elongate basin, and because the sediments lack calcareous microfossils, deposition probably occurred below the calcite compensation depth. Primary transport direction was parallel to depositional strike, but, unlike the Flysch Zone, there is considerable evidence for lateral transport of sediment into the trench.

The Te Anau Assemblage of New Zealand is part of an arc trench–ocean basin complex (Carter and others, 1978), of which the shelf and slope components were described previously. The Te Anau Assemblage consists of mud turbidites, hemipel-

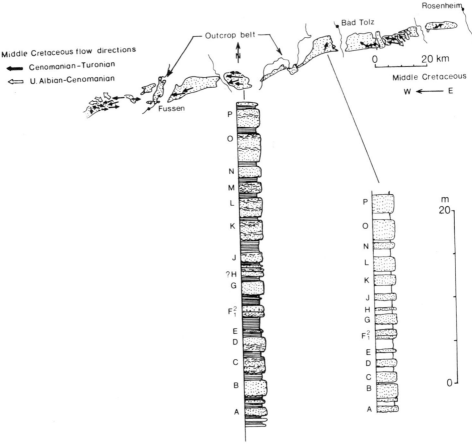

Figure 16.6 Correlation of sandstone beds in the Middle Cretaceous trench sediments of the eastern Alps (after Hesse, 1974).

agic mudstones and siltstones, pelagic chert, and silty contourites that represent the deepest-water deposits of the complex. Interspersed among the fine-grained rocks are channelized sandy flysch and conglomerates up to several hundred meters thick. These coarse beds were probably deposited in axial channels and on submarine fans, and the finer-grained turbidites with which they are associated probably represent levee and overbank turbidites.

The longitudinal shape of the basin could be confidently established in the previous examples, and primary transport directions were observed to occur parallel to the elongation of the troughs. These were the primary criteria used to identify these deposits as arc-trench complexes, but these criteria could not be established for the Te Anau Assemblage because of structural complications in the terrain. The Assemblage, instead, was interpreted as an arc-trench deposit on the basis of its association of trench-type deposits with pelagic sediments (Fig. 14.9). Identification of such an association is especially important in many subduction complexes where the sediments may consist of a chaotic mix of stacked trench fills, pelagic sediments, and slope and intraslope basin deposits (Fig. 16.7). The sedimentary features of some of these facies are not radically different, and many of the criteria used to define trench sequences, such as basin shape and transport direction, are rendered difficult to interpret by structural complexities.

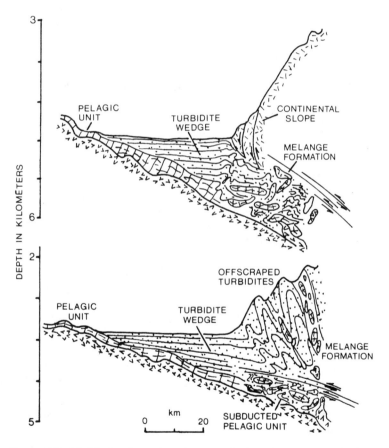

Figure 16.7 Models for formation of a heterolithic melange composed of off-scraped remnants of oceanic plate and trench-wedge facies (upper), and an accretionary wedge composed mainly of trench-wedge sediments after selective offscraping of pelagic sediments (lower) (after Scholle and Marlow, 1974).

Unfortunately, pelagic and hemipelagic sediments may not be common constituents of accretionary wedges built during plate subduction (Scholle and Marlow, 1974). Thrust slices in the wedge may, instead, consist solely of a selectively off-scraped terrigenous component because the plate facies were subducted beneath the wedge during its construction (Fig. 16.7).

Continental Rise–Abyssal Plain System

The continental rise is a broad, gently sloping wedge of sediment bounded on the landward side by the continental slope and on the basinward side by the abyssal plain (Fig. 13.1). The rise is up to 1000 km wide, and the sediments under it are up to 10 km thick and are best developed along passive continental margins.

Most of the rise surface is remarkably smooth, and it normally lacks any marked channeling perpendicular to contours (Stanley and others, 1971). The upper limit of the rise occurs at the break where steep gradients of the slope give way to much gentler ones of the rise (Fig. 13.1), and the lower limit is marked also by a slope break at the abyssal plain. In some places the boundary is also marked by lower rise hills that have been interpreted either as the toes of large slides or as giant bedforms or current-molded topography (Emery and others, 1970; Lancelot and others, 1972).

Sedimentary Processes and Products

The nature of the depositional processes involved in the emplacement of rises is still controversial. Earlier hypotheses stressed the importance of turbidites that supposedly flowed down submarine canyons and onto fans that eventually coalesced into the rise. Later studies suggested that thermohaline currents flowing perpendicular to contours played the largest role in shaping the rise by transporting sediment parallel to regional gradient. Detailed seismic work has also suggested the presence of large masses of slumped sediments in the rise.

Although their effect on sedimentation throughout the column has not yet been fully assessed, the role of contour currents on modern continental rises is well-established. Except in a few places where the presence of channels indicates the downslope movement of sediment, flow directions indicated by the orientation of ripple marks and current lineations on the slope are parallel to bathymetric contours.

Off the eastern North America rise, the presence of structures formed by traction transport of relatively coarse sediment has been attributed to the action of the Western Boundary Undercurrent (Heezen and others, 1966). South-flowing deep-ocean currents are deflected against the western margin of the Atlantic Ocean north of the equator by the Coriolis force. Current velocities of 2 to 20 cm/s are sufficient to transport clay and silt but not to erode it, and as long as these currents do not exceed the threshold velocity necessary to erode the substrate, aggradation of the rise occurs.

The seaward thinning and basinward fining of the sediment wedge under the rise might be explained as simple responses of the sediments to increasing distance from source, but the peculiarities of contour currents offer an alternative hypothesis. The degree of deflection of deep geostrophic currents is velocity-dependent, so currents capable of transporting the coarsest size and the highest volume of sediment are those nearest the slope, whereas those flowing near the toe of the rise are slower moving and less competent (Fig. 16.8). Thickest accumulation of the coarsest sediment would occur under the faster-velocity streams.

The landward increase in flow velocities within deep geostrophic flows may also explain the distribution of sedimentary processes along the continental margin. Current velocities of flows highest on the rise and on the slope may occasionally exceed the threshold of erosion of muds, suggesting that the boundary between slope and rise marks the transition between a depositional and essentially erosional regime. Currents flowing in the basin beyond the toe of the rise may be too slow to carry significant amount of sediment, explaining why turbidity flows dominate depositional processes on abyssal plains (Fig. 16.8).

Not all sediment transport is along strike (Stanley and others, 1971). Submarine fans are neither numerous nor morphologically very distinct, but they do occur where abundant sediment is transported down the rise gradient. Although evidence of their passage is obliterated by reworking, dilute turbidity flows probably are the major agents of sediment delivery to the rise. Both the Laurentian Fan and the Hatteras Fan, for example, occur on the continental rise off eastern North America. During the Pleistocene, they were principal sources of coarse sediment to the Hatteras and Sohm Abyssal plains (Fig. 16.9), and coarse sediment continues to be delivered to the Laurentian Fan (Stow and others, 1983–1984).

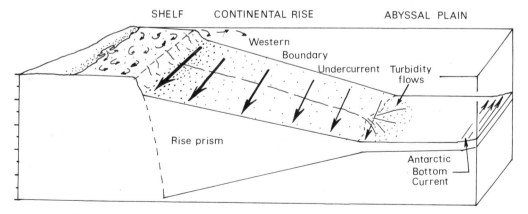

Figure 16.8 Diagrammatic representation of the mechanisms involved in the transport and deposition of sediments on the continental rise by geostrophic currents (after Heezen and others, 1966; Copyright 1966 by the American Association for the Advancement of Science).

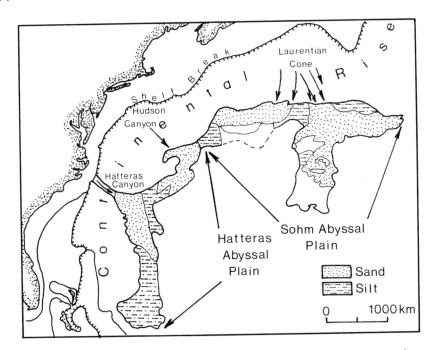

Figure 16.9   Distribution of turbidites in relation to source on the Hatteras and Sohm abyssal plains (after Horn and others, 1971).

The Hatteras Fan, off the middle Atlantic coast of North America, however, is now largely inactive. It is fed by the Hatteras Canyon, which is the trunk channel for a tributary system of large extent on the slope and upper rise (Fig. 16.10). The canyon was rendered inactive by the postglacial sea-level rise, and the fan surface is covered by up to 100 cm of mud that is periodically remolded by bottom currents (Cleary and Conolly, 1974). The morphology of the Hatteras Fan, however, is not greatly different from that of fans on the west coast of the United States, suggesting that at one time considerable amounts of coarse material must have been deposited on it.

Other features of rise fans are also comparable to those fans on active margins. Levees on the Hatteras Fan, for instance, border channels of the distributary system and consist of interbedded muds and thin sands (Cleary and Conolly, 1974) (Fig. 16.10), and the channels are underlain by massive or graded beds of medium- to fine-grained sand. The suprafan area is underlain by alternating beds of fine to medium quartz sand and clay, with thickest sand beds occurring near channels on the suprafan. Thinner beds of finer-grained sand occur in more distal areas and on topographic highs only infrequently affected by turbidity flows.

In addition to these modes of downslope transport, evidence exists for movement of large masses of sediment down the slope and onto the rise. About half of the lower portion of the rise wedge consists of sediments with weak and distorted

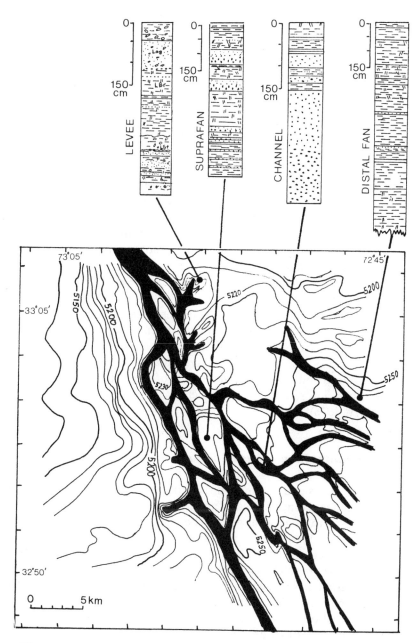

Figure 16.10 Physiography and typical sedimentary sequences of the Hatteras deep-sea fan (after Cleary and Conolly, 1974).

reflectors on echograms that are believed to represent slumps or gravitational slides (Emery and others, 1970) (Fig. 16.11). The slumps occur preferentially low in the sequence under the lower rise and may represent a period when sediments accumulated rapidly and/or tectonic activity was pervasive. Other slump deposits have been identified in the region of the Laurentian Channel, where a slump occurred in 1929 that triggered a turbidity flow of large magnitude out onto the Sohm Abyssal Plain.

Contour currents presently appear to be the main transport agent on continental rises, and the products of these currents are called contourites. They consist principally of silt and clay, although cross-laminated fine sand with placer layers are also known to occur. Sand beds may be rhythmically bedded and superficially resemble turbidites, but unlike turbidites, both upper and lower bounding surfaces are sharp and the sediments tend to be relatively well sorted (Bouma and Hollister, 1973; Lovell and Stow, 1981).

Contourites are characteristically uniform along strike. Contributions from different mineralogical terrains may be thoroughly mixed over the long transport paths that the currents follow, and their capacity to transport relatively dilute suspensions over very long distances tends to reduce the potential for variations in primary features along strike.

Contour currents occur only on the west side of oceans in the northern hemisphere (east side in the southern hemisphere), because the deep return flow of the thermohaline circulation system is southward and is deflected by the Coriolis force. On the east side of oceans, no equivalent current is observed, and rise sediments are fine-grained and poorly sorted deposits of low-density turbidity currents, hemipelagic suspension settling, and sediment mass movement (Lancelot and others, 1972; Embley, 1976).

Continental Rise-Abyssal Plain Sequences

Abyssal plains and continental rises together form sediment accumulations that are the largest and longest lived of any sedimentary depositional system. They have been interpreted to be nascent geosynclines (Dietz and Holden, 1974), and it is beyond the scope of this section to attempt a stratigraphic analysis of features of such magnitude and complexity. The continental rises bordering the North Atlantic Ocean, for example, began construction during the initial stages of continental rifting, and DSDP drill holes along the western and eastern margins provide some indication of the major events that occurred during their evolution.

Initial deposits on the western margin date from the Jurassic and consist of carbonate oozes that accumulated slowly on topographic highs in a pelagic setting and argillaceous red limestones that filled depressions (Lancelot and others, 1972). Differential accumulation rates between highs and lows, as well as transport by turbidity currents of sediment from topographic highs into basins, served to smooth the original, irregular basement topography (Fig. 16.12).

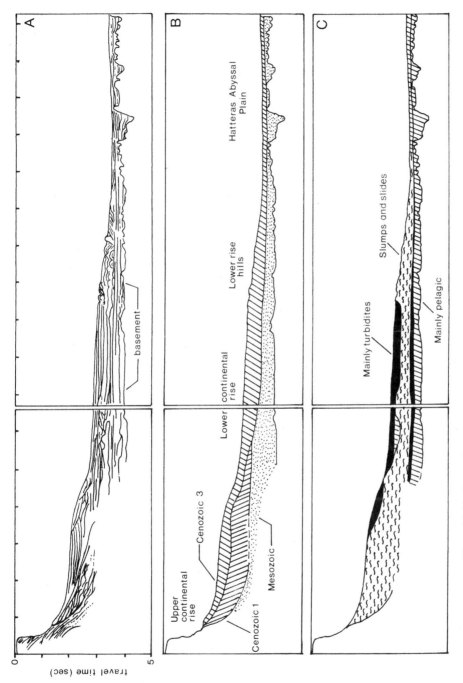

Figure 16.11 Internal architecture of the continental rise off New York City shown in (A) transverse seismic profile, (B) interpretative cross section of the chronology of the deposits, and (C) distribution of facies (after Emery and others, 1970, Figs. 11, 13, 22, reprinted with permission of the American Association of Petroleum Geologists; and Hollister and others, 1972).

407

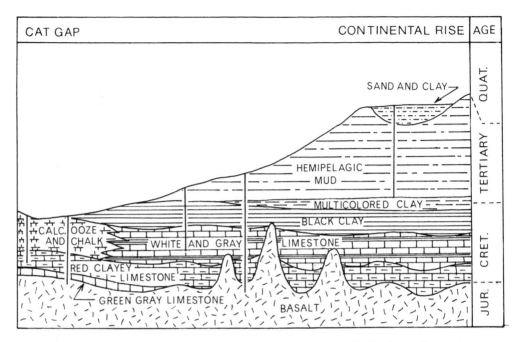

Figure 16.12 Diagrammatic transverse cross section showing distribution and vertical sequence of sediments on the continental margin on the east coast of North America (after Lancelot and others, 1972).

Organic clay deposited during the early Cretaceous represents a period of stagnation that was apparently oceanwide. Stagnant conditions were not temporally continuous, however, and periods of oxygenated conditions, probably caused by fluctuations in circulation patterns, occurred when pelagic white limestones were deposited.

A change in the relative motions of the American and African plates resulted in a period of intense tectonic and volcanic activity. The late Cretaceous section is greatly condensed, and terrigenous sediments form only a small portion of a sequence dominated by volcanogenic material.

Beginning in the Eocene, however, abundant terrigenous detritus was shed into the basin, initiating the main phase of rise construction. The sediments consisted initially of hemipelagic muds, but during the Quaternary, large amounts of coarse material accumulated not only on the rise but on the abyssal plain as well (Hollister and others, 1972).

The sequence of events on the eastern side of the Atlantic is broadly similar to that on the west. At Site 368 on the Cape Verde Rise, the hole bottomed in Cretaceous black shales, but at nearby sites off the rise black shales overlie Jurassic carbonates (Gardner and others, 1978; Lancelot and others, 1977) (Fig. 16.13). The shales are overlain gradationally by terrigenous clays and silty clays that were prob-

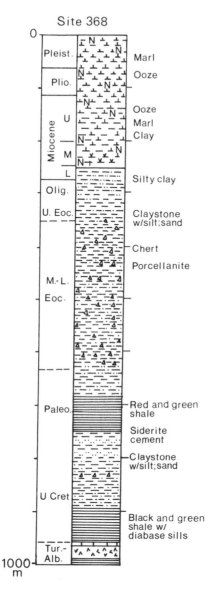

Figure 16.13 Vertical sequence of sediments in the Cape Verde Rise off the coast of western Africa (after Lancelot and others, 1977).

ably deposited by low-density turbidity currents. Cycles of various periodicities occur within this sequence. Short-term cycles are represented by rhythmic interbedding of silts and muds in turbidites. Cycles with periods of tens of thousands of years probably resulted from variation in organic productivity versus terrigenous sediment yield, redox potential at the sea floor, and degree of $CaCO_3$ saturation of sea-floor waters (Dean and others, 1978).

The upper part of the sequence is dominated by pelagic nannofossil oozes. Uplift under the rise cut off the delivery of terrigenous material and redirected it into the Cape Verde Basin. At the same time, the calcite compensation depth increased, allowing the deposition of pelagic marls and calcareous oozes (Fig. 16.13). The quantity of terrigenous sediment provided to the rise increased substantially during the Pleistocene, however, in response to climatic changes in source areas and eustatic sea-level lowering.

Conclusion: Event Stratigraphy in Clastic Depositional Sequences

Ager (1973) coined the term "event stratigraphy" to correlate rocks according to the events that produced them, rather than on their intrinsic petrologic character or fossil content. This is one logical extension of the facies concept, wherein laterally contiguous depositional environments experience the same extrinsic factors but respond differently to them.

The concept was developed by Ager when he recognzied from detailed correlations of Triassic rocks by Audley-Charles (1970) and others (see *Quart. Jour. Geol. Soc. London,* v. 126, pts. 1 and 2) that, although the lower part of the Triassic in Britain is almost completely lacking in marine fossils, correlations with the Triassic of the European continent could be made by comparing responses in various areas to marine transgressions and regressions.

For the most part, widespread transgressions of the epicontinental seas of the continent did not reach the British Isles as such, but they were made manifest in the rocks indirectly by their effect on local climate and base level. During the Early Triassic, for instance, continental environments were widespread over most of northwestern Europe and marine conditions prevailed only in inner zones of the Alps. A transgression occurred, however, that expanded marine environments into France and changed the climate in Britain. During this transgression, the arid climate that had produced a prolonged period of mainly eolian sedimentation gave way to more humid conditions, which resulted in a period of widespread alluviation.

This brief transgression was followed by a regression. Eolian sedimentation was reestablished in Great Britain, marginal marine evaporites were deposited in the Jura and in Germany, and normal marine conditions again were confined to the inner zone of the Alps.

This was followed by a second, more profound transgression, which spread normal marine conditions to Germany and Luxembourg, where the Hauptmuschelkalk was deposited. Equivalent strata in Britain consist of the Waterstone Beds, which were deposited in a variety of paralic, brackish-water, and marginal marine settings.

Similar responses are recorded in the overlying Triassic strata during succeeding cycles of sea-level rise and fall, further establishing the principle that the indirect effects of extrinsic mechanisms can be a useful tool in correlating sections where fossil evidence is nondiagnostic and facies equivalency cannot physically be established.

The concept, although not the term, had been applied earlier to coeval Devonian rocks in New York by McCave (1967, 1969). Sandstones, shales, and conglomerates of the Catskill redbeds were deposited during a major regression as a westward-prograding clastic wedge of predominantly alluvial sediments.

The regression was interrupted periodically by transgressions, however. To the west, these transgressive episodes are marked by the occurrence of the Tully, Portland Point, and Centerfield limestones, which grade eastward, toward the Catskill front, into nearshore and marginal marine rocks (Friedman and Johnson, 1966; McCave, 1967) (Fig. C.1). The direct effect of the transgression, therefore, was to decrease the yield of clastic sediment to the marine basin and allow the deposition of carbonate rocks.

The transgression, however, also influenced the clastic wedge indirectly where correlative rocks include anomalously thick sequences of alluvial overbank sediments (Fig. C.1). These apparently were deposited when base level of the streams on the coastal plain rose during transgression.

The concept of event stratigraphy has been successfully applied to the Tertiary basins of the California Borderland. These basins responded rapidly to relative changes in sea level because they were small with a large sediment influx that created abrupt lateral and vertical facies changes. Most importantly, rapid subsidence rates allowed preservation of the deposits. Coeval strata commonly change within 5 to 10 km downdip from subaerial alluvial fan to submarine fan facies at abyssal depths (May and others, 1983).

During transgressions, coarse-grained sediment was trapped in nearshore zones, and a lag of broken shell fragments, phosphatic and glauconitic pebbles, and winnowed coarse-grained sand was deposited on the shelf (Clifton, 1981a; May and others, 1983). Fine-grained sediment did escape into the deep basins, but sedimentation rates were slow, and hemipelagic sedimentation predominated on submarine fans and on basin plains.

Progradational pulses may occur during high stands when deltas and shorelines migrate basinward. They may also occur during eustatic low levels where subaerial fan and fan delta facies prograde over a narrow shelf and deliver coarse sediment directly to slope canyons. Three major progradational cycles, for instance, were deposited during eustatic low levels on the Eocene Ferrelo Fan off southern California (Howell and Vedder, 1983–1984). Alluvial fan and coastal plain fan delta

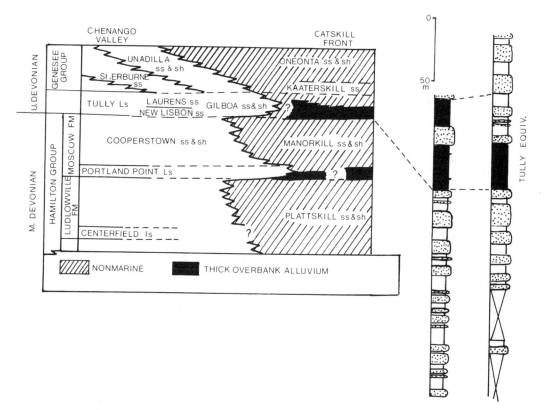

Figure C.1   Correlation chart for the upper part of the Middle Devonian and the lower part of the Upper Devonian in New York State showing position within the sequence of two exceptionally thick sections of overbank alluvium (after McCave, 1969, Figs. 1, 5 reprinted with permission of the American Association of Petroleum Geologists).

facies were deposited in onshore areas (Howell and Link, 1979), and paralic facies pass abruptly downdip into shelf and slope canyon deposits (Fig. C.2). The cycles are manifested in the submarine fan by formation of sand-filled channels in inner and mid-fan areas and by coarsening-upward progradational lobes in outer fan settings.

Rapid responses in these basins to changes in sea level also permit interbasinal correlation of events because they respond in a similar manner to worldwide changes in sea level. A major cycle was initiated in them when sea level dropped and a basal unconformity was formed during the late Early Eocene (May and others, 1983) (Fig. C.3). This was followed by a rapid transgression, which deposited a thin retrogradational sequence. During the subsequent stillstand, a thick progradational sequence accumulated through the Middle and Late Eocene. The stillstand was interrupted during the medial Middle Eocene when a slight regression resulted in increased sediment yield to the basin and initiation of a period of fan progradation.

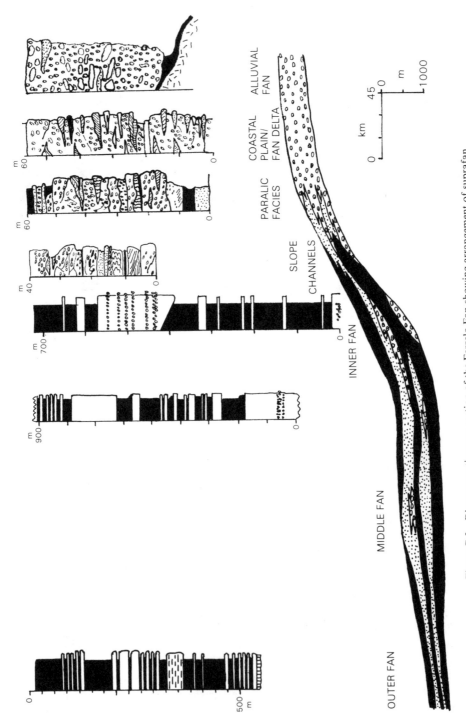

Figure C.2 Diagrammatic cross section of the Ferrelo Fan showing arrangement of suprafan bulges, distribution of conglomeratic sediments deposited in pulses during low sea-level stands, and representative sequences of the conglomeratic sediments in various parts of the basin (after Howell and Link, 1979; Howell and Vedder, 1983–1984).

414

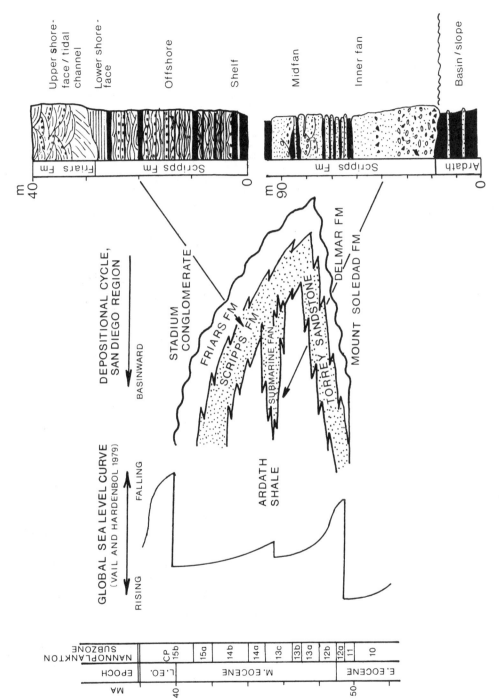

Figure C.3 Correlation of middle Eocene strata in the San Diego Basin showing control of facies distributions by a eustatically controlled onlap event related to a global sea-level rise. Graphic logs show typical stratigraphic sequences developed during the event in the basin and on basin margins (after May and others, 1983).

415

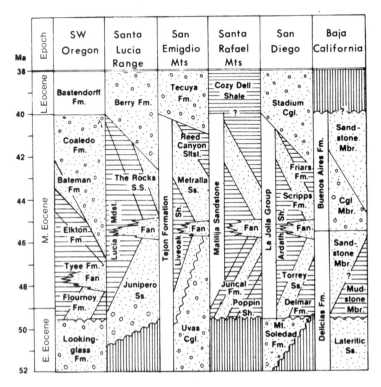

Figure C.4 Stratigraphic correlation chart for Eocene depocenters along the Pacific coast showing synchronous development of similar facies in response to worldwide changes in sea level (after May and others, 1983).

This progradational event, along with the major depositional cycle on which it is superimposed, has correlatives in many of the basins along the California Borderland (Fig. C.4). Not all components are present everywhere, but the synchronous development of depositional patterns, recognizable by their response to changing sea level, suggests a primary eustatic control on stratigraphic development.

It is apparent from these examples that the ultimate goals of stratigraphy are attainable on a much finer scale than was previously possible. Major events in earth history have long been recognized. Mountain-building episodes, basin formation, and profound transgressions and regressions have resulted in radical changes in the depositional milieu.

Sedimentology has now advanced to the point where the interpretation of various depositional settings can be made with such confidence that the effects of minor tectonic episodes, eustatic changes in sea level, and climatic fluctuations can be recognized. With this enhanced ability to interpret the significance of such events, it is now possible to make correlations that are directly related to the more subtle rhythms of earth evolution.

References

ADAMS, J., and J. PATTON, 1979, Sebkha-dune deposition in the Lyons Formation (Permian) Northern Front Range, Colorado: *Mountain Geologist,* v. 16, pp. 47–57.

AGER, D. V., 1970, The Triassic system in Britain and its stratigraphic nomenclature: *Quart. J. Geol. Soc. London,* v. 126, pp. 3–17.

———, 1973, *The Nature of the Stratigraphical Record:* MacMillan Press Ltd., London, 122 p.

AHLBRANDT, T. S., 1975, Comparison of textures and structures to distinguish eolian environments, Killpecker dune field, Wyoming: *Mountain Geologist,* v. 12, pp. 61–73.

———, and S. G. FRYBERGER, 1979, Eolian deposits in the Nebraska Sand Hills: *U.S. Geol. Survey Prof. Paper 1120A,* 24 p.

———, and ———, 1983, Sedimentary features and significance of interdune deposits: in ETHERIDGE, F. G., and R. M. FLORES, eds., *Non-marine Depositional Environments: Models for Exploration,* SEPM Spec. Pub. 31, pp. 293–314.

———, S. ANDREWS, and D. T. GWYNNE, 1978, Bioturbation in eolian deposits: *J. Sed. Petrology,* v. 48, pp. 839–848.

AIGNER, T., 1982, Calcareous tempestites: storm-dominated stratification in Upper Muschelkalk limestones (Middle Triassic, S.W. Germany): in EINSELE, G., and A. SEILACHER, eds., *Cyclic and Event Stratification,* Springer-Verlag, New York, pp. 248–261.

———, and H. E. REINECK, 1982, Proximality trends in modern storm sands from the Helgoland Bight (North Sea) and their implications for basin analysis: *Senckenbergiana Maritima,* v. 14, pp. 183–215.

ALEXANDER, J., and M. R. LEEDER, 1987, Active tectonic control on alluvial architecture: in ETHERIDGE, F. G., R. M. FLORES, and D. M. HARVEY, eds., *Recent Developments in Fluvial Sedimentolgy,* Soc. Econ. Paleontologists and Mineralogists, Spec. Pub. 39, Tulsa, Okla., pp. 243–252.

417

ALLEN, J. R. L., 1965a, A review of the origin and characteristics of Recent alluvial sediments: *Sedimentology,* v. 5, p. 89–191.

——, 1965b, The sedimentation and paleogeography of the Old Red Sandstone of Anglesey, Northern Wales: *Proc. Yorkshire Geol. Soc.,* v. 35, pp. 139–185.

——, 1965c, A review of the origin and characteristics of recent alluvial sediments: *Sedimentology,* v. 5, pp. 89–191.

——, 1970, *Physical Processes of Sedimentation:* American Elsevier, New York, 248 p.

——, 1978, Studies in fluviatile sedimentation: an exploratory quantitative model for the architecture of avulsion-controlled alluvial sites: *Sedim. Geol.,* v. 21, pp. 129–148.

——, 1980, Sandwaves: a model of origin and internal structure: *Sedim. Geol.,* v. 26, pp. 281–328.

ALLEN, P. A., and R. HOMEWOOD, 1984, Evolution and mechanics of a Miocene tidal sandwave: *Sedimentology,* v. 31, pp. 63–81.

ANDERTON, R., 1976, Tidal-shelf sedimentation: an example from the Scottish Dalradian: *Sedimentology,* v. 23:4, pp. 429–458.

ANDREWS, P. B., 1970, Facies and genesis of a hurricane washover fan, St. Joseph Island, central Texas coast: *Texas Bur. Econ. Geol. Rep. Invest.,* 67, 147 p.

ARTHUR, M. A., R. VON HUENE, and C. G. ADELSECK, JR., 1980, Sedimentary evolution of Japan fore-arc region off northern Honshu, Legs 56 and 57, Deep-sea Drilling Project: *Initial Reports of the Deep-sea Drilling Project,* v. 56, 57, pt. 1, pp. 521–568.

ASQUITH, D. O., 1970, Depositional topography and major marine environments, late Cretaceous, Wyoming: *Am. Assoc. Petroleum Geologists Bull.,* v. 54, pp. 1184–1224.

——, 1974, Sedimentary models, cycles, and deltas, Upper Cretaceous, Wyoming: *Am. Assoc. Petroleum Geologists Bull.,* v. 58, pp. 2274–2283.

AUDLEY-CHARLES, M. G., 1970, Stratigraphical correlation of the Triassic rocks of the British Isles: *Quart. J. Geol. Soc. London,* v. 126, pts. 1 and 2, pp. 19–48.

AUGUSTINUS, P. G. E. F., 1980, Actual development of the chenier coast of Surinam (South America): *Sedim. Geol.,* v. 26, pp. 91–114.

BAGANZ, B. P., J. C. HORNE, and J. C. FERM, 1975, Carboniferous and Recent lower delta plains—a comparison: *Gulf Coast Assoc. Geol. Socs. Trans.,* v. 37, pp. 556–591.

BAGNOLD, R. A., 1941, *The Physics of Blown Sand and Desert Dunes:* Methuen, London, 265 p.

——, 1954, Experiments on a gravity free dispersion of large solid spheres in a Newtonian fluid under shear: *Roy. Soc. London Proc.* Ser. A, v. 225, pp. 49–63.

BART, H. A., 1977, Sedimentology of cross-bedded sandstones in Arikaree Group, Miocene, southeastern Wyoming: *Sedim. Geol.,* v. 19, pp. 165–184.

BARTSCH-WINKLER, S., and H. R. SCHMOLL, 1984, Bedding types in Holocene tidal channel sequences, Knik Arm, Upper Cook Inlet, Alaska: *J. Sed. Petrology,* v. 54:4, pp. 1239–1250.

BARWIS, J. H., 1978, Sedimentology of some South Carolina tidal-creek point bars, and a comparison with their fluvial counterparts: in MIALL, A. D., ed., *Fluvial Sedimentology,* Can. Soc. Petrol. Geologists, Mem. 5, pp. 129–160.

——, and M. O. HAYES, 1979, Regional patterns of modern barrier island and tidal inlet deposits as applied to paleoenvironmental studies: in FERM, J. C., and J. C. HORNE, eds.,

Carboniferous Depositional Environments in the Appalachian Region, University of South Carolina, Carolina Coal Group, pp. 472-498.

BATES, C. C., 1953, Rational theory of delta formation: *Am. Assoc. Petroleum Geologists Bull.,* v. 37, p. 2119-2162.

BEHRENSMEYER, A. K., and L. TAUXE, 1982, Isochronous fluvial systems in Miocene deposits of northern Pakistan: *Sedimentology,* v. 29, pp. 331-353.

BELDERSON, R. H., M. A. JOHNSON, and N. H. KENYON, 1982, Bedforms: in, STRIDE, A. H., ed., *Offshore Tidal Sands,* Chapman and Hall, New York, pp. 27-57.

———, and others, 1984, A "braided" distributary system on the Orinoco deep-sea fan: *Marine Geology,* v. 56, pp. 195-206.

BERGER, A., and others, eds., 1984, *Milankovitch and Climate:* D. Reidel Publishing Co., Dordrecht, Holland, 510 p.

BERNARD, H. A., R. J. LeBLANC, and C. F. MAJOR, JR., 1962, Recent and Pleistocene geology of southeast Texas: *Geol. Gulf Coast and Central Texas, Houston Geol. Soc. Guidebook of Excursions,* p. 175-225.

BERRYHILL, H. L., JR., and others, 1976, Environmental studies, South Texas Outer Continental Shelf, 1975: Geology: Natl. Tech. Information Service, Springfield, Va., Pub. PB251-341, 353 p.

———, and others, 1977, Environmental studies, South Texas Outer Continental Shelf, 1976: Geology: Natl. Tech. Information Service, Springfield, Va., Pub. PB289-144, 306 p.

BIGARELLA, J. J., R. D. BECKER, and G. M. DUARTE, 1969, Coastal dune structures from Parana (Brazil): *Marine Geol.,* v. 7, pp. 5-55.

BLAKEY, R. C., and R. GUBITOSA, 1984, Controls of sandstone body geometry and architecture in the Chinle Formation (Upper Triassic), Colorado Plateau: *Sedim. Geol.,* v. 38, pp. 51-86.

BLISSENBACH, E., 1954, Geology of alluvial fans in semi-arid regions: *Geol. Soc. America Bull.,* v. 65:1, pp. 175-190.

BLUCK, B. J., 1967, Deposition of some Upper Old Red Sandstone conglomerates in the Clyde area: a study in the significance of bedding: *Scottish J. Geol.,* v. 3, pp. 139-167.

BOND, G., 1967, River valley morphology, stratigraphy, and paleoclimatology in southern Africa: in BISHOP, W. W., and J. D. CLARK, eds., *Background to Evolution in Africa:* University of Chicago Press, Chicago, pp. 303-311.

BOOTHROYD, J. C., 1978, Mesotidal inlets and estuaries: in DAVIS, R. A., JR., ed., *Coastal and Sedimentary Environments,* Springer-Verlag, New York, pp. 287-360.

———, and G. M. ASHLEY, 1975, Process, bar morphology, and sedimentary structures on braided outwash fans, northeastern Gulf of Alaska: in JOPLING, A. V., and B. C. McDONALD, eds., *Glaciofluvial and Glaciolacustrine Sedimentation,* SEPM Spec. Pub. 23, pp. 193-222.

———, and D. K. HUBBARD, 1973, Genesis of bedforms in mesotidal estuaries: in CRONIN, L. E., ed., *Estuarine Research,* Vol. 2, *Geology and Engineering,* Academic Press, New York, pp. 365-380.

———, and D. NUMMEDAL, 1978, Proglacial braided outwash: a model for humid alluvial fan deposits: in MIALL, A. D., ed., *Fluvial Sedimentology,* Can. Soc. Petroleum Geologists, Mem. 5, pp. 641-668.

BOUMA, A. H., 1962, *Sedimentology of Some Flysch Deposits:* Elsevier, Amsterdam, 168 p.

———, 1979, Continental slopes: in DOYLE, L. J., and O. H. PILKEY, eds., *Geology of Continental Slopes,* SEPM Spec. Pub. 27, pp. 1–16.

———, and C. D. HOLLISTER, 1973, Deep ocean basin sedimentation: in *Turbidites and Deepwater Sedimentation,* Pacific Sec. SEPM Short Course Lecture Notes, pp. 79–118.

———, and others, 1978, Intraslope basin in northwest Gulf of Mexico: in BOUMA, A. H., G. T. MOORE, and J. M. COLEMAN, eds., *Framework, Facies, and Oil-trapping Characteristics of the Upper Continental Margin,* Am. Assoc. Petroleum Geologists, Studies in Geol. 7, pp. 289–302.

———, and others, 1982, Continental shelf and epicontinental seaways: in SCHOLLE, P. A., and D. SPEARING, eds., *Sandstone Depositional Environments,* Am. Assoc. Petrol. Geologists Mem. 31, pp. 281–328.

BOURGEOIS, J., 1980, A transgressive shelf sequence exhibiting hummocky stratification: the Cape Sebastian Sandstone (Upper Cretaceous), southwestern Oregon: *J. Sed. Petrology,* v. 50, pp. 681–702.

BOYD, D. R., and B. F. DYER, 1964, Frio barrier bar system of south Texas: *Gulf Coast Assoc. Geol. Soc. Trans.,* v. 14, pp. 309–322.

BRADLEY, W. H., 1973, Oil shale formed in desert environment: Green River Formation, Wyoming: *Geol. Soc. America Bull.,* v. 84, pp. 1121–1124.

———, and H. P. EUGSTER, 1969, Geochemistry and paleolimnology of the trona deposits and associated authegenic minerals of the Green River Formation of Wyoming: *U.S. Geol. Survey Prof. Paper 496-B,* 71 p.

BRENNER, R. L., 1980, Construction of process-response models for ancient epicontinental seaway depositional systems using partial analogs: *Am. Assoc. Petroleum Geologists Bull.,* v. 64, pp. 1223–1243.

———, and D. J. P. SWIFT, and G. C. GAYNOR, 1985, Reevaluation of coquinoid sandstone depositional model, Upper Jurassic of central Wyoming and south-central Montana: *Sedimentology,* v. 32:3, pp. 363–372.

BRETZ, J. H., A. T. U. SMITH, and G. E. NEFF, 1956, Channeled Scablands of Washington: New data and interpretations: *Geol. Soc. America Bull.,* v. 67, pp. 957–1049.

BRIDGE, J. S., 1984, Large-scale facies sequences in alluvial overbank environments: *J. Sed. Petrology,* v. 54:2, pp. 583–588.

———, and M. R. LEEDER, 1979, A simulation model of alluvial stratigraphy: *Sedimentology,* v. 26, pp. 617–644.

———, and others, 1986, Sedimentology and morphology of a low-sinuosity river: Calamus River, Nebraska Sand Hills: *Sedimentology,* v. 33, pp. 851–870.

BRIDGES, P. H., and M. R. LEEDER, 1976, Sedimentary model for intertidal mudflat channels with examples from the Solway Firth, Scotland: *Sedimentology,* v. 23, pp. 533–552.

BROOKFIELD, M. E., 1977, The origin of bounding surfaces in ancient aeolian sandstones: *Sedimentology,* v. 24, pp. 303–332.

BROWN, L. F., JR., and W. L. FISHER, 1980, Seismic stratigraphic interpretation and petroleum exploration: *Am. Assoc. Petroleum Geologists, Continuing Education Course Note Series No. 16,* 181 p.

———, A. W. CLEAVES, II, and A. W. ERXLEBEN, 1973, Pennsylvanian depositional systems in north-central Texas: *Texas Bur. Econ. Geol. Guidebook No. 14,* 122 p.

BRUUN, P., 1962, Sea level rise as a cause of shore erosion: *ASCE, J. Waterways and Harbors Div. Proc.,* v. 88, pp. 117–130.

BULL, W. B., 1964, Geomorphology of segmented alluvial fans in western Fresno County, California: *U.S. Geological Survey Prof. Paper 352-E,* pp. 89–129.

———, 1972, Recognition of alluvial fan deposits in the stratigraphic record: in RIGBY, K. J., and W. K. HAMBLIN, Recognition of ancient sedimentary environments, *SEPM Spec. Pub. 16,* pp. 68–83.

BYRNE, J. V., D. O. LEROY, and C. M. RILEY, 1959, The chenier plain and its stratigraphy, southwestern Louisiana: *Gulf Coast Assoc. Geol. Socs.,* pp. 237–260.

CACCHIONE, D. A., and J. B. SOUTHARD, 1974, Incipient sediment movement by shoaling internal gravity waves: *J. Geophys. Res.,* v. 79, pp. 2237–2242.

———, G. T. ROWE, and A. MALAHOFF, 1978, Submersible investigation of outer Hudson submarine canyon: in STANLEY, D. J., and G. KELLING, *Sedimentation in Submarine Canyons, Fans, and Trenches,* Dowden, Hutchinson, and Ross, Inc., Stroudsburg, Pa., pp. 42–50.

CAMPBELL, C. V., 1976, Reservoir geometry of a fluvial sheet sandstone: *Am. Assoc. Petroleum Geologists Bull.,* v. 60, pp. 1009–1020.

———, and R. Q. OAKS, JR., 1973, Estuarine sandstone filling tidal scours, Lower Cretaceous, Fall River Formation, Wyoming: *J. Sed. Petrology,* v. 43, pp. 765–778.

CANT, D. J., 1978, Development of a facies model for sandy braided river sedimentation: comparison of the South Saskatchewan River and the Battery Point Formation: in MIALL, A. D., ed., *Fluvial Sedimentology,* Can. Soc. Petroleum Geologists, Mem. 5, pp. 627–639.

———, and R. G. WALKER, 1976, Development of a braided fluvial facies model for the Devonian Battery Point Sandstone, Quebec: *Can. J. Earth Sci.,* v. 13, pp. 102–119.

———, and ———, 1978, Fluvial processes and facies sequences in the sandy braided South Saskatchewan River, Canada: *Sedimentology,* v. 25, pp. 625–648.

CARTER, C. H., 1978, Regressive barrier and barrier-protected deposit, depositional environments, and geographic setting of the late Tertiary Cohansey Sand: *J. Sed. Petrology,* v. 48, pp. 933–950.

CARTER, R. M., and J. K. LINDQVIST, 1977, Balleny Group, Chalky Island, southern New Zealand: an inferred Oligocene submarine canyon and fan complex: *Pacific Geol.,* v. 12, pp. 1–46.

———, and others, 1978, Sedimentation patterns in an ancient arc-trench-ocean basin complex: Carboniferous to Jurassic Rangitata orogen, New Zealand: in STANLEY, D. J., and G. KELLING, eds., *Sedimentation in Submarine Canyons, Fans and Trenches,* Dowden, Hutchinson, and Ross, Inc., Stroudsburg, Pa., pp. 340–361.

CHAPLIN, J. R., 1982, *Field Guidebook to the Paleoenvironments and Biostratigraphy of the Borden and Parts of the Newman and Breathitt Formations (Mississippian–Pennsylvanian) in Northeastern Kentucky:* Guidebook for 12th Ann. Field Conf., Great Lakes Section-SEPM, 196 p.

CHORLEY, R. J., and B. A. KENNEDY, 1971, *Physical Geography: A Systems Approach:* Prentice-Hall, London.

CHOUGH, S., and HESSE, R., 1980, Submarine meandering thalweg and turbidity currents flowing for 4,000 km in the northwest Atlantic Mid-ocean Channel, Labrador Sea: *Geology,* v. 4, pp. 529–533.

CHURCH, M., 1978, Paleohydrological reconstructions from a Holocene valley fill: in MIALL, A. D., ed., *Fluvial Sedimentology*, Can. Soc. Petrol. Geologists, Mem. 5., pp. 743-772.

CLEARY, W. J., and J. R. CONOLLY, 1974, Hatteras deep-sea fan: *J. Sed. Petrology*, v. 44, pp. 1140-1154.

CLEMMENSEN, L. B., and K. ABRAHAMSEN, 1983, Aeolian stratification and facies association in desert sediments, Arran basin (Permian), Scotland: *Sedimentology*, v. 30, pp. 311-339.

CLIFTON, H. E., 1969, Beach lamination-nature and origin: *Marine Geol.*, v. 7, pp. 553-559.

——, 1981a, Progradational sequences in Miocene shoreline deposits, southeastern Caliente Range, California: *J. Sed. Petrology*, v. 51:1, pp. 165-184.

——, 1981b, Submarine canyon deposits, Point Lobos, California: in FRIZZEL, V., ed., *Upper Cretaceous and Paleocene Turbidites, Central California Coast*, Pacific Sec.-SEPM, Field Trip 6, pp. 79-92.

——, 1982, Estuarine deposits: in SCHOLLE, P. A., and D. SPEARING, eds., Sandstone Depositional Environments, *Am. Assoc. Petroleum Geologists Mem. 31*, pp. 179-190.

——, and R. L. PHILLIPS, 1981, Lateral trends and vertical sequences in estuarine sediments, Willapa Bay, Washington: in FIELD, M. E., ed., *Quaternary Depositional Environments of the Pacific Coast:* Pacific Coast Paleogeography Symposium No. 4, SEPM Pac. Sec., pp. 55-71.

——, R. E. HUNTER, and R. L. PHILLIPS, 1971, Depositional structures and processes in the non-barred, high energy nearshore: *J. Sed. Petrology*, v. 41, pp. 651-670.

CLUFF, R. M., 1980, Paleoenvironments of the New Albany Shale Group (Devonian–Mississippian) of Illinois: *J. Sed. Petrology*, v. 50, pp. 767-780.

COLEMAN, J. M., and S. M. GAGLIANO, 1964, Cyclic sedimentation in the Mississippi river delta plain: *Gulf Coast Assoc. Petroleum Geologists Trans.*, v. 14, pp. 67-80.

——, and L. D. WRIGHT, 1975, Modern river deltas: variability of process and sand bodies: in BROUSSARD, M. L., ed., *Deltas, Models for Exploration*, Houston Geol. Soc., pp. 99-149.

——, D. B. PRIOR, and J. F. LINDSAY, 1983, Deltaic influences on shelf edge instability processes: in STANLEY, D. J., and G. T. MOORE, eds., *The Shelfbreak, Critical Interface on Continental Margins*, SEPM Spec. Pub. 33, pp. 121-138.

——, and others, 1981, Morphology and dynamic sedimentology of the eastern Nile delta shelf: in NITTROUER, C. A., ed., *Sedimentary Dynamics of Continental Shelves*, Developments in Sedimentology, v. 32, Elsevier, New York, pp. 301-326.

COLQUHOUN, D. J., T. A. BOND, and D. CHAPPEL, 1972, Santee submergence, example of cyclic submerged and emerged sequences: in NELSON, B. W., ed., *Environmental Framework of Coastal Plain Estuaries*, Geol. Soc. America Mem. 133, pp. 475-498.

COOK, D. O., and D. S. GORSLINE, 1972, Field observations of sand transport by shoaling waves: *Marine Geol.*, v. 13, pp. 31-55.

COOK, H. E., M. E. FIELD, and J. V. GARDNER, 1982, Characteristics of sediments on modern and ancient continental slopes: in SCHOLLE, P. A., and D. SPEARING, eds., *Sandstone Depositional Environments:* Am. Assoc. Petrol. Geologists, Tulsa, Okla., pp. 329-364.

COSTELLO, W. R., and R. G. WALKER, 1972, Pleistocene sedimentology: Credit River, southern Ontario: a new component of the braided river model: *J. Sed. Petrology*, v. 42, pp. 389-400.

COTTER, M. R., 1975, Late Cretaceous sedimentation in a low energy coastal zone: The Ferron Sandstone of Utah: *J. Sed. Petrology,* v. 95, pp. 664–685.

CRAFT, J. H., and J. S. BRIDGE, 1987, Shallow-marine sedimentary processes in the Late Devonian Catskill Sea, New York State: *Geol. Soc. America Bull.,* v. 98, pp. 338–355.

CRAM, J. M., 1979, The influence of continental shelf width on tidal range: Paleo-oceanographic implication: *J. Geol.,* v. 87, pp. 175–228.

CREAGER, J. S., and R. W. STERNBERG, 1972, Some specific problems in understanding bottom sediment distribution and dispersal on the continental shelf: in SWIFT, D. J. P., D. B. DUANE, and O. H. PILKEY, eds., *Shelf Sediment Transport: Process and Pattern,* Dowden, Hutchinson and Ross, Inc., Stroudsburg, Pa., pp. 347–362.

CURRAY, J. R., 1969, Shallow structure of the continental margin: in STANLEY, D. J., ed., *New Concepts of Continental Margin Sedimentation:* Am. Geol. Inst., Washington, D.C., pp. 1–22.

——, and D. G. MOORE, 1964, Holocene regressive littoral sand, Costa de Nayarit, Mexico: in VAN STRAATEN, L. M. J. U., ed., *Deltaic and Shallow Marine Deposits,* Developments in Sedimentology, Elsevier, Amsterdam, pp. 76–82.

——, F. J. EMMEL, and P. J. S. CRAMPTON, 1969, Holocene history of a strandplain, lagoonal coast, Nayarit, Mexico: in CASTANARES, A. A., and F. B. PHLEGER, eds., *Coastal Lagoons—a Symposium,* Universidad Nacional Autonoma, Mexico, pp. 63–100.

CURTIS, D. M., 1970, Miocene deltaic sedimentation, Louisiana Gulf Coast: in MORGAN, J. P., ed., *Deltaic Sedimentation, Modern and Ancient,* SEPM Spec. Pub. 15, pp. 293–308.

DAMUTH, J. E., and others, 1983, Age relationships of distributary channels on Amazon deep-sea fan: implications for fan growth pattern: *Geology,* v. 11, pp. 470–473.

DAVIDSON-ARNOTT, R. G. D., and B. GREENWOOD, 1976, Facies relationships on a barred coast, Kouchibouguac Bay, New Brunswick, Canada: in DAVIS, R. A., JR., and R. L. ETHINGTON, eds., *Beach and Nearshore Sedimentation,* SEPM Spec. Pub. 24, pp. 49–168.

DAVIES, D. K., 1972, Deep sea sediments and their sedimentation, Gulf of Mexico: *Am. Assoc. Petroleum Geologists Bull.,* v. 56, pp. 2212–2239.

——, F. G. ETHRIDGE, and R. R. BERG, 1971, Recognition of barrier environments: *Am. Assoc. Petroleum Geologists Bull.,* v. 55, pp. 550–565.

DAVIS, R. A., JR., 1983, *Depositional Systems:* Prentice-Hall, Englewood Cliffs, N.J., 669 p.

——, and M. O. HAYES, 1984, What is a wave-dominated coast: *Marine Geol.,* v. 60:1/4, pp. 313–330.

——, and others, 1972, Comparison of ridge and runnel systems in tidal and non-tidal environments: *J. Sed. Petrology,* v. 32, pp. 413–421.

DAVIS, W. M., 1899, The geographical cycle: *Geog. Jour.,* v. 14, pp. 481–504.

DEAN, W. E., and others, 1978, Cyclic sedimentation along the continental margin of northwest Africa: in *Initial Reports of the Deep-Sea Drilling Project,* v. 41, pp. 965–990.

DEERY, J. R., and J. D. HOWARD, 1977, Origin and character of washover fans on the Georgia coast, U.S.A.: *Trans. Gulf Coast Assoc. Geol. Socs.,* v. 27, pp. 259–271.

DEMICCO, R. V., and E. G. KORDESCH, 1986, Facies sequences of a semi-arid closed basin: the Lower Jurassic East Berlin Formation of the Hartford Basin, New England, U.S.A.: *Sedimentology,* v. 33, pp. 107–118.

DENNY, C. S., 1967, Fans and pediments: *Am. J. Sci.,* v. 265, pp. 81–105.

DE PRATTER, C. B., and J. D. HOWARD, 1977, History of shoreline changes determined by archeological dating: Georgia coast: *Gulf Coast Assoc. Geol. Socs.,* v. 27, pp. 252–258.

DE RAAF, J. F. M., J. R. BOERSMA, and A. VAN GELDER, 1977, Wave generated structures and sequences from a shallow marine succession, Lower Carboniferous, County Cork, Ireland: *Sedimentology,* v. 24, pp. 451–483.

DICKINSON, K. A., H. L. BERRYHILL, JR., and C. W. HOLMES, 1972, Criteria for recognizing ancient barrier coastlines: in RIGBY, J. K., and W. K. HAMBLIN, eds., *Recognition of Ancient Sedimentary Environments,* SEPM Spec. Pub. 16, pp. 192–214.

DIETZ, R. S., and J. C. HOLDEN, 1974, Collapsing continental rises: actualistic concept of geosynclines—a review: in DOTT, R. H., JR., and R. H. SHAVER, eds., *Modern and Ancient Geosynclinal Sedimentation,* SEPM Spec. Pub. 19, pp. 14–25.

DILL, R. F., 1964, Sedimentation and erosion in Scripps submarine canyon head: in MITTER, R. L., ed., *Papers in Marine Geology* (Shepard Commemorative Volume), Macmillan, New York, pp. 23–41.

——, 1981, Role of multiple-headed submarine canyons, river mouth migration, and episodic activity in generation of basin-filling turbidity currents (abs): *Am. Assoc. Petroleum Geologists Bull.,* v. 65, pp. 918.

DILLON, W. P., 1970, Submergence effects on a Rhode Island barrier lagoon and inferences on migration of barriers: *J. Geol.,* v. 78, pp. 94–106.

DONALDSON, A. C., R. H. MARTIN, and W. H. KANES, 1970, Holocene Guadalupe delta of Texas Gulf Coast: in MORGAN, J. P., ed., *Deltaic Sedimentation, Modern and Ancient,* SEPM, Spec. Pub. 15, pp. 107–137.

DONOVAN, R. N., 1975, Devonian lacustrine limestones at the margin of the Orcadian Basin, Scotland: *J. Geol. Soc. London,* v. 131, pp. 489–510.

DORJES, J., and J. D. HOWARD, 1975, Fluvial-marine transition indicators in an estuarine environment, Ogeechee River-Ossabaw Sound: *Senckenbergiana Marit.,* v. 7, pp. 137–180.

——, and others, 1970, Sedimentologie uns Makrobenthos der Nordergrunde und der Aussenjade (Nordsee): *Senkenbergiana Marit.,* v. 2, pp. 31–59.

DOTT, R. H., JR., and J. BOURGEOIS, 1982, Hummocky stratification: significance of its variable bedding sequences: *Geol. Soc. Am. Bull.,* v. 93, pp. 663–680.

——, and C. W. BYERS, 1981, SEPM Research Conference on Modern and Ancient Cratonic Sedimentation: *J. Sed. Petrology,* v. 51:1, pp. 330–347.

DOYLE, L. J., and O. H. PILKEY, eds., 1979, *Geology of Continental Slopes:* SEPM Spec. Pub. 27, 374 p.

DRAKE, D. E., P. G. HATCHER, and G. H. KELLER, 1978, Suspended particulate matter and mud deposition in upper Hudson Submarine Canyon: in STANLEY, D. J., and G. KELLING, eds., *Sedimentation in Submarine Canyons, Fans, and Trenches,* Dowden, Hutchinson and Ross, Stroudburg, Pa., pp. 33–41.

——, R. L. KOLPACK, and P. J. FISCHER, 1972, Sediment transport on the Santa Barbara-Oxnard shelf, Santa Barbara Channel, California: in SWIFT, D. J. P., D. B. DUANE, and O. H. PILKEY, eds., *Shelf Sediment Transport: Process and Pattern,* Dowden, Hutchinson and Ross, Stroudsburg, Pa., pp. 307–331.

DUANE, D. B., and others, 1972, Linear shoals on the Atlantic inner continental shelf, Florida to Long Island: in SWIFT, D. J. P., D. B. DUANE, and O. H. PILKEY, eds., *Shelf Sediment*

Transport: Process and Pattern, Dowden, Hutchinson, and Ross, Stroudsburg, Pa., pp. 447–498.

Duc, A. W., and R. S. Tye, 1987, Evolution and stratigraphy of a regressive barrier/back-barrier complex: Kiawah Island, South Carolina: *Sedimentology,* v. 34, pp. 237–251.

Eittreim, S., and M. Ewing, 1972, Suspended particulate matter in the deep waters of the North American basin: in Gordon, A. L., ed., *Studies in Physical Oceanography 2,* Gordon & Breach, New York, pp. 123–168.

Elliott, T., 1978a, Clastic shorelines: in Reading, H. G., ed., *Sedimentary Environments and Facies,* Elsevier, New York, pp. 143–177.

———, 1978b, Deltas: in Reading, H. G., ed., *Sedimentary Environments and Facies,* Elsevier, New York, pp. 97–142.

Embley, R. W., 1976, New evidence for occurrence of debris flow deposits in the deep sea: *Geology,* v. 4:6, pp. 371–374.

Emery, K. O., 1968, Relict sediments on continental shelves of the world: *Am. Assoc. Petroleum Geologists Bull.,* v. 52, pp. 445–464.

———, 1977, Stratigraphy and structure of pull-apart margins: in McFarlan, E., C. L. Drake, and L. S. Pittmann, eds., *Geology of Continental Margins,* Am. Assoc. Petroleum Geologists, Cont. Education Course Note Series No. 5, pp. B1–B20.

———, and others, 1970, Continental rise off eastern North America: *Am. Assoc. Petroleum Geologists Bull.,* v. 54, pp. 44–108.

Ettensohn, F. R., 1980, An alternative to the barrier-shoreline model for deposition of Mississippi and Pennsylvania rocks in northeastern Kentucky: *Geol. Soc. Am. Bull.,* v. 91:3, pt. I, pp. 130–135.

Eugster, H. P., and L. A. Hardie, 1975, Sedimentation in an ancient playa-lake complex: the Wilkins Peak Member of the Green River Formation of Wyoming: *Geol. Soc. Am. Bull.,* v. 86, pp. 319–334.

Evans, G., 1975, Intertidal flat deposits of the Wash, western margin of the North Sea: in Ginsburg, R. N., ed., *Tidal Deposits: A Casebook of Recent Examples and Fossil Counterparts,* Springer-Verlag, Berlin, pp. 13–20.

Ewing, J. A., 1973, Wave-induced bottom currents on the outer shelf: *Marine Geol.,* v. 15, pp. M31–M35.

Farre, J. A., and others, 1983, Breaching the shelf break: Passage from youthful to mature phase in submarine canyon evolution: in Stanley, D. J., and G. T. Moore, eds., *The Shelf Break: Critical Interface on Continental Margins,* SEPM Spec. Pub. 33, pp. 25–39.

Farrell, K. M., 1987, Sedimentology and facies architecture of overbank deposits of the Mississippi River, False River region, Louisiana: in Etheridge, F. G., R. M., Flores, and M. D. Harvey, eds., *Recent Developments in Fluvial Sedimentology,* Soc. Econ. Paleontologists and Mineralogists Spec. Pub. 39, Tulsa, Okla., pp. 111–120.

Ferm, J. C., and J. C. Horne, 1979, *Carboniferous Depositional Environments in the Appalachian Region:* University of South Carolina, Carolina Coal Group, 763 p.

———, and others, 1971, Carboniferous depositional environments in northeastern Kentucky: *Kentucky Geol. Soc. Guidebook,* Lexington, Ky., 30 p.

Field, M. E., 1980, Sand bodies on coastal plain shelves: Holocene record of the U.S. Atlantic inner shelf off Maryland: *J. Sed. Petrology,* v. 50, pp. 505–528.

———, 1981, Sediment mass transport in basins: controls and patterns: in DOUGLAS, R. G., I. P. COLBURN, and D. S. GORSLINE, eds., *Depositional Systems of Active Continental Margin Basins,* Short Course Notes: Pac. Sec.-SEPM, pp. 61-84.

———, and D. B. DUANE, 1976, Post-Pleistocene history of the United States inner continental shelf: significance to the origin of barrier islands: *Geol. Soc. Am. Bull.,* v. 87, pp. 691-702.

———, and B. D. EDWARDS, 1980, Slopes of the southern California Continental Borderland: a regime of mass transport: in FIELD, M. E., and others, eds., *Quaternary Depositional Environments of the Pacific Coast:* Pacific Coast Paleogeography Symposium No. 4, SEPM Pac. Sec., pp. 169-184.

———, and O. H. PILKEY, 1971, Deposition of deep-sea sands: Comparison of two areas of the Carolina continental rise: *J. Sed. Petrology,* v. 41:2, pp. 526-536.

———, and others, 1981, Sand waves on an epicontinental shelf, northern Bering Sea: *Marine Geol.,* v. 42, pp. 233-258.

FISCHER, A. G., and M. A. ARTHUR, 1977, Secular variations in the pelagic realm: in COOK, H. E., and P. ENOS, eds., *Deep-Water Carbonate Environments:* SEPM Spec. Pub. 25, pp. 19-50.

FISCHER, W. L., and J. H. MCGOWEN, 1967, Depositional systems in the Wilcox Group of Texas and their relationship to occurrence of oil and gas: *Gulf Coast Assoc. Geol. Socs. Trans.,* v. 17, pp. 105-125.

———, and ———, 1969, Depositional systems in Wilcox Group (Eocene) of Texas and their relation to occurrence of oil and gas: *Am. Assoc. Petrol. Geologists Bull.,* v. 53, pp. 30-54.

———, and others, 1969, Delta systems in the exploration for oil and gas: *Texas Bur. Econ. Geol. Res. Colloquium,* 78 p.

FITZGERALD, D. M., S. PENLAND, and D. NUMMEDAL, 1984, Control of barrier island shape by inlet sediment bypassing: East Frisian Islands, West Germany: *Marine Geol.,* v. 60:1/4, pp. 355-376.

FLEMMING, B. W., 1980, Sand transport and bedform patterns on the continental shelf between Durban and Port Elizabeth: *Sedim. Geol.,* v. 26, pp. 179-206.

FOLK, R. L., 1971, Longitudinal dunes of the northwestern edge of the Simpson Desert, Australia, 1, Geomorphology and grain size relationships: *Sedimentology,* v. 16, pp. 5-54.

FRASER, G. S., 1976, Sedimentology of a Middle Ordovician quartz arenite-carbonate transition in the Upper Mississippi Valley: *Geol. Soc. Am. Bull.,* v. 86, pp. 833-845.

———, 1982, Upper Triassic shelf sedimentation, Prudhoe Bay, Alaska (Abs): *11th Internatl. Cong. on Sedimentology,* p. 98.

———, and R. H. CLARKE, 1976, Transgressive-regressive shelf deposition, Shublik Formation, Prudhoe Bay area, Alaska (abs): *Am. Assoc. Petroleum Geologists Bull.,* v. 60, p. 672.

———, and J. C. COBB, 1982, Late Wisconsinian proglacial sedimentation along the West Chicago Moraine in northeastern Illinois: *J. Sed. Petrology,* v. 52, pp. 473-491.

———, and N. C. HESTER, 1977, Sediments and sedimentary structures of a beach ridge complex, southwestern shore of Lake Michigan: *J. Sed. Petrology,* v. 47, pp. 1187-1200.

———, and L. J. SUTTNER, 1986, *Alluvial Fans and Fan Deltas:* International Human Resources Development Corporation Boston, 199 pp.

——, N. K. BLEUER, and N. D. SMITH, 1983, History of Pleistocene alluviation of the middle and upper Wabash Valley: in SHAVER, R. H., and J. A. SUNDERMAN, eds., *Field Trips in Midwestern Geology,* vol. 1, Geol. Soc. America and Indiana Geol. Survey, pp. 200-224.

FRAZIER, D. E., 1974, Depositional episodes: their relationship to the Quarternary stratigraphic framework in the northwestern portion of the Gulf Basin: *Texas Bur. Econ. Geol. Circ. 74-1,* 28 p.

FREY, R. W., and T. V. MAYOU, 1971, Decapod burrows in Holocene barrier island beaches and washover fans: *Senckenbergiana Marit.,* v. 3, pp. 53-77.

FRIEDMAN, G. M., and K. G. JOHNSON, 1966, The Devonian Catskill deltaic complex of New York, type example of a "tectonic delta complex": in SHIRLEY, M. L., ed., *Deltas in Their Geologic Framework:* Houston Geol. Soc., Houston, Tex., pp. 171-188.

——, and J. E. SANDERS, 1978, *Principles of Sedimentology:* John Wiley & Sons, New York, 792 p.

FRYBERGER, S. G., T. S. AHLBRANDT, and S. ANDREWS, 1979, Origin, sedimentary features and significance of low-angle eolian "sand sheet" deposits, Great Sand Dunes National Monument and vicinity, Colorado: *J. Sed. Petrology,* v. 49, pp. 733-746.

GALLOWAY, W. E., 1975, Process framework for describing the morphologic and stratigraphic evolution of deltaic depositional systems: in BROUSSARD, M. L., ed., *Deltas, Models for Exploration,* Houston Geol. Soc., pp. 87-98.

——, 1981, Depositional architecture of Cenozoic Gulf Coastal Plain fluvial systems: in ETHERIDGE, F. G., and R. M. FLORES, eds., *Non-marine Depositional Environments: Models for Exploration,* SEPM Spec. Pub. 31, pp. 127-155.

——, and D. K. HOBDAY, 1983, *Terrigenous Clastic Depositional Systems:* Springer-Verlag, New York, 423 p.

GARDNER, J. V., W. E. DEAN, and L. JANSA, 1978, Sediments recovered from the northwest African continental margin, Leg 41, Deep Sea Drilling Project: in *Initial Reports of the Deep Sea Drilling Project,* v. 41, pp. 1121-1134.

GARNER, H. F., 1959, Stratigraphic–sedimentary significance of contemporary climate and relief in four regions of the Andes Mountains: *Geol. Soc. Am. Bull.,* v. 70, pp. 1327-1368.

GARRISON, L. E., N. E. KENYON, and A. H. BOUMA, 1982, Channel systems and lobe construction in the Mississippi Fan: *Geomarine Letters,* v. 2, pp. 31-39.

GILBERT, G. K., 1885, The topographic features of lake shores: *Fifth Ann. Rept. U.S. Geol. Survey,* Washington, D.C., pp. 75-123.

GLENNIE, K. W., 1970, Desert sedimentary environments: *Developments in Sedimentology 14,* Elsevier, Amsterdam, 222 p.

——, 1972, Permian Rotliegendes of Northwest Europe interpreted in the light of modern desert sedimentation studies: *Am. Assoc. Petrol. Geol. Bull.,* v. 56, pp. 1048-1071.

——, 1983, Lower Permian Rotliegend desert sedimentation in the North Sea area: in BROOKFIELD, M. E., and T. S. AHLBRANDT, eds., *Eolian Sediments and Processes,* Elsevier, Amsterdam, pp. 521-542.

GOLDSMITH, V., 1973, Internal geometry and origin of vegetated coastal sand dunes: *J. Sed. Petrology,* v. 43, pp. 1128-1142.

GORDON, E. A., and J. S. BRIDGE, 1987, Evolution of Catskill (Upper Devonian) river systems: Intra- and extrabasinal controls: *J. Sed. Petrology,* v. 57:2, pp. 234-249.

GORSLINE, D. S., 1981, Fine sediment transport and deposition in active margin basins: in DOUGLAS, R. G., I. P. COLBURN, and D. S. GORSLINE, eds., *Depositional Systems of Active Continental Margin Basins:* Short Course Notes: Pac. Sec. SEPM, pp. 39–60.

GOULD, H. R., and E. McFARLAN, JR., 1959, Geologic history of the chenier plain, south-western Louisiana: *Gulf Coast Assoc. Geol. Socs.,* v. 9, pp. 261–272.

GRAHAM, S. A., and S. B. BACHMAN, 1983, Structural controls on submarine fan geometry and internal architecture: Upper La Jolla fan system, offshore Southern California: *Am. Assoc. Petroleum Geologists Bull.,* v. 67:1, pp. 83–96.

GREENWOOD, B., 1984, Hummocky lamination in the surf zone (abs): *1st Ann. Midyear Mtg.,* SEPM, p. 36.

——, and D. J. SHERMAN, 1986, Hummocky cross-stratification in the surf zone: flow parameters and bedding genesis: *Sedimentology,* v. 33, pp. 33–45.

GREER, S. A., 1975, Sandbody geometry and sedimentary facies at the estuary–marine transi-tion zone, Ossabaw Sound, Georgia: *Senckenbergiana Marit.,* v. 7, pp. 105–136.

GROW, J. A., R. E. MATTICK, and J. S. SCHLEE, 1979, Multichannel seismic depth sections and internal velocities over outer continental slope between Cape Hatteras and Cape Cod: in WATKINS, J. R., and others, eds., *Geological and Geophysical Investigations of Conti-nental Margins:* Am. Assoc. Petroleum Geologists Mem. 29, pp. 65–83.

GUINASSO, N. H., and D. R. SCHINK, 1975, Quantitative estimates of biological mixing rates in abyssal sediments: *Geophys. Res.,* v. 80, pp. 3032–3043.

HAQ, B. U., J. HARDENBOL, and P. R. VAIL, 1987, Chronology of fluctuating sea levels since the Triassic: *Science,* v. 235, pp. 1156–1165.

HALLAM, A., 1984, Pre-Quaternary sea-level changes: *Ann. Rev. Earth Planet Sci.,* v. 12, pp. 205–243.

HALSEY, S. D., 1979, Nexus: new model of barrier island development: in LEATHERMAN, S. P., ed., *Barrier Islands from the Gulf of St. Lawrence to the Gulf of Mexico,* Academic Press, New York, pp. 185–210.

HAMBLIN, A. P., and R. G. WALKER, 1979, Storm-dominated shallow marine deposits: the Fernie-Koutenay (Jurassic) transition, southern Rocky Mountains: *Can. J. Earth Sci.,* v. 16, pp. 1673–1690.

HARBAUGH, J. W., and G. BONHAM-CARTER, 1977, Computer simulation of continental mar-gin sedimentation: in GOLDBERG, E. D., and others, eds., *The Sea,* vol. 6, John Wiley & Sons, New York, pp. 623–649.

HARDIE, L. A., and H. P. EUGSTER, 1970, The evolution of closed-basin brines: *Mineralog. Soc. Am. Spec. Pub. 3,* pp. 273–290.

——, J. P. SMOOT, and H. P. EUGSTER, 1978, Saline lakes and their deposits: a sedimento-logical approach: in MATTER, A., and M. TUCKER, eds., *Modern and Ancient Lake Sedi-ments,* Spec. Pub. 2, Internatl. Assoc. Sedimentologists, pp. 7–41.

HARMS, J. C., and others, 1975, *Depositional Environments as Interpreted from Primary Sedimentary Structures and Stratification Sequences:* SEPM, Short Course Notes No. 2, 161 p.

HARRISON, S. C., 1975, Tidal-flat complex, Delmarva Peninsula, Virginia: in GINSBURG, R. N., ed., *Tidal Deposits, a Casebook of Recent Examples and Fossil Counterparts,* Springer-Verlag, pp. 31–38.

HAYES, M. O., 1967, Hurricanes as geological agents: case studies of Hurricane Carla, 1961, and Cindy, 1963: *Texas Bur. Econ. Geol. Rept. Inv. 61,* 54 p.

——, 1976, Morphology of sand accumulation in estuaries: an introduction to the symposium: in CRONIN, L. E., ed., *Estuarine Res. v. II, Geology and Engineering,* Academic Press, London, pp. 3–22.

HAYES, M. O., 1979, Barrier island morphology as a function of tidal and wave regime: in LEATHERMAN, S. P., ed., *Barrier Islands from the Gulf of St. Lawrence to the Gulf of Mexico,* Academic Press, New York, pp. 1–27.

HECKEL, P. H., 1972, Recognition of ancient shallow marine environments: in RIGBY, J. K., and W. K. HAMBLIN, eds., *Recognition of Ancient Sedimentary Environments,* SEPM Spec. Pub. 16, pp. 226–286.

HEEZEN, B. C., and C. HOLLISTER, 1964, Deep-sea current evidence from abyssal sediments: *Marine Geology,* v. 1:2, pp. 141–174.

——, and ——, 1971, *The Face of the Deep:* Oxford University Press, New York, 659 p.

——, C. D. HOLLISTER, and W. F. RUDDIMAN, 1966, Shaping of the continental rise by deep geostrophic contour currents: *Science,* v. 152, pp. 502–508.

HEIN, F. J., and R. G. WALKER, 1977, Bar evolution and development of stratification in the gravelly braided Kicking Horse River, British Columbia: *Can. J. Earth Sci.,* v. 14, pp. 562–570.

HELLER, P. L., and W. R. DICKINSON, 1985, Submarine ramp facies model for delta-fed, sand-rich turbidite systems: *Am. Assoc. Petroleum Geologists Bull.,* v. 69:6, pp. 960–979.

HERZER, R. M., and D. W. LEWIS, 1979, Growth and burial of a submarine canyon off Motunau, north Canterbury, New Zealand: *Sedim. Geology,* v. 24:1, pp. 69–83.

HESSE, R., 1974, Long-distance continuity of turbidites: possible evidence for an Early Cretaceous trench–abyssal plain in the East Alps: *Geol. Soc. Am. Bull.,* v. 85, pp. 859–870.

HEWARD, A. P., 1978, Alluvial fan and lacustrine sediments from the Stephanian A and B (La Magdalena, Cinera-Matallana and Sabero coalfields, northern Spain: *Sedimentology,* v. 25, pp. 523–544.

——, 1981, A review of wave-dominated clastic shoreline deposits: *Earth Sci. Rev.,* v. 17, pp. 223–276.

HINE, A. C., 1976, Bedform distribution and migration patterns on tidal deltas in the Chatham Harbor estuary, Cape Cod, Massachusetts: in CRONIN, L. E., ed., *Estuarine Res. Vol. II, Geology and Engineering,* Academic Press, London, pp. 235–252.

HOBDAY, D. K., and A. J. TANKARD, 1978, Transgressive barrier and shallow-shelf interpretation of the lower Paleozoic Peninsula Formation, South Africa: *Geol. Soc. Am. Bull.,* v. 89, pp. 1733–1744.

HOLLISTER, C. D., and others, 1972, *Initial Reports of the Deep Sea Drilling Project:* U.S. Government Printing Office, Washington, D.C., v. 11, 1077 p.

HOOKE, R. L., 1967, Processes on arid-region alluvial fans: *J. Geol.,* v. 75, pp. 438–460.

HORN, D. R., J. I. EWING, and M. EWING, 1972, Graded bed sequences emplaced by turbidity currents north of 2° N in the Pacific, Atlantic, and Mediterranean: *Sedimentology,* v. 18, pp. 247–276.

——, and others, 1971, Turbidites of the Hatteras and Sohm abyssal plains, western North Atlantic: *Marine Geol.,* v. 11, pp. 287–324.

HORNE, J. C., 1979, Estuarine deposits in the Carboniferous of the Pocahantas Basin: in FERM, J. C., and J. C. HORNE, eds., *Carboniferous Depositional Environments in the Appalachian Region,* University of South Carolina Coal Group, pp. 692–706.

——, and others, 1978, Depositional models in coal exploration and mine planning in Appalachian region: *Am. Assoc. Petroleum Geologists Bull.,* v. 62, pp. 2379–2411.

HOUBOLT, J. J., 1968, Recent sediments in the southern bight of the North Sea: *Geol. Mijnbouw,* v. 47:4, pp. 245–273.

——, and J. D. M. JONKER, 1968, Recent sediments in the eastern part of the Lake of Geneve (Lac Leman): *Geol. Mijnbouw,* v. 47, pp. 131–148.

HOWARD, J. D., 1972, Trace fossils as criteria for recognizing shorelines in stratigraphic record: in RIGBY, J. K., and W. K. HAMBLIN, eds., *Recognition of Ancient Sedimentary Environments:* SEPM Spec. Pub. 16, pp. 215–225.

——, and H. E. REINECK, 1972, Physical and biogenic sedimentary structures of the nearshore shelf: *Senckenbergiana Marit.,* v. 4, pp. 81–123.

——, and ——, 1981, Depositional facies of high-energy beach-to-offshore sequence: a comparison with low-energy sequence: *Am. Assoc. Petrol. Geologists Bull.,* v. 65, pp. 807–830.

——, and R. M. SCOTT, 1983, Comparison of Pleistocene and Holocene barrier island beach-to-offshore sequences, Georgia and northeast Florida coasts, U.S.A.: *Sedim. Geology,* v. 34, pp. 167–183.

——, G. H. REMMER, and J. L. JEWITT, 1975, Hydrography and sediment of the Duplin River, Sapelo Island, Georgia: *Senckenbergiana Marit.,* No. 7, pp. 237–256.

HOWARTH, M. J., 1982, Tidal currents of the continental shelf: in STRIDE, A. H., ed., *Offshore Tidal Sands,* Chapman and Hall, New York, pp. 10–26.

HOWELL, D. G., and J. E. JOYCE, 1981, Field guide to the Upper Cretaceous Pigeon Point Formation: in FRIZZEL, V., ed., *Upper Cretaceous and Paleocene Turbidites, Central California Coast,* Pacific Sec.-SEPM Field Trip 6, pp. 61–70.

——, and M. H. LINK, 1979, Eocene conglomerate sedimentology and basin analysis, San Diego and southern California borderland: *J. Sed. Petrology,* v. 49, pp. 517–540.

——, and W. R. NORMARK, 1982, Sedimentology of submarine fans: in SCHOLLE, P. A., and D. R. SPEARING, eds., *Sandstone Depositional Environments:* Am. Assoc. Petroleum Geologists, Mem., 31, Tulsa, Okla., pp. 365–404.

——, and J. G. VEDDER, 1983–1984, Ferrelo Fan, California: Depositional system influenced by eustatic sea level changes: *Geomarine Letters,* v. 3:2–4, pp. 187–192.

——, and R. VON HUENE, 1981, Tectonics and sediment along active continental margins: in DOUGLAS, R. G., and others, eds., *Depositional Systems of Active Continental Margin Basins,* SEPM Pacific Sec., Los Angeles, pp. 1–13.

HOYT, J. H., 1967, Barrier island formation: *Geol. Soc. Am. Bull.,* v. 78, pp. 1125–1136.

——, 1972, Erosional and depositional estuarine "Terraces," southeastern United States: in NELSON, B. W., ed., *Environmental Framework of Coastal Plain Estuaries,* Geol. Soc. Am. Mem. 133, pp. 465–474.

——, and J. V. HENRY, 1967, Influence of island migration on barrier island sedimentation: *Geol. Soc. Am. Bull.,* v. 78, pp. 77–86.

HUBBARD, D. K., and J. H. BARWIS, 1976, Discussion of tidal inlet sand deposits: examples from the South Carolina coast: in HAYES, M. O., and T. W. KANA, eds., *Terrigenous*

Clastic Depositional Environments, University of South Carolina, Coastal Res. Div., Tech. Rep. II-CRD, pp. II-128–II-142.

——, G. OERTEL, and D. NUMMEDAL, 1979, The role of waves and tidal currents in the development of tidal-inlet sedimentary structures and sand-body geometry: Examples from North Carolina, South Carolina, and Georgia: *J. Sed. Petrology,* v. 49:4, pp. 1073–1091.

HUBERT, J. F., and M. G. HYDE, 1982, Sheet-flow deposits of graded beds and mudstones on an alluvial sandflat–playa system: Upper Triassic Blomidon redbeds, St. Marys Bay, Nova Scotia: *Sedimentology,* v. 28, pp. 457–474.

——, A. A. REED, and P. J. CAREY, 1976, Paleogeography of the East Berlin Formation, Newark Group, Connecticut Valley: *Am. J. Sci.,* v. 276, pp. 1183–1207.

HUNT, C. B., and D. R. MABEY, 1966, Stratigraphy and structure of Death Valley, California: *U.S. Geol. Survey Prof. Paper 494A.*

——, and others, 1966, Hydrologic Basin, Death Valley, California, *U.S. Geol. Survey Prof. Paper 494B,* 138 p.

HUNT, R. E., D. J. P. SWIFT, and H. PALMER, 1977, Constructional shelf topography, Diamond Shoals, North Carolina: *Geol. Soc. America Bull.,* v. 88, pp. 299–311.

HUNTER, R. E., 1977, Basic types of stratification in small eolian dunes: *Sedimentology,* v. 24, pp. 361–388.

——, and D. M. RUBIN, 1983, Interpreting cyclic crossbedding, with an example from the Navajo Sandstone: in BROOKFIELD, M. E., and T. S. AHLBRANDT, eds., *Eolian Sediments and Processes,* Elsevier, Amsterdam, pp. 429–454.

INGERSOLL, R. V., 1978, Submarine fan facies of the Upper Cretaceous Great Valley Sequence, northern and southern central California: *Sedim. Geol.,* v. 21, pp. 205–230.

——, and S. A. GRAHAM, 1983, Recognition of the shelf-slope break along ancient, tectonically active continental margins: in STANLEY, D. J., and others, eds., *The Shelf Break: Critical Interface on Continental Margins:* SEPM Spec. Pub. 33, pp. 107–117.

INGLE, J. C., 1966, *The Movement of Beach Sand: Developments in Sedimentology,* v. 5, Elsevier, New York, 221 p.

——, 1981, Origins of Neogene diatomites around the North Pacific rim: in GARRISON, R. E., and R. G. DOUGLAS, eds., *Monterey Formation and Related Siliceous Rocks of California,* Pacific Sec. SEPM, pp. 159–180.

——, D. E. KARIG, and A. H. BOUMA, 1973, Leg 31, Western Pacific floor: *Geotimes,* v. 18, pp. 22–25.

INMAN, D. L., and C. E. NORDSTROM, 1971, On the tectonic and morphologic classification of coasts: *J. Geol.,* v. 79, pp. 1–21.

JACKA, A. D., and others, 1968, Permian deep-sea fans of the Delaware Mountain Group (Guadalupian) Delaware Basin: in SILVER, B. A., ed., *Guadalupian Facies, Apache Mountain Area, West Texas,* Permian Basin Sec. SEPM Pub. 68-11, pp. 49–67.

JACKSON, R. G., II, 1975, Velocity-bedform-texture patterns of meander bends in the lower Wabash River of Illinois and Indiana: *Geol. Soc. Am. Bull.,* v. 86, pp. 1511–1522.

——, 1976, Depositional model of point bars in the lower Wabash River: *J. Sed. Petrology,* v. 46, pp. 579–594.

JOHNSON, D. W., 1919, *Shore Processes and Shoreline Development:* John Wiley & Sons, New York, 584 p.

JOHNSON, H. D., 1977, Shallow marine sand bar sequences: an example from the late Precambrian of North Norway: *Sedimentology,* v. 24, pp. 245–270.

JOHNSON, K. G., and G. M. FRIEDMAN, 1969, The Tully clastic correlatives (Upper Devonian) of New York State: A model for recognition of alluvial, dune (?), tidal, nearshore (bar and lagoon) and offshore sedimentary environments: *J. Sed. Petrology,* v. 39, pp. 451–485.

KARCZ, I., 1972, Sedimentary structures formed by flash floods in southern Israel: *Sedim. Geol.,* v. 7, pp. 161–182.

KARL, H. A., 1976, Processes influencing transportation and deposition of sediment on the continental shelf, southern California: unpublished Ph.D. thesis, University of Southern California, 331 p.

KELLER, G. H., and A. F. RICHARDS, 1967, Sediments of the Malacca Straits, southeast Asia: *J. Sed. Petrology,* v. 37, pp. 102–127.

KELLING, G., and D. J. STANLEY, 1976, Sedimentation in canyon, slope, and base-of-slope environments: in STANLEY, D. J., and D. J. P. SWIFT, eds., *Marine Sediment Transport and Environmental Management,* John Wiley & Sons, New York, pp. 379–436.

KENYON, N. H., 1970, Sand ribbons of European tidal seas: *Marine Geol.,* v. 9, pp. 25–39.

———, and others, 1981, Offshore tidal sandbanks as indicators of net sand transport and as potential deposits: in NIO, S. D., and others, eds., *Holocene Marine Sedimentation in the North Sea Basin,* Internatl. Assoc. Sedimentologists Spec. Pub. 5, pp. 257–268.

KEPFERLE, R. C., 1977, Stratigraphy, petrology, and depositional environments of the Kenwood Siltstone Member, Borden Formation (Mississippian) Kentucky and Indiana: *U.S. Geol. Survey Prof. Paper 1007,* 49 p.

KLEIN, G. DEV., 1970, Tidal origin of a Precambrian quartzite—the lower fine-grained quartzite (Dalradian) of Islay, Scotland: *J. Sed. Petrology,* v. 40, pp. 973–985.

———, 1971, A sedimentary model for determining paleotidal range: *Geol. Soc. Am. Bull.,* v. 82, pp. 2585–2592.

———, 1977, Tidal circulation model for deposition of clastic sediment in epeiric and mioclinal shelf seas: *Sedim. Geol.,* v. 18, pp. 1–12.

———, 1982, Probable sequential arrangement of depositional systems on cratons: *Geology,* v. 10, pp. 17–22.

———, 1984, Relative rates of tectonic uplift as determined from episodic turbidite deposition in marine basins: *Geology,* v. 12, pp. 48–50.

———, and T. A. RYER, 1978, Tidal circulation patterns in Precambrian, Paleozoic, and Cretaceous epeiric and mioclinal shelf seas: *Geol. Soc. Am. Bull.,* v. 89, pp. 1050–1058.

———, and others, 1982, Sedimentology of a subtidal, tide-dominated sand body in the Yellow Sea, southwest Korea: *Marine Geology,* v. 50, pp. 221–240.

KNOX, J. C., 1972, Valley alluviation in southwestern Wisconsin: *Assoc. Am. Geogr. Ann.,* v. 62, pp. 401–410.

KOCUREK, G., 1981a, Erg reconstruction: the Entrada Sandstone (Jurassic) of northern Utah and Colorado: *Paleog. Paleoclim. Paleoecology,* v. 36, pp. 125–153.

———, 1981b, Significance of interdune deposits and bounding surfaces in aeolian dune sands: *Sedimentology,* v. 28, pp. 753–780.

———, 1984, Origin of first-order bounding surfaces in aeolian sandstones: *Sedimentology,* v. 31, pp. 123–127.

———, and G. FIELDER, 1982, Adhesion structures: *J. Sed. Petrology,* v. 52:4, pp. 1229–1242.

———, and J. NIELSON, 1986, Conditions favourable for the formation of warm-climate aeolian sand sheets: *Sedimentology,* v. 33, pp. 795–816.

KOLB, C. R., 1963, Sediments forming the bed and banks of the lower Mississippi River and their effect on river migration: *Sedimentology,* v. 2, pp. 227–234.

———, and J. R. VAN LOPIK, 1966, Depositional environments of the Mississippi River deltaic plain—southeastern Louisiana: in SHIRLEY, M. L., ed., *Deltas in Their Geologic Framework,* Houston Geol. Soc., Houston, Tex., pp. 17–61.

KOMAR, P. D., 1976, *Beach Processes and Sedimentation:* Prentice-Hall, Englewood Cliffs, N.J., 429 p.

———, R. H. NEWDECK, and L. D. KULM, 1972, Observations and significance of deep-water oscillatory ripple marks on the Oregon continental shelf: in SWIFT, D. J. P., D. B. DUANE, and O. H. PILKEY, eds., *Shelf Sediment Transport: Process and Patterns,* Dowden, Hutchinson, and Ross, Stroudsburg, Pa., pp. 601–619.

KRAFT, J. C., 1971, Sedimentary facies patterns and geologic history of a Holocene marine transgression: *Geol. Soc. America Bull.,* v. 82, pp. 2131–2158.

———, 1978, Coastal stratigraphic sequences: in DAVIS, R. A., JR., ed., *Coastal Sedimentary Environments,* Springer-Verlag, New York, pp. 361–384.

———, R. B. BIGGS, and S. D. HALSEY, 1979, Morphology and vertical sedimentary sequence models in Holocene transgressive barrier systems: in COATES, D. R., ed., *Coastal Geomorphology,* Publications in Geomorphology, State University of New York, Binghamton, pp. 321–354.

KRUIT, C., and others, 1975, Une excursion aux cones al' alluvions en eau profonde d'age tertiare pres de San Sebastian (Province de Guipuzcoa, Espagne): *9th Internatl. Sed. Cong. Guidebook No. 23,* Nice, France, 75 p.

KULM, L. D., and G. A. FOWLER, 1974, Oregon continental margin structure and stratigraphy: in BURK, C. A., and C. L. DRAKE, eds., *Geology of Continental Margins:* Springer-Verlag, New York, pp. 261–283.

———, and K. F. SCHEIDEGGER, 1979, Quaternary sedimentation on the tectonically active Oregon continental slope: in DOYLE, L. J., and O. H. PILKEY, eds., *Geology of Continental Slopes,* SEPM Spec. Pub. 27, pp. 247–264.

———, and others, 1973, *Initial Reports of the Deep Sea Drilling Project,* v. 18, U.S. Government Printing Office, Washington, D.C., 1077 p.

———, and others, 1975, Oregon continental shelf sedimentation: interrelationships of facies distribution and sedimentary processes: *J. Geol.,* v. 83, pp. 145–176.

KUMAR, N., and J. E. SANDERS, 1972, Sand body created by migration of Fire Island Inlet, Long Island, New York (abs): *Am. Assoc. Petroleum Geologists Bull.,* v. 56, p. 634.

———, and ———, 1974, Inlet sequence: a vertical succession of sedimentary structures and textures created by the lateral migration of tidal inlets: *Sedimentology,* v. 21, pp. 491–532.

LANCELOT, Y., 1977, Site 369, continental slope off Cape Bojador, Spanish Sahara: *Initial Reports of the Deep Sea Drilling Project,* Vol. 41, U.S. Government Printing Office, Washington, D.C., pp. 327–420.

———, J. C. HATHAWAY, and C. D. HOLLISTER, 1972, Lithology of sediments from the western North Atlantic, Leg XI, Deep Sea Drilling Project: in *Initial Reports of the Deep-Sea Drilling Project,* Vol. 9, U.S. Government Printing Office, Washington, D.C., pp. 901–950.

——, and others, 1977, Site 368: Cape Verde Rise: in *Initial Reports of the Deep Sea Drilling Project,* Vol. 12, U.S. Government Printing Office, Washington, D.C., pp. 233–326.

LAND, C. B., JR., 1972, Stratigraphy of Fox Hills Sandstone and associated formations, Rock Springs Uplift and Wamsutter Arch area, Sweetwater County, Wyoming: a shoreline–estuary sandstone model for the Late Cretaceous: *Colo. School of Mines Quart.,* v. 67, 69 p.

LANGOZKY, Y., and A. SNEH, 1966, The Dead Sea–Arana Rift Valley project: *Israel Inst. Petroleum Res. Geophys. Report 1018,* pp. 5–10.

LARSONNEUR, C., 1975, Tidal deposits, Mont Saint-Michel Bay, France: in GINSBURG, R. N., ed. *Tidal Deposits, a Casebook of Recent Examples and Fossil Counterparts,* Springer-Verlag, New York, pp. 21–30.

LEEDER, M. R., 1982, *Sedimentology:* George Allen and Unwin, London, 344 p.

LEETMA, A., 1976, Some simple mechanisms for steady shelf circulation: in STANLEY, D. J., ed., *Marine Sediment Transport and Environmental Management,* John Wiley & Sons, New York, pp. 23–28.

LEOPOLD, L. B., and M. G. WOLMAN, 1957, River channel patterns: braided, meandering, and straight: *U.S. Geol. Survey Prof. Paper 282-B,* 85 p.

LEPICHON, X., and RENARD, V., 1982, Avalanching: a major process of erosion and transport in deep-sea canyons: evidence from submersible and multi-narrow beam surveys: in SCRUTTON, R. A., and N. TALWANI, *The Ocean Floor,* John Wiley & Sons, New York, pp. 113–128.

LINK, M. H., and J. E. WELTON, 1982, Sedimentology and reservoir potential of Matilija Sandstone: An Eocene sand-rich deep-sea fan and shallow marine complex, California: *Am. Assoc. Petroleum Geologists Bull.,* v. 60, pp. 1514–1534.

LOOPE, D. B., 1984, Origin of extensive bedding planes in aeolian sandstones: A defense of Stokes' hypothesis: *Sedimentology,* v. 31, pp. 123–125.

LOVE, D. W., 1979, Quaternary fluvial geomorphic adjustments in Chaco Canyon, New Mexico: in RHOADES, D. E., ed., *Ann. Geomorph. Symp. 10,* State University of New York, Binghamton, pp. 277–308.

LOVELL, J. P. B., and D. A. V. STOW, 1981, Identification of ancient sandy contourites: *Geology,* v. 9, pp. 347–349.

LOWE, D. R., 1972, Submarine canyon and slope channel sedimentation model as inferred from Upper Cretaceous deposits, western California, USA: in *Internatl. Geol. Cong. 24,* Sec. 6, pp. 75–81.

LUDWICK, J. C., 1973, Tidal currents, sediment transport, and sand banks in Chesapeake Bay entrance, Virginia: in CRONIN, L. E., ed., *Estuarine Research, Vol. 2, Geology and Engineering,* Academic Press, New York, pp. 365–380.

LUPE, R., and T. S. AHLBRANDT, 1979, Sediments of the ancient eolian environment—reservoir inhomogeneity: in MCKEE, E. D., ed., *A Study of Global Sand Seas:* U.S. Geol. Survey Prof. Paper 1052, pp. 241–252.

MACKENZIE, D. B., 1972, Tidal sand flat deposits in Lower Cretaceous Dakota Group near Denver Colorado: *Mountain Geologist,* v. 8, pp. 141–150.

————, 1975, Tidal sand flat deposits in Lower Cretaceous Dakota Group near Denver, Colorado: in GINSBURG, R. N., *Tidal Deposits: A Casebook of Recent Examples and Fossil Counterparts,* Springer-Verlag, New York, pp. 117-126.

MADER, D., 1981, Genesis of the Bundsandstein (Lower Triassic) in the Western Eifel (Germany): *Sedim. Geol.,* v. 29, pp. 1-29.

MAINGUET, N., and Y. CALLOT, 1974, Air photo study of typology and interrelation between the texture and structure of dune patterns in the Fachi-Bilma Erg, Sahara: *Zeitschrift Geomorph.,* Suppl. Band 20, pp. 62-69.

————, and M. C. CHEMIN, 1983, Sand seas of the Sahara and Sahel: an explanation of their thickness and sand dune type by the sand budget principle: in BROOKFIELD, M. E., and T. S. AHLBRANDT, eds., *Eolian Sediments and Processes,* Elsevier, Amsterdam, pp. 353-364.

MALDONADO, A., and others, 1985, Valencia Fan (northwestern Mediterranean): Distal deposition fan variant: *Marine Geology,* v. 62, pp. 295-320.

MANOHAR, M., 1955, *Mechanics of Bottom Sediment Movement due to Wave Action:* U.S. Army Corps of Engineers, Beach Erosion Control Board Tech. Mem. 75, p. 121.

MAY, J. A., J. E. WARME, and R. A. SLATER, 1983, Role of submarine canyons on shelfbreak erosion and sedimentation: modern and ancient examples: in STANLEY, D. J., and G. T. MOORE, eds., *The Shelfbreak: Critical Interface on Continental Margins,* SEPM Spec. Pub. 33, pp. 315-332.

————, R. K. YEO, and J. E. WARME, 1984, Eustatic control on synchronous stratigraphic development: Cretaceous and Eocene coastal basins along an active margin: *Sedim. Geol.,* v. 40, pp. 131-150.

McCAVE, I. N., 1967, Shallow and marginal marine sediments associated with the Catskill complex in the Middle Devonian of New York: in KLEIN, G. DEV., ed., *Late Paleozoic and Mesozoic Continental Sedimentation, Northeastern North America:* Geol. Soc. America Spec. Paper 106, pp. 75-108.

————, 1969, Correlation of marine and non-marine strata with example from the Devonian of New York State: *Am. Assoc. Petroleum Geologists Bull.,* v. 53:1, pp. 155-162.

————, 1971, Sandwaves in the North Sea off the coast of Holland: *Marine Geol.,* v. 10, pp. 199-225.

McCUBBIN, D. G., 1972, Facies and paleocurrents of Gallup Sandstone, model for alternating deltaic and strandplain progradation (abs): *Am. Assoc. Petrol. Geologists Bull.,* v. 56, p. 638.

————, 1982, Barrier island and strand-plain facies: in SHOLLE, P. A., and D. R. SPEARING, eds., *Sandstone Depositional Environments,* Am. Assoc. Petroleum Geologists Mem. 31, Tulsa, Okla., pp. 247-279.

————, and M. J. BRADY, 1969, Depositional environment of the Almond reservoirs, Patrick Draw field, Wyoming: *Mountain Geologist,* v. 6, pp. 3-26.

McGOWEN, J. H., and L. E. GARNER, 1970, Physiographic features and stratification types of coarse-grained point bars: modern and ancient examples: *Sedimentology,* v. 14, pp. 77-111.

————, and C. G. GROAT, 1971, Van Horn Sandstone, West Texas: an alluvial fan model for mineral exploration: *Texas Bur. Econ. Geol.,* Rept. Inv. 72, 57 p.

McGREGOR, B., and others, 1982, Wilmington Submarine Canyon: A marine fluvial-like system: *Geology,* v. 10, pp. 27-30.

McKee, E. D., 1966, Structures of dunes at White Sands National Monument, New Mexico (and comparison with structures of dunes from other selected areas): *Sedimentology,* v. 7, pp. 1–69.

——, 1978, Sedimentary structures in dunes: in McKee, E. D., ed., *A Study of Global Sand Seas,* U.S. Geol. Surv. Prof. Paper 1052, pp. 83–134.

——, 1982, Sedimentary structures in dunes of the Namib Desert, Southwest Africa: *Geol. Soc. Am. Spec. Paper 188,* 64 p.

——, 1983, Eolian sand bodies of the world: in Brookfield, M. E., and T. S. Ahlbrandt, eds., *Eolian Sediments and Processes:* Elsevier, Amsterdam, pp. 1–25.

——, and R. J. Moiola, 1975, Geometry and growth of the White Sands, New Mexico dune field: *U.S. Geol. Survey J. Res.,* v. 3, pp. 59–66.

——, and T. S. Sterrett, 1961, Laboratory experiments on form and structure of long-shore bars and beaches: in *Geometry of Sandstone Bodies—A Symposium:* Am. Assoc. Petroleum Geologists Ann. Mtg., Tulsa, Okla., pp. 13–28.

——, and G. C. Tibbitts, Jr., 1964, Primary structure of a seif dune and associated deposits in Libya: *J. Sed. Petrology,* v. 34. pp. 5–17.

——, E. J. Crosby, and H. L. Berryhill, Jr., 1967, Flood deposits, Bijou Creek, Colorado, June, 1965: *J. Sed. Petrology,* v. 37, pp. 829–851.

Merk, G. P., 1960, *Great Sand Dunes of Colorado: Guide to Geology of Colorado,* Rocky Mountain Assoc. Geologists, pp. 127–129.

Miall, A. D., 1977, A review of the braided-river depositional environment: *Earth Sci. Rev.,* v. 13, pp. 1–62.

——, 1981, Alluvial sedimentary basins: tectonic setting and basin architecture: in Miall, A. D., ed., *Sedimentation and Tectonics in Alluvial Basins:* Geol. Assoc. Canada Spec. Paper 23, pp. 1–33.

——, 1986, Eustatic sea level changes interpreted from seismic stratigraphy: A critique of the methodology with particular reference to the North Sea Jurassic record: *Am. Assoc. Petroleum Geologists Bull.,* v. 70:2, pp. 131–137.

Middleton, G. V., 1973, Johannes Walther's law of the correlation of facies: *Geol. Soc. Am. Bull.,* v. 84, pp. 979–988.

——, 1978, Facies: in Fairbridge, R. W., and J. Bourgeois, eds., *Encyclopedia of Sedimentology,* Dowden, Hutchinson and Ross, Stroudsburg, Pa., pp. 323–325.

——, and M. A. Hampton, 1973, Sediment gravity flows: mechanics of flow and deposition: in Middleton, G. V., and A. H. Bouma, eds., *Turbidites and Deep Water Sedimentation:* SEPM Pac. Sec., Short Course Lect. Notes, pp. 1–38.

Miller, J. A., 1975, Facies characteristics of Laguna Madre wind-tidal flats: in Ginsburg, R. N., ed., *Tidal Deposits,* Springer-Verlag, New York, pp. 67–74.

Mitchum, R. M., Jr., P. R. Vail, and S. Thompson, III, 1977, Seismic stratigraphy and global changes of sea level, Part 2: The depositional sequence as a basic unit for stratigraphic analysis: in Payton, C. E., ed., *Seismic Stratigraphy—Applications to Hydrocarbon Exploration,* Am. Assoc. Petroleum Geologists Mem. 26, Tulsa, Okla., pp. 53–62.

Monteleone, P. H., and G. S. Fraser, 1978, Pleistocene sedimentation in the San Andreas Trough—Millerton Formation of Tomales Bay, California (abs): *Geol. Soc. America, Cordilleran Sec., 74th Ann. Mtg.,* p. 138.

MOODY, D. W., 1964, Coastal geomorphology and processes in relation to the development of submarine sand ridges off Bethany Beach, Delaware: Ph.D. Thesis, Johns Hopkins University, Baltimore, Md., 167 p.

MOORE, D. G., 1969, Reflector profiling studies of the California continental borderland: structure and Quaternary turbidite basins: *Geol. Soc. Am. Spec. Paper 107,* 142 p.

MOORE, G. F., and D. E. KARIG, 1976, Development of sedimentary basins on the lower trench slope: *Geology,* v. 4, pp. 693–697.

MOORE, J. C., 1973, Cretaceous continental margin sedimentation, southwestern Alaska: *Geol. Soc. Am. Bull.,* v. 84, pp. 595–614.

MORTON, R. A., 1981, Formation of storm deposits by wind-forced currents in the Gulf of Mexico and North Sea: in NIO, S. D., and others, eds., *Holocene Marine Sedimentation in the North Sea Basin,* Internatl. Assoc. Sedimentologists Spec. Pub. 5, pp. 303–396.

——, and A. C. DONALDSON, 1973, Sediment distribution and evolution of tidal deltas along a tide-dominated shoreline, Wachapreague, Virginia: *Sedim. Geol.* v. 10, pp. 285–299.

MOSLOW, T. F., and S. D. HERON, JR., 1978, Relict inlets: preservation and occurrence in the Holocene stratigraphy of southern Core Banks, North Carolina: *J. Sed. Petrology,* v. 48, pp. 1275–1286.

——, and ——, 1979, Quaternary evolution of Core Banks, North Carolina: Cape Lookout to New Drum Inlet: in LEATHERMAN, S. P., ed., *Barrier Islands,* Academic Press, New York, pp. 211–236.

MOUGENOT, D., G. BOILOT, and J. P. REHAULT, 1983, Prograding shelfbreak types on passive continental margins: some European examples: in STANLEY, D. J., and G. T. MOORE, eds., *The Shelfbreak: Critical Interface on Continental Margins,* SEPM Spec. Pub. 33, pp. 61–78.

MUTTI, E., 1977, Thin-bedded turbidite facies and related depositional environments in the Paleogene Hecho Group System (south-central Pyrenees, Spain): *Sedimentology,* v. 24, pp. 107–131.

——, 1979, Turbidites et cones sous-marine profonds: in HOMEWOOD, P., ed., *Sedimentation Detritique (fluviatile, littorale, marine),* Institute de Geologie, Universite de Fribourg, Fribourg, pp. 353–419.

——, 1983–1984, The Hecho Eocene submarine fan system, south-central Pyrenees, Spain: *Geomarine Letters,* v. 3:2–4, pp. 199–202.

——, and F. RICCI-LUCCHI, 1972, Le torbiditi, dell'Appennino settenrionale: introduzione all'analisi di facies: *Soc. Geol. Ital. Mem.,* v. 11, pp. 161–199.

MUTTI, E., and F. RICCI-LUCCHI, 1978, Turbidites of the northern Apennines: introduction to facies analysis: *Internat. Geology Review,* v. 20, pp. 125–166.

NARDIN, T. R., 1983, Late Quaternary depositional systems and sea level change—Santa Monica and San Pedro Basins, California Continental Borderland: *Am. Assoc. Petroleum Geologists Bull.,* v. 67:7, pp. 1104–1124.

NEAL, J. T., 1965, Environmental occurrence and general hydrology of U.S. playas, ch. 1, Environ. Res. Paper No. 96, Air Force Cambridge Res. Lab. Rept. 65-266, 176 p.

NELSON, B. W., 1970, Hydrography, sediment dispersal and recent historical development of the Po River delta, Italy: in MORGAN, J. P., and R. H. SHAVER, eds., *Deltaic Sedimentation Modern and Ancient,* SEPM Spec. Pub. 15, pp. 152–184.

NELSON, C. H., 1982, Modern shallow water graded sand layers from storm surges, Bering Shelf: A mimic of Bouma sequences and turbidite systems: *J. Sed. Petrology,* v. 52, pp. 537–545.

———, 1983, Modern submarine fans and debris aprons: an update of the first half century: in BOARDMAN, S. J., ed., *Revolution in the Earth Sciences, Advances in the Past Half-Century,* Kendall/Hunt, Dubuque, Iowa, pp. 148–166.

———, and L. D. KULM, 1973, Submarine fans and channels: in MIDDLETON, G. V., and A. H. BOUMA, eds., *Turbidites and Deep-Water Sedimentation,* Pac. Sec. SEPM Short Course, Anaheim, Calif., pp. 39–78.

———, and T. H. NILSEN, 1974, Depositional trends of modern and ancient deep-sea fans: in DOTT, R. H., JR., and R. H. SHAVER, eds., *Modern and Ancient Geosynclinal Sedimentation,* SEPM Spec. Pub. 19, pp. 54–76.

———, and ———, 1984, *Modern and Ancient Deep-Sea Fan Sedimentation:* SEPM Short Course 14, Tulsa, Okla., 404 p.

———, and others, 1978, Thin-bedded turbidites in modern submarine canyons and fans: in STANLEY, D. J., and G. KELLING, eds., *Sedimentation in Submarine Canyons, Fans, and Trenches,* Dowden, Hutchinson, and Ross, Stroudsburg, Pa., pp. 177–189.

NIEDORODA, A. W., and others, 1984, Shoreface morphodynamics on wave-dominated coasts: *Marine Geol.,* v. 60:1/4, pp. 331–354.

———, and others, 1985, Barrier island evolution, middle Atlantic shelf, U.S.A., Part II: Evidence from the shelf floor: *Marine Geology,* v. 63, pp. 363–396.

NIJMAN, W., and C. PUIGDEFABREGAS, 1978, Coarse-grained point bar structure in a molasse-type fluvial system, Eocene Castisent Sandstone Formation, South Pyrenean Basin: in MIALL, A. D., ed., *Fluvial Sedimentology,* Can. Soc. Petroleum Geologists Mem. 5, pp. 487–510.

NILSEN, T. H., 1969, Old Red sedimentation in the Beulandet–Vaerlandet Devonian district, Western Norway: *Sedim. Geol.,* v. 3, pp. 35–57.

———, and G. W. MOORE, 1979, Reconnaisance study of Upper Cretaceous to Miocene stratigraphic units and sedimentary facies, Kodiak and adjacent islands Alaska, with a section on sedimentary petrography by G. R. WINKLER: *U.S. Geol. Survey Prof. Paper 1093,* 34 p.

NIO, S. D., 1976, Marine transgressions as a factor in the formation of sand wave complexes: *Geol. Mijnbouw,* v. 55, pp. 18–40.

———, and C. H. NELSON, 1982, The North Sea and northeastern Bering Sea: a comparative study of the occurrence and geometry of sand bodies of two shallow epicontinental shelves: in NELSON, C. H., and others, eds., *The Northeastern Bering Shelf: New Perspectives of Epicontinental Shelf Processes and Depositional Products,* Geol. en Mijnbouw, v. 61:1, pp. 105–113.

NITTROUER, C. A., and R. W. STERNBERG, 1981, The formation of sedimentary strata in an allochthonous shelf environment: the Washington continental shelf: *Marine Geology,* v. 42, pp. 201–232.

NORMARK, W. R., 1970, Growth patterns of deep sea fans: *Geol. Soc. Am. Bull.,* v. 54, pp. 2170–2195.

———, 1978, Fan valleys, channels, and depositional lobes on modern submarine fans: characteristics for recognition of sandy turbidite environments: *Am. Assoc. Petroleum Geologists Bull.,* v. 62, pp. 912–931.

———, and G. R. Hess, 1980, Quaternary growth patterns of California submarine fans: in FIELD, M. E., and others, eds., *Quaternary Depositional Environments of the Pacific Coast,* Pacific Sec.-SEPM, pp. 201–210.

———, and D. J. W. PIPER, 1972, Sediments and growth pattern of Navy deep-sea fan, San Clemente Basin, California borderland: *J. Geol.,* v. 80, pp. 198–223.

———, and ———, 1983–1984, Navy Fan, California Borderland; Growth pattern and depositional processes: *Geomarine Letters,* v. 3:2–4, pp. 101–108.

NORRIS, R. M., and K. S. NORRIS, 1961, Algodones Dunes of southeastern California: *Geol. Soc. Am. Bull.,* v. 72, pp. 605–620.

OERTEL, G. F., 1979, Barrier island development during the Holocene recession, southeastern United States: in LEATHERMAN, S. P., ed., *Barrier Islands,* Academic Press, New York, pp. 291–320.

OLSEN, P. E., 1987, A 40-million-year lake record of early Mesozoic orbital climatic forcing: *Science,* v. 234, pp. 842–848.

OOMKENS, E., 1970, Depositional sequences and sand distribution in the post-glacial Rhone delta complex: in MORGAN, J. P., ed., *Deltaic Sedimentation, Modern and Ancient,* SEPM Spec. Pub. 15, pp. 198–212.

———, 1974, Lithofacies relationships in the late Quaternary Niger delta complex: *Sedimentology,* v. 21, pp. 195–222.

OTVOS, E. G., JR., 1970, Development and migration of barrier islands, northern Gulf of Mexico: *Geol. Soc. Am. Bull.,* v. 81, pp. 241–246.

———, 1977, Post-Pleistocene history of the United States inner continental shelf: significance to origin of barrier islands: discussion: *Geol. Soc. Am. Bull.,* v. 88, pp. 734–736.

———, 1978, New Orleans–South Hancock Holocene barrier trends and origins of Lake Pontchartrain: *Gulf Coast Assoc. Geo. Socs. Trans.,* v. 28, pp. 337–355.

———, 1979, Barrier island evolution and history of migration, north central Gulf Coast: in LEATHERMAN, S. P., ed., *Barrier Islands,* Academic Press, New York, pp. 273–290.

———, and W. A. PRICE, 1979, Problems of chenier genesis and terminology—an overview: *Marine Geology,* v. 31, pp. 251–264.

PALMER, J. J., and A. J. SCOTT, 1984, Stacked shoreline and shelf sandstone of LaVentana Tongue (Campanian), northwestern New Mexico: *Am. Assoc. Petroleum Geologists Bull.,* v. 68:1, pp. 74–91.

PANAGEOTOU, W., and S. P. LEATHERMAN, 1986, Holocene–Pleistocene stratigraphy of the inner shelf off Fire Island, New York: Implications for barrier-island migration: *J. Sed. Petrology,* v. 56:4, pp. 528–537.

PARKER, R. S., 1976, Experimental study of drainage system evolution: unpublished report, Colorado State University, Fort Collins.

PHILLIPS, S., 1987, Shelf sedimentation and depositional sequence stratigraphy of the Upper Cretaceous Woodbine–Eagle Ford Groups, East Texas: Ph.D. thesis, Cornell University, Ithaca, New York.

PICARD, M. D., and L. R. HIGH, JR., 1972, Criteria for recognizing lacustrine rocks: in RIGBY, J. K., and W. K. HAMBLIN, eds., *Recognition of Ancient Sedimentary Environments,* SEPM Spec. Pub. 16, pp. 108–145.

———, and ———, 1973, Sedimentary structures of ephemeral streams: *Developments in Sedimentology,* vol. 17, Elsevier, New York, 223 p.

PICKERING, K. T., 1981a, Konigsfjord Formation—a late Precambrian submarine fan in northeast Finmark, North Norway: *Norges Geologiske Undersokelse,* v. 367, pp. 77–104.

——, 1981b, Two types of outer fan lobe sequences from the late Precambrian Konigsfjord Formation submarine fan, Finmark, North Norway: *J. Sed. Petrology,* v. 51, pp. 1277–1286.

PIERCE, J. W., and D. J. COLQUHOUN, 1970, Holocene evolution of a portion of the North Carolina coast: *Geol. Soc. Am. Bull.,* v. 81, pp. 3697–3714.

PILKEY, O. H., S. D. LOCKER, and W. J. CLEARY, 1980, Comparison of sand-layer geometry on flat floors of 10 modern depositional basins: *Am. Assoc. Petroleum Geologists Bull.,* v. 64:6, pp. 841–856.

PINET, P. R., and P. POPENOE, 1982, Blake Plateau: Control of Miocene sedimentation patterns by large-scale shifts in the Gulf Stream axis: *Geology,* v. 10, pp. 257–259.

PIPER, D. J. W., and W. R. NORMARK, 1983, Turbidite depositional patterns and flow characteristics, Navy Submarine Fan, California Borderland: *Sedimentology,* v. 30, pp. 681–694.

——, W. R. NORMARK, and J. C. INGLE, 1976, The Rio Dell Formation: a Plio-Pleistocene basin slope deposit in northern California: *Sedimentology,* v. 23, pp. 307–328.

——, R. VON HUENE, and J. R. DUNCAN, 1973, Late Quaternary sedimentation in the active eastern Aleutian Trench: *Geology,* v. 1, pp. 19–22.

PISCIOTTO, K. A., and R. E. GARRISON, 1981, Lithofacies and depositional environments of the Monterey Formation, California: in GARRISON, R. E., and R. G. DOUGLAS, eds., *The Monterey Formation and Related Siliceous Rocks of California,* Pacific Sec.-SEPM, pp. 97–122.

PITTMAN, W. C., III, 1978, Relationship between eustacy and stratigraphic sequences of passive margins: *Geol. Soc. Am. Bull.,* v. 89, pp. 1389–1403.

——, and X. GOLOVCHENKO, 1983, The effect of sealevel change on the shelfedge and slope of passive margins: in, STANLEY, D. J., and G. T. MOORE, eds., *The Shelfbreak: Critical Interface on Continental Margins,* SEPM Spec. Pub. 33, pp. 41–58.

POSTMA, H., 1961, Transport and accumulation of suspended matter in the Dutch Wadden Sea: *Neth. J. Sea Res.,* v. 1, pp. 148–190.

——, 1967, Sediment transport and sedimentation in the estuarine environment: in LAUFF, G. H., ed., *Estuaries,* Am. Assoc. Advancement of Sci., Washington, D.C., pp. 158–179.

PRICE, W. A., and R. H. PARKER, 1979, Origins of permanent inlets separating barrier islands and influence of drowned valleys on tidal records along the Gulf Coast of Texas: *Gulf Coast Assoc. Geol. Socs. Trans.,* v. 29, pp. 371–385.

PRITCHARD, D. W., 1967, What is an estuary: physical viewpoint: in LAUFF, G. H., ed., *Estuaries,* Am. Assoc. Advancement Sci., Washington, D.C., pp. 3–5.

——, and H. H. CARTER, 1971, Estuarine circulation patterns: in SCHUBEL, J. R., ed., *The Estuarine Environment,* Am. Geol. Inst., Washington, D.C., p. IV 1–17.

PUIGDEFABREGAS, C., 1973, Miocene point bar deposits in the Ebro Basin, Northern Spain: *Sedimentology,* v. 20, pp. 133–144.

——, and A. VAN VLIET, 1978, Meandering stream deposits from the Tertiary of the southern Pyrenees: in MIALL, A. D., ed., *Fluvial Sedimentology,* Can. Soc. Petroleum Geologists Mem. 5, pp. 469–485.

RAMPINO, M. R., and J. E. SANDERS, 1980, Holocene transgression in south central Long Island, New York: *J. Sed. Petrology,* v. 50, pp. 1063–1080.

———, 1981, Evolution of the barrier islands of southern Long Island, New York: *Sedimentology,* v. 28, p. 37–47.

———, 1982, Holocene transgression in south central Long Island—reply: *J. Sed. Petrology,* v. 52, pp. 1020–1025.

REIMNITZ, E., 1971, Surf-beat origin for pulsating bottom currents in the Rio Balsas submarine canyon, Mexico: *Geol. Soc. Am. Bull.,* v. 82, pp. 81–89.

———, and M. GUTIERREZ-ESTRADA, 1970, Rapid changes in the head of the Rio Balsas submarine canyon system, Mexico: *Marine Geology,* v. 8, pp. 245–258.

REINECK, H. E., 1958, Longitudinale Schrag-schichten in Watt: *Geol. Rundschau,* v. 47, pp. 73–82.

———, 1963, Sedimentgefuge im Bereich der sudlichen Nordsee: *Abh. Senckenbergische naturforsch. Ges. 505,* 138 p.

———, 1967, Layered sediments of tidal flats, beaches, and shelf bottoms of the North Sea: in LAUFF, G. H., ed., *Estuaries,* Am. Assoc. Advanc. Sci., Washington, D.C., pp. 191–206.

———, 1972, Tidal flats: in RIGBY, J. K., and W. K. HAMBLIN, eds., *Recognition of Ancient Sedimentary Environments,* SEPM Spec. Pub. 16, pp. 146–159.

———, and I. B. SINGH, 1971, Der Golf von Gaeta/Tyrrhenisches Meer. 3. Die Gefuge von Vorstrand und Schelfsedimenten: *Senckenbergiana Marit.,* v. 3, pp. 185–201.

———, and ———, 1972, Genesis of laminated sand and graded rhythmites in storm-sand layers of shelf mud: *Sedimentology,* v. 18, pp. 123–128.

———, and ———, 1980, *Depositional Sedimentary Environments:* Springer-Verlag, New York, 549 p.

———, W. F. GUTMANN, and G. HERTWECK, 1967, Das Schickgebiet sudlich Helgoland als Beispiel rezenter Schelfablagerungen. *Senckenbergiana Lethaea,* v. 48, pp. 219–275.

———, and others, 1968, Sedimentologie, Faunenzonierung, und Faziesabfolge von der Ostkuste der inneren Deutschen Bucht: *Senckenbergiana Lethaea,* v. 49, pp. 261–309.

REINSON, G. E., 1977, Tidal current control of submarine morphology at the mouth of the Miramichi estuary, New Brunswick: *Can. J. Earth Sci.,* v. 14, pp. 2524–2532.

RICCI-LUCCHI, F., and E. VALMORI, 1980, Basin-wide turbidites in a Miocene oversupplied deep-sea plain: a geometrical analysis: *Sedimentology,* v. 27, pp. 241–270.

RICE, C. L., 1985, Fluvial model for deposition of basal Pennsylvanian quartzarenite in eastern Kentucky and southwestern Virginia: *Am. Assoc. Petroleum Geologists Bull.,* v. 69:9, p. 1446.

RICOY, J. U., and L. F. BROWN, JR., 1977, Depositional systems in the Sparta Formation (Eocene) Gulf Coast basin of Texas: *Trans. Gulf Coast Assoc. Geol. Socs.,* v. 27, pp. 139–154.

ROSS, G. M., 1983, Bigbear Erg: A Proterozoic intermontane eolian sand sea in the Hornby Group, Northwest Territories, Canada: in BROOKFIELD, M. E., and T. S. AHLBRANDT, eds., *Eolian Sediments and Processes,* Elsevier, Amsterdam, pp. 483–520.

ROY, P. S., B. G. THOM, and L. D. WRIGHT, 1980, Holocene sequences on an embayed high-energy coast: An evolutionary model: *Sedim. Geology,* v. 26, pp. 1–19.

RUBIN, D. M., and R. E. HUNTER, 1984, Bedding planes in aeolian sandstones: another reply: *Sedimentology*, v. 31, pp. 128–132.

RUPKE, N. A., 1975, Deposition of fine-grained sediments in the abyssal environment of the Algero-Balearic Basin, western Mediterranean Sea: *Sedimentology*, v. 22, pp. 95–109.

———, 1978, Deep clastic seas: in READING, H. G., ed., *Sedimentary Environments and Facies*, Elsevier, New York, pp. 372–415.

———, and D. J. STANLEY, 1974, Distinctive properties of turbiditic and hemipelagic mud layers in Alegro-Balearic Basin, western Mediterranean Sea: *Smithsonian Contrib. Earth Sci. 13*, 40 p.

———, D. J. STANLEY, and R. STUCKENRATH, 1974, Late Quaternary rates of abyssal mud deposition in the western Mediterranean Sea: *Mar. Geol.*, v. 17:2, pp. m9–m16.

RUST, B. R., 1978, Depositional models for braided alluvium: in MIALL, A. D., ed., *Fluvial Sedimentology*, Can. Soc. Petroleum Geologists Mem. 5, pp. 605–626.

RYDER, R. T., T. D. FOUCH, and J. H. ELISON, 1976, Early Tertiary sedimentation in the western Uinta Basin, Utah: *Geol. Soc. Am. Bull.*, v. 87, pp. 496–512.

RYER, T. A., 1977, Patterns of Cretaceous shallow-marine sedimentation, Coalville and Rockport areas, Utah: *Geol. Soc. Am. Bull.*, v. 88, pp. 177–188.

SALAMUNI, R., and J. J. BIGARELLA, 1967, The Botucatu Formation: in BIGARELLA, J. J., R. D. BECKER, and I. D. PINTO, eds., *Problems in Brazilian Gondwana Geology*, 1st Internatl. Symp. on Gondwana Strat. and Paleontol., 344 p.

SANDERS, J. E., and N. KUMAR, 1975, Holocene shoestring sand on inner continental shelf off Long Island, New York: *Am. Assoc. Petroleum Geologists Bull.*, v. 59, pp. 997–1009.

SANGREE, J. B., and others, 1978, Recognition of continental slope seismic facies, offshore Texas–Louisiana: in BOUMA, A. H., G. T. MOORE, and J. M. COLEMAN, eds., *Framework, Facies and Oil-trapping Characteristics of the Upper Continental Margin*, Am. Assoc. Petroleum Geologists, Studies in Geol. 7, pp. 87–116.

SANGREY, D. A., 1977, Marine geotechnology—state of the art: *Marine Geotechnol.*, v. 2, pp. 45–80.

SAXENA, R. S., 1976, Modern Mississippi Delta—depositional environments and processes: *Guidebook*, AAPG/SEPM Ann. Mtg., New Orleans, 125 p.

SCHOLLE, D. W., and M. S. MARLOW, 1974, Sedimentary sequence in modern Pacific trenches and the deformed circum-Pacific eugeosyncline: in DOTT, R. H., JR., and R. H. SHAVER, eds., *Modern and Ancient Geosynclinal Sedimentation*, SEPM Spec. Pub. 19, pp. 193–211.

SCHUMM, S. A., 1960, The effect of sediment type on the shape and stratification of some modern fluvial deposits: *Am. Jour. Sci.*, v. 258, pp. 177–184.

———, 1968, Speculations concerning paleohydrologic controls of terrestrial sedimentation: *Geol. Soc. Am. Bull.*, v. 79, pp. 1573–1588.

———, 1977, *The Fluvial System:* Wiley & Sons, New York, 338 p.

———, and R. M. BEATHARD, 1976, Geomorphic thresholds: an approach to river management: in *Rivers 76*, v. 1, 3rd Symposium of the Waterways: Harbors and Coastal Eng. Div. of Am. Soc. Civil Engineers, pp. 707–724.

SCHWARTZ, R. K., 1975, Nature and genesis of some storm washover deposits: *U.S. Coastal Eng. Res. Center Tech. Mem. 61*, 69 p.

SCHWELLER, W. J., and L. D. KULM, 1978, Depositional patterns and channelized sedimentation in active eastern Pacific trenches: in STANLEY, D. J., and G. KELLING, eds., *Sedimentation in Submarine Canyons, Fans, and Trenches,* Dowden, Hutchinson, and Ross, Stroudsburg, Pa., pp. 311–324.

SCOTT, R. M., and B. C. BIRDSALL, 1978, Physical and biogenic characteristics of sediments from Hueneme submarine canyon, California coast: in STANLEY, D. J., and G. KELLING, eds., *Sedimentation in Submarine Canyons, Fans, and Trenches,* Dowden, Hutchinson, and Ross, Stroudsburg, Pa., pp. 51–64.

SEELY, D. R., and W. R. DICKINSON, 1977, Stratigraphy and structure of compressional margins: in *Geology of Continental Margins,* Am. Assoc. Petroleum Geologists Contin. Educ. Course Note Ser. No. 5, pp. C1–C23.

SHARP, R. P., 1966, Kelso Dunes, Mojave Desert, California: *Geol. Soc. Am. Bull.,* v. 77, pp. 1045–1074.

SHELTON, J. W., 1973, Models of sand and sandstone deposits: a methodology for determining sand genesis and trend: *Oklahoma Geol. Survey Bull. 118,* 122 p.

SHEPARD, F. P., 1973, *Submarine Geology,* 3rd ed.: Harper & Row, New York, 517 p.

———, 1981, Submarine canyons: Multiple causes and long-time persistence: *Am. Assoc. Petroleum Geologists Bull.,* v. 65, pp. 1062–1078.

———, and R. F. DILL, 1966, *Submarine Canyons and Other Sea Valleys:* Rand-McNally, Chicago, 381 p.

———, N. F. MARSHALL, and P. A. MCLOUGHLIN, 1974, Currents in submarine canyons: *Deep-Sea Res.,* v. 21, pp. 691–706.

———, and others, 1977, Current-meter recordings of low-speed turbidity currents: *Geology,* v. 5, pp. 297–301.

SHEPARD, R. G., 1978, Distinction of aggradational and degradational fluvial regimes in valley fill alluvium, Tapia Canyon, New Mexico: in MIALL, A. D., ed., *Fluvial Sedimentology,* Can. Soc. Petroleum Geologists Mem. 5, pp. 277–286.

SILVESTER, R., 1964, *Coastal Engineering II: Sedimentation, Estuaries, Tides, Effluents, and Modelling:* Elsevier, Amsterdam, 378 p.

SIMONS, D. B., and E. V. RICHARDSON, 1961, Forms of bed roughness in alluvial channels: *Am. Soc. Civil Engineers Proc.,* v. 87:HY3, pp. 87–105.

SLATER, R. A., 1981, Submarine observations of the sea floor near the proposed Georges Bank lease sites along the North Atlantic outer continental shelf and upper slope: *U.S. Geol. Survey Open File Rept.,* 81-742, 65 p.

———, 1984, A numerical model of tides in the Cretaceous seaway of North America: *J. Geol.,* v. 93, pp. 333–345.

SLATT, R. M., 1984, Continental shelf topography: key to understanding distribution of shelf sand-ridge deposits from Cretaceous Western Interior Seaway: *Am. Assoc. Petroleum Geologists Bull.,* v. 68:9, pp. 1107–1120.

SLINGERLAND, R., 1986, Numerical computation of co-oscillating paleotides in the Catskill epeiric sea of eastern North America: *Sedimentology,* v. 33, pp. 487–497.

SMITH, D. G., and P. E. PUTNAM, 1980, Anastamosed river deposits: modern and ancient examples in Alberta, Canada: *Can. J. Earth Sci.,* v. 17, pp. 1396–1406.

———, and N. D. SMITH, 1980, Sedimentation in anastomosed river systems: examples from alluvial valleys near Banff, Alberta: *J. Sed. Petrology,* v. 50, pp. 157–164.

SMITH, G. I., 1962, Subsurface stratigraphy and geochemistry of late Quaternary evaporites, Searles Lake, California: *U.S. Geol. Survey Prof. Paper.* 450-C, p. 65-68.

SMITH, J. D., 1977, Modelling of sediment transport on continental shelves: in GOLDBERG, E. D., and others, eds., *The Sea*, 6, John Wiley & Sons, New York, pp. 539-577.

———, and T. S. HOPKINS, 1972, Sediment transport on the continental shelf of Washington and Oregon in light of recent current measurements: in SWIFT, D. J. P., D. B. DUANE, and O. H. PILKEY, eds., *Shelf Sediment Transport: Process and Pattern,* Dowden, Hutchinson, and Ross, Stroudsburg, Pa., pp. 143-180.

SMITH, N. D., 1970, The braided stream depositional environment: comparison of the Platte River with some Silurian clastic rocks, north central Appalachians: *Geol. Soc. Am. Bull.,* v. 81, pp. 2993-3014.

———, 1972, Some sedimentological aspects of planar cross-stratification in a sandy braided river: *J. Sed. Petrology,* v. 42, pp. 624-634.

———, 1974, Sedimentology and bar formation in the Upper Kicking Horse River, a braided outwash stream: *J. Geol.,* v. 82, pp. 205-224.

———, 1978, Some comments on terminology for bars in shallow rivers: in MIALL, A. D., ed., *Fluvial Sedimentology,* Can. Soc. Petroleum Geologists Mem. 5, pp. 85-88.

SMOOT, J. P., 1983, Depositional subenvironments in an arid closed basin; the Wilkins Peak Member of the Green River Formation (Eocene), Wyoming, U.S.A.: *Sedimentology,* v. 30, pp. 801-827.

SNEH, A., 1983, Desert stream sequences in the Sinai Peninsula: *J. Sed. Petrology,* v. 53:4, pp. 1271-1279.

SOUTHARD, J. B., and D. A. CACCHIONE, 1972, Experiments on bottom sediment movement by internal breaking waves: in SWIFT, D. J. P., D. B. DUANE, and O. H. PILKEY, eds., *Shelf Sediment Transport: Process and Pattern,* Dowden, Hutchinson, and Ross, Stroudsburg, Pa., pp. 83-98.

———, and D. J. STANLEY, 1976, Shelf-break processes and sedimentation: in STANLEY, D. J., and D. J. P. SWIFT, eds., *Marine Sediment Transport and Environmental Management,* John Wiley & Sons, New York, pp. 351-378.

STANLEY, D. J., 1967, Comparing patterns of sedimentation in some modern and ancient submarine canyons: *Earth Planetary Sci. Letters,* v. 3, pp. 371-380.

———, 1975, Submarine canyon and slope sedimentation (Gres d'Annot) in the French Maritime Alps: *Proc. IX Internatl. Sed. Cong.,* Nice, 129 p.

———, and G. KELLING, 1978, *Sedimentation in Submarine Canyons, Fans, and Trenches:* Dowden, Hutchinson, and Ross, Stroudsburg, Pa., 395 p.

———, and G. T. MOORE, eds., 1983, *The Shelfbreak: Critical Interface on Continental Margins:* SEPM Spec. Pub. 33, 467 p.

———, and R. UNRUG, 1972, Submarine channel deposits, fluxoturbidites and other indicators of slope and base-of-slope environments in modern and ancient marine basins: in RIGBY, J. K., and W. K. HAMBLIN, eds., *Recognition of Ancient Sedimentary Environments,* SEPM Spec. Pub. 16, pp. 287-340.

———, S. K. ADDY, and E. W. BEHRENS, 1983, The mudline: variability of its position relative to shelfbreak: in STANLEY, D. J., and G. T. MOORE, eds., *The Shelfbreak: Critical Interface on Continental Margins,* SEPM Spec. Pub. 33, pp. 279-298.

———, P. FENNER, and G. KELLING, 1972a, Currents and sediment transport at Wilmington

Canyon shelfbreak, as observed by underwater television: in SWIFT, D. J. P., D. B. DUANE, and O. H. PILKEY, eds., *Shelf Sediment Transport, Process and Pattern,* Dowden, Hutchinson and Ross, Stroudsburg, Pa., pp. 621–644.

——, H. SHENG, and C. P. PEDRAZA, 1971, Lower continental rise east of the middle Atlantic states: predominant sediment dispersal perpendicular to isobaths: *Geol. Soc. Am. Bull.,* v. 82, pp. 1831–1839.

——, and others, 1972b, Late Quaternary progradation and sand spillover on the outer continental margin off Nova Scotia, southeast Canada: *Smithsonian Inst. Contrib. Earth Sci.,* No. 8, 88 p.

STEEL, R. J., 1974, New Red Sandstone floodplain and piedmont sedimentation in the Hebridean Province: *J. Sed. Petrology,* v. 44, pp. 336–357.

——, 1976, Devonian basins of western Norway, sedimentary response to tectonism and varying tectonic context: *Tectonophysics,* v. 36, pp. 207–224.

——, and others, 1977, Coarsening upward cycles in the alluvium of Hornelen Basin (Devonian, Norway)—sedimentary response to tectonic events: *Geol. Soc. Am. Bull.,* v. 88, pp. 1124–1134.

STEWART, D. J., 1981, A meander-belt sandstone of the Lower Cretaceous of southern England: *Sedimentology,* v. 28, pp. 1–20.

STOKES, W. L., 1968, Multiple parallel-truncation bedding planes—a feature of wind deposited sandstone: *J. Sed. Petrology,* v. 38, pp. 510–515.

STOW, D. A. V., 1981, Laurentian Fan: Morphology, sediments, processes, and growth pattern: *Am. Assoc. Petroleum Geologists Bull.,* v. 65, pp. 375–393.

——, D. G. HOWELL, and C. HANS NELSON, 1983–1984, Sedimentary, tectonic, and sea level controls on submarine fan and slope-apron turbidite systems: *Geomarine Letters,* v. 3:2–4, pp. 57–64.

STUBBLEFIELD, W. L., and D. J. P. SWIFT, 1976, Ridge development as revealed by sub-bottom profiles on the central New Jersey shelf: *Marine Geology,* v. 20, pp. 315–334.

SURDAM, R. C., and K. O. STANLEY, 1979, Lacustrine sedimentation during the culminating phase of Eocene Lake Gosuite, Wyoming (Green River Formation): *Geol. Soc. Am. Bull.,* v. 80, pp. 93–110.

——, and C. A. WOLFBAUER, 1975, Green River Formation, Wyoming: a playa-lake complex: *Geol. Soc. Am. Bull.,* v. 86, pp. 335–345.

SWIFT, D. J. P., 1969a, Inner shelf sedimentation: processes and products: in STANLEY, D. J., ed., *The New Concepts of Continental Margin Sedimentation: Application to the Geological Record:* Am. Geol. Inst., Washington, D.C., pp. DS4-1–DS4-46.

——, 1969b, Outer shelf sedimentation: processes and products: in STANLEY, D. J. P., ed., *The New Concepts of Continental Margin Sedimentation: Application to the Geological Record,* Am. Geol. Inst., Washington, D.C., pp. DS5-1–DS5-26.

——, 1975, Barrier island genesis: evidence from the central Atlantic Shelf, eastern USA: *Sedim. Geol.,* v. 14, pp. 1–43.

——, 1976, Coastal sedimentation: in STANLEY, D. J., and D. J. P. SWIFT, eds., *Marine Sediment Transport and Environmental Management,* John Wiley & Sons, New York, pp. 255–310.

——, and M. E. FIELD, 1981, Evolution of a classic sand ridge field: Maryland sector, North American inner shelf: *Sedimentology,* v. 28, pp. 461–482.

——, and S. D. HERON, JR., 1969, Stratigraphy of the Carolina Cretaceous: *Southeastern Geol.,* v. 10, pp. 201–245.

——, and T. F. MOSLOW, 1982, Holocene transgression in south central Long Island, New York—discussion: *J. Sed. Petrology,* v. 52, pp. 1014–1019.

——, and A. W. NIEDORODA, 1985, Fluid and sediment dynamics on continental shelves: in TILLMAN, R. W., and others, eds., *Shelf Sands and Sandstone Reservoirs,* SEPM Short Course Notes 13, Tulsa, Okla., pp. 47–134.

——, and P. SEARS, 1974, Estuarine and littoral depositional patterns in the surficial sand sheet, central and southern Atlantic shelf of North America: in *International Symposium on Interrelationships of Estuarine and Continental Shelf Sedimentation,* Inst. Geol. du Bassin d'Aquitaine, Mem. 7, pp. 171–189.

——, D. B. DUANE, and T. F. MCKINNEY, 1973, Ridge and swale topography of the Middle Atlantic Bight, North America: secular response to the Holocene hydraulic regime: *Marine Geology,* v. 15, pp. 227–247.

——, G. L. FREELAND, and R. A. YOUNG, 1979, Time and space distribution of megaripples and associated bedforms, Middle Atlantic Bight, North American Atlantic shelf: *Sedimentology,* v. 26, pp. 384–406.

——, D. J. STANLEY, and J. R. CURRAY, 1971, Relict sediments on continental shelves: a reconsideration: *J. Geol.,* v. 79, pp. 322–346.

——, and others, 1972, Holocene evolution of the shelf surface, central and southern Atlantic shelf of North America: in SWIFT, D. J. P., D. B. DUANE, and O. H. PILKEY, eds., *Shelf Sediment Transport Process and Pattern,* Dowden, Hutchinson, and Ross, Stroudsburg, Pa., pp. 499–574.

——, and others, 1978, Evolution of a shoal retreat massif, North Carolina shelf: Inferences from areal geology: *Marine Geology,* v. 27, pp. 19–42.

——, and others, 1983, Hummocky cross-stratification and megaripples: A geological double standard: *J. Sed. Petrology,* v. 53:4, pp. 1295–1317.

——, and others, 1985, Barrier island evolution, middle Atlantic shelf, U.S.A., Part I: Shoreface dynamics: *Marine Geology,* v. 63, pp. 331–363.

——, and others, 1987, Shelf construction in a foreland basin: Storm beds, shelf sandbodies, and shelf-slope depositional sequences in the Upper Cretaceous Mesaverde Group, Book Cliffs, Utah: *Sedimentology,* v. 34, pp. 423–457.

TALBOT, M. R., 1985, Major bounding surfaces in aeolian sandstones—a climatic model: *Sedimentology,* v. 32, pp. 257–265.

——, and M. A. S. WILLIAMS, 1979, Cyclic alluvial fan sedimentation on the flanks of fixed dunes, Janjari, Central Niger: *Catena,* v. 6, pp. 43–62.

TANNER, W. F., 1974, The incomplete floodplain: *Geology,* v. 2:2, pp. 105–106.

THOM, B. G., and P. S. ROY, 1985, Relative sea levels and coastal sedimentation in southeast Australia in the Holocene: *J. Sed. Petrology,* v. 55:2, pp. 257–264.

THOMPSON, R. W., 1968, Tidal flat sedimentation on the Colorado River delta: *Geol. Soc. Am. Mem. 107,* 133 p.

——, 1975, Tidal flat sediments of the Colorado River delta, northwestern Gulf of California: in GINSBURG, R. N., ed., *Tidal Deposits: A Casebook of Recent Examples and Fossil Counterparts,* Springer-Verlag, New York, pp. 57–65.

THORNBURG, T. M., and L. D. KULM, 1987, Sedimentation in the Chile Trench: Depositional morphologies, lithofacies, and stratigraphy: *Geol. Soc. Am. Bull.,* v. 98, pp. 33–52.

TUNBRIDGE, I. P., 1984, Facies model for a sandy ephemeral stream and clay playa complex; the Middle Devonian Trentishoe Formation of North Devon, U.K.: *Sedimentology,* v. 31, pp. 697–715.

TWICHELL, D. C., and D. G. ROBERTS, 1982, Morphology, distribution, and development of submarine canyons on the United States Atlantic continental slope between Hudson and Baltimore canyons: *Geology,* v. 10, pp. 408–412.

TWIDALE, C. R., 1972, Landform development in the Lake Eyre region, Australia: *Geogr. Rev.,* v. 62, pp. 40–70.

TYE, R. S., 1984, Geomorphic evolution and stratigraphy of Price and Capers Inlets, South Carolina: *Sedimentology,* v. 31, pp. 655–674.

UNDERWOOD, M. B., and D. E. KARIG, 1980, Role of submarine canyons in trench and trench-slope sedimentation: *Geology,* v. 8, pp. 432–436.

——, S. B. BACHMANN, and W. J. SCHWELLER, 1980, Sedimentary processes and facies associations within trench and trench-slope settings: in FIELD, M. E., and others, eds., *Quaternary Depositional Environments of the Pacific Coast,* Pacific Sec.-SEPM, pp. 211–230.

URRIEN, C. M., 1972, Rio de la Plata estuary environments: in NELSON, B. W., ed., *Environmental Framework of Coastal Plain Estuaries,* Geol. Soc. Am. Mem. 133, pp. 213–236.

VAIL, P. R., and R. G. TODD, 1981, Northern North Sea Jurassic unconformities, chronostratigraphy and sea-level changes from seismic stratigraphy: in ILLING, L. V., and G. D. HOBSON, eds., *Petroleum Geology of the Continental Shelf of Northwest Europe,* Inst. of Petroleum, London, Heyden, pp. 216–235.

——, J. HARDENBOL, and R. G. TODD, 1984, Jurassic unconformities, chronostratigraphy, and sea level changes from seismic stratigraphy: in *Interregional Unconformities and Hydrocarbon Accumulation,* Am. Assoc. Petroleum Geologists Mem. 36, pp. 129–144.

——, R. M. MITCHUM, JR., and S. THOMPSON, III, 1977a, Seismic stratigraphy and global changes of sea level, Part 3: Relative changes of sea level from coastal onlap: in PAYTON, C. E., ed., *Seismic Stratigraphy—Applications to Hydrocarbon Exploration,* Am. Assoc. Petroleum Geologists Mem. 26, Tulsa, Okla., pp. 63–81.

——, ——, and ——, 1977b, Seismic stratigraphy and global changes of sea level, Part 4: Global cycles of relative changes of sea level: in PAYTON, C. E., ed., *Seismic Stratigraphy—Applications to Hydrocarbon Exploration,* Am. Assoc. Petroleum Geologists Mem. 26, Tulsa, Okla., pp. 83–97.

VALENTINE, P. C., R. A. COOER, and J. R. UZMANN, 1984, Submarine sand dunes and sedimentary environments in Oceanographer Canyon: *J. Sed. Petrology,* v. 54:3, pp. 704–715.

VAN DE GRAAF, F. R., 1972, Fluvial-deltaic facies of the Castlegate Sandstone (Cretaceous), east-central Utah: *J. Sed. Petrology,* v. 42, pp. 558–571.

VAN HORN, M. D., 1979, Stratigraphy of the Almond Formation, east-central flank, Rock Springs uplift, Sweetwater County, Wyoming—a mesotidal-shoreline model for the Late Cretaceous: unpublished M.S. thesis, Colorado School of Mines, Golden, 150 p.

VAN HOUTEN, F. B., 1964, Cyclic lacustrine sedimentation, Upper Triassic Lockatong Formation, central New Jersey and adjacent Pennsylvania: *Kansas Geol. Survey Bull. 169,* pp. 497–531.

VON STACKELBURG, U. V., 1972, Faziesverteilung in Sedimenten des indisch-pakistanischen Kontinentalrandes (Arabisches Meer): *Meteor Forshungsergebnisse,* Reiche C, No. 9, pp. 1–173.

VAN VLIET, A., 1978, Early Tertiary deepwater fans of Guipuzcoa, northern Spain: in STANLEY, D. J., and G. KELLING, eds., *Sedimentation in Submarine Canyons, Fans, and Trenches,* Dowden, Hutchinson, and Ross, Stroudsburg, Pa., pp. 190–209.

VANNEY, J. R., and D. J. STANLEY, 1983, Shelfbreak physiography: an overview: in STANLEY, D. J., and G. T. MOORE, eds., *The Shelfbreak: Critical Interface on Continental Margins,* SEPM Spec. Pub. 33, pp. 1–24.

VISHER, G. S., 1965, Fluvial processes as interpreted from ancient and recent fluvial deposits: in MIDDLETON, G. V., ed., *Primary Sedimentary Structures and Their Hydrodynamic Interpretation,* SEPM Spec. Pub. 12, pp. 116–132.

——, 1968, *A Guidebook to the Geology of the Blue-Jacket–Bartlesville Sandstone, Oklahoma:* Oklahoma City Geol. Soc., *72 p.*

——, 1975, A Pennsylvanian interdistributary tidal flat deposit: in GINSBURG, R. N., ed., *Tidal Deposits, a Casebook of Recent Examples and Fossil Counterparts,* Springer-Verlag, New York, pp. 179–186.

VISSER, M. J., 1980, Neap-spring cycles reflected in Holocene subtidal large-scale bedform deposits: a preliminary note: *Geology,* v. 8, pp. 543–546.

VON STRAATEN, L. M. J. U., 1964, De Boden der Waddenzee: *Nederl. Geol. Ver.,* pp. 75–151.

VOS, R. G., 1977, Sedimentology of an upper Paleozoic river-, wave-, and tide-influenced delta system: *J. Sed. Petrology:* v. 47, pp. 1242–1260.

WAGNER, G., 1950, *Einfuhrung in die Erd-und Landschaftgesichte:* Verlag der Hohenlohe'-schen Ruchhandlung F. Rau, Ohringen, 664 p.

WALKER, R. G., 1976, Facies models 2, turbidites and associated coarse clastic deposits: *Geosci. Canada,* v. 3:1, pp. 25–36.

——, 1978, Deep-water sandstone facies and ancient submarine fans: models for exploration for stratigraphic traps: *Am. Assoc. Petrol. Geologists Bull.,* v. 62, pp. 932–966.

——, ed., 1979, *Facies Models:* Geoscience Canada Reprint Ser. 1, Geol. Soc. Canada, 211 p.

——, and D. J. CANT, 1979, Sandy fluvial systems: in WALKER, R. G., ed., *Facies Models:* Geoscience Canada Reprint Ser. 1, Geol. Soc. Canada, pp. 23–32.

——, and J. C. HARMS, 1971, The "Catskill Delta": a prograding muddy shoreline in central Pennsylvania: *J. Geol.,* v. 79, pp. 381–399.

——, and G. V. MIDDLETON, 1979, Facies models 4. Eolian sands: in WALKER, R. G., ed., *Facies Models:* Geoscience Canada Reprint Ser. 1, Geol. Soc. Canada, pp. 33–42.

——, and E. MUTTI, 1973, Turbidite facies and facies associations: in *Turbidites and Deep Water Sedimentation:* Pacific Sec.-SEPM Short Course, pp. 119–157.

WALKER, T. R., and J. C. HARMS, 1972, Eolian origin of Flagstone beds, Lyons Sandstone (Permian), type area, Boulder County Colorado: *Mountain Geologist,* v. 9, pp. 279–288.

WANLESS, H. R., and others, 1970, Late Paleozoic deltas in the central and eastern United States: in MORGAN, J. P., *Deltaic Sedimentation, Modern and Ancient,* SEPM Spec. Pub. 15, pp. 215–245.

WATTS, A. B., 1981, The U.S. Atlantic continental margin: Subsidence history, crustal structure and thermal evolution: in BALLY, A. B., and B. C. SCHREIBER, eds., *Geology of Passive Continental Margins: History, Structure and Sedimentologic Record,* Am. Assoc. Petroleum Geologists, Education Course Notes No. 19, pp. 2-1-2.75.

WEIMER, R. J., J. D. HOWARD, and D. R. LINDSAY, 1982, Tidal flats: in SCHOLLE, P. A., and D. R. SPEARING, eds., *Sandstone Depositional Environments:* Am. Assoc. Petroleum Geologists, Tulsa, Okla., pp. 191–246.

WELLS, J. T., and J. M. COLEMAN, 1981, Periodic mudflat progradation, northeastern coast of South America: A hypothesis: *J. Sed. Petrology,* v. 51:4, pp. 1069–1076.

WILKINSON, B. H., and J. R. Byrne, 1977, Lavaca Bay—transgressive deltaic sedimentation in central Texas estuary: *Am. Assoc. Petroleum Geologists Bull.,* v. 61, pp. 527–545.

WILSON, I. G., 1971, Desert sandflow basins and a model for the development of ergs: *Geog. Jour.,* v. 137, pp. 180–199.

——, 1972, Aeolian bedforms—their development and origins: *Sedimentology,* v. 19, pp. 173–210.

——, 1973, Ergs: *Sedim. Geol.,* v. 10, pp. 77–106.

WOODBURY, H. D., J. H. SPOTTS, and W. H. AKERS, 1978, Gulf of Mexico continental slope sediments and sedimentation: in BOUMA, A. H., G. T. MOORE, and J. M. COLEMAN, eds., *Framework Facies and Oil-Trapping Characteristics of the Upper Continental Margin,* Am. Assoc. Petroleum Geologists, Studies in Geol. 7, pp. 117–138.

WRIGHT, L. D., 1977, Sediment transport and deposition at river mouths: a synthesis: *Geol. Soc. Am. Bull.,* v. 88, pp. 857–868.

——, J. M. COLEMAN, and B. G. THOM, 1973, Processes of channel development in a high-tide-range environment: Cambridge Gulf, Ord River Delta: *J. Geology,* v. 81, pp. 15–41.

WRIGHT, M. E., and R. G. WALKER, 1981, Cardium Formation (U. Cretaceous), at Seebe, Alberta, storm-transported sandstones and conglomerates in shallow marine depositional environments below fair-weather wave base: *Can. J. Earth Sci.,* v. 18, pp. 795–809.

WUNDERLICH, F., 1972, Beach dynamics and beach development: *Senckenbergiana Marit.,* v. 4, pp. 47–79.

Index

Abyssal plains
    classification of, 385
    controls on sediment distribution, 383
    controls on sequence evolution, 386
    in the rock record, 388
    processes on, 383
    response to climate change, 386
    response to sea level fluctuations, 388
    sediment sources of, 385, 386
    shapes of, 383
Aghullas current, 257
Aleutian Abyssal Plain, 385, 388
Aleutian Trench, 388, 394
Allochthonous shelf
    definition, 255
    facies distribution on, 270
    sediment delivery to, 262, 270
Alluvial fan, arid
    characteristics of, 32
    controls on sequence development, 37
    defined, 32
    examples in the rock record, 38
    facies distributions of, 35
    processes on, 33
Alluvial fan, humid
    channel development on, 77
    controls on sequence evolution, 80
    cyclic sedimentation on, 80
    facies of, 77
    in the rock record, 82
    processes on, 77
Alluvial plains
    channel types in, 88
    characteristics of, 87
    coastal, 103
    controls on sequence evolution, 100
    sequence simulation, 101, 103
    sequences in the rock record, 105
Almond formation estuarine deposits, 240
Anastomosing channels
    characteristics of, 97
    controls on sequence evolution, 99
    facies of, 97
Antidunes, origin, 7
Appalachian carboniforous deltaic sequences, 222
Arc trenches, defined, 386, 390
Arikaree Group intermittent streams, 45
Astoria Canyon, 348
Authigenic minerals on shelves, 267, 281
Autochthonous shelf
    definition, 255
    facies distribution on, 268, 303
    sediment delivery to, 261
Avulsion
    on arid alluvial fans, 33, 77
    effects on delta evolution, 217

Avulsion (*cont.*)
  processes in meandering channels, 98
  on submarine fans, 373, 376

Backshore, 163
Balearic Abyssal Plain, 386
Barrier island migration
  continuous retreat, 186
  effects of preexisting topography, 180
  effects of sediment supply, 188
  overstepping, 185
  stepwise retreat, 186
Barrier islands
  characteristics of, 169
  controls on sequence evolution, 183
  on delta margins, 210
  in estuaries, 232, 238
  origin of, 178
  response to sea level change, 178, 185
  sediments of, 169
  sequences in, 170, 187
  sequences in the rock record, 190
Battery Point Sandstone alluvial sequences,
    110
Beach ridges
  on strandplains, 192
  on wave-dominated deltas, 210, 214
Bedform stability diagrams, 7
Bigbear erg, 67
Bioturbation
  in deltas, 213
  in estuaries, 232, 243
  on shelves, 267, 283, 310
  on wave-dominated coasts, 161, 164, 165,
    169
Bluejacket Sandstone tidal deposits, 149
Borden Group deltaic sediments, 313
Braided channels
  bar types in, 95
  characteristics of, 94
  controls on sequence evolution, 99
  cyclic deposition in, 99
  sequences in the rock record, 109

Calcium compensation depth, 363
Caliente Range, Miocene shelf sequences in,
    281
California Borderland, 328, 330, 385, 386, 412
Cannes de Roche Formation alluvial se-
    quences, 109

Cape Sebastian Sandstone shelf sediments,
    297, 318
Cape Verde Rise, 408
Cardium Formation shelf sediments, 303
Carmello Formation slope deposits, 358
Castinet Formation alluvial sequence, 105
Catskill Delta, 412
Catskill Delta chenier plain, 155
Chandeleur Islands, origin of, 216
Cheniers
  characteristics of, 152
  in Colorado River Delta, 147
  deposits in Wilcox Group, 246
  origin of, 152
Chezy equation, 7
Chinle Formation alluvial sequences, 105
Chute cutoff, 93
Chutes in meandering rivers, 90
Cliff House Formation barrier island deposits,
    192
Climate, effects of
  on abyssal plains, 386
  on arid alluvial fans, 37
  on arid lakes, 50
  on continental shelves, 277
  on cyclic sedimentation, 132
  on ergs, 66
  on humid lakes, 123
Coastal alluvial fan, 243
Coastal dunes, 163
Coastal onlap, 291
Cohansey Sand barrier island deposits, 174,
    190
Complex response defined, 17
Continental rise
  controls on sequence evolution, 406
  defined, 402
  morphology of, 402
  processes on, 402
Continental shelf
  classification, 253
  controls on sequence evolution, 272, 289
  definition, 253
  effects of deep basinal currents, 257
  effects of sea level change, 279, 289
  facies on, 262
  facies distribution, 268
  mud deposition on, 267, 270, 285
  processes, 257, 286
  progressive sorting on, 284
  response to storm flow, 274, 276, 278
  sand bodies on, 263

sediment delivery to, 261, 270
sequences in the rock record, 293
structural features, 253
subdivisions of, 286
Continental slope
    active margin sequences, 344
    classification, 326
    controls on sequence evolution, 339
    currents on, 331
    cyclic sedimentation on, 359
    definition, 321
    effects of sea level fluctuations, 340, 360
    effects of tides, 332
    facies on, 333
    internal structure, 327
    mass movement on, 329, 335
    morphologic elements, 323
    organic sediments on, 334
    passive margin sequences, 342
    sedimentation near deltas, 342
    seismic character of, 340
    sequences in the rock record, 351
    shape, 323, 327
    suspension settling on, 332
Continuity principle defined, 22
Contour currents
    on continental rises, 402, 406
    in trenches, 393
Contourites, 393, 402, 406
Convergent margins, 328
Coriolis Force, 402
    effect on shelf currents, 257
Costa de Nayaret strandplain deposits, 196
Cotulla Barrier/Indio Lagoon system, 246
Crevasse splays, on delta plains, 208
Critical shear stress, 6
Cyclic sedimentation
    on arid alluvial fans, 37
    in arid lakes, 50
    climatic controls on, 132
    East Berlin Formation, 124
    Green River Formation, 52
    in braided channels, 99
    on deltas, 217
    on humid alluvial fans, 80
    in humid lakes, 123
    in intermittent streams, 43
    Lockatong Formation, 123
    Monterey Formation, 355
    on submarine fans, 374

Dakota Group tidal deposits, 147
Debris flows
    on arid alluvial fans, 33
    mechanics of, 12
Delmarva Peninsula, tidal flats on, 145
Delta front
    facies on, 210
    processes along, 210
Delta lobe evolution, 216
Delta plain
    facies on, 207
    processes on, 207
Delta plumes, 272, 296
    in rock record, 296, 315
Delta sequences, balance between subsidence
    and deposition, 220
Deltas
    characteristics of, 202
    classification of, 205
    controls on sequence evolution, 213, 220,
      247
    cyclic deposition on, 217
    effects of jet-flow, 204
    effects of sea level change, 217
    facies preservation during evolution, 214
    in lagoons, 173
    occurrence of, 202
    processes affecting morphology, 204
    sequences in the rock record, 222
Density currents, on shelves, 260
Deposition from suspension, 14
Depositional components defined, 27
Depositional system
    classified, 26
    defined, 2
Deserts
    characteristics of, 31
    components of, 32
    defined, 31
Diffusive transport on shelves, 261
Dikaka, 48
Distal facies
    in abyssal plains, 385
    on arid alluvial fans, 36
    in ergs, 65
    on humid alluvial fans, 77
    in intermittent streams, 45
    on submarine fans, 371
    in submarine trenches, 393
Downlap sequence, 292
Dunes, eolian
    bounding surfaces in, 58

Dunes (*cont.*)
    defined, 54
    stratification in, 56
    types of, 56
Dunes, subaqueous, origin, 7

East Berlin Formation
    intermittent streams, 47
    lake deposits, 124
Ebb tidal delta
    occurrence, 170
    sequences in, 170
Eckman spiral, 258
Elongate submarine fans, 366
Enclosed basin plains
    sediment sources, 386
    shapes, 386
Entrada Sandstone erg, 67
Environments of deposition
    classified, 25
    defined, 24
Eolian dunes, on coasts, 163
Eolian plains
    defined, 63
    deposits in, 64
Epeiric seas
    characteristics of, 306
    modern counterparts, 306, 309, 310
    processes in, 307
Epicontinental seaways, 313
    sand bodies in, 314
Ergs
    characteristics of, 53
    controls on sequence evolution, 65
    cyclic deposition in, 66
    defined, 53
    facies distributions of, 64
    facies of, 54
    processes in, 54
    sequences in the rock record, 67
Estuarine coasts
    classification of, 231
    controls on sequence evolution, 238
    definition of, 231
    deposits in the rock record, 240
    facies in, 233, 237
    processes in, 233, 235
Event stratigraphy
    application to California Borderland, 413
    defined, 2
    defined, 411
Extrinsic stress defined, 17

Facies concept, 22
Fall River Formation tidal deposits, 147
Feedback response defined, 17
Ferrelo Fan, 412
Flood tidal deltas
    characteristics of, 174
    in estuaries, 240
    processes on, 174
    sequences of, 175
Floodbasins in meandering rivers, 93
Flow regimes, 7
Fluidization, mechanics of, 13
Foreshore, 161
    on barriers, 169
Frio Barrier System, 188
Froude number, 6

Gallup Sandstone
    coastal deposits, 166
    strandplain deposits, 197
Galveston Island barrier deposits, 189
Gault Formation trench deposits, 399
Geomorphic cycles, 20
Georgia coastal deposits, 164
Geostrophic flow, 259
Graded storm layers
    characteristics of, 276
    origin of, 267
    sequences in, 277
Graded time defined, 19
Grain flows
    mechanics of, 13
    products of, 13
Green River Formation saline lakes, 52
Gulf Coast Eocene coastal zone deposits, 247
Gulf Coast Miocene deltaic deposits, 221
Gulf Coastal Plain Tertiary alluvial sequences, 117
Gulf of California, tidal flats in, 145
Gulf of Gaeta, shelf sediments in, 285
Gulf of Mexico, continental shelf, 310
Gulf stream, 257
Gully Canyon (submarine), 347
Gyre Basin (submarine), 344

Hatteras Abyssal Plain, 385, 403
Hatteras Fan (submarine), 403
Hecho Group submarine fan deposits, 378
High tidal flat, 140
Hornelen Basin alluvial fans, 82
Hudson Canyon (submarine), 351

Hueneme Canyon (submarine), 348
Humid alluvial fans
    channel development on, 77
    controls on sequence development, 80
    cyclic sedimentation on, 80
    facies of, 78
    in the rock record, 82
    processes on, 77
Humid system
    characteristics of, 76
    components of, 77
    defined, 76
    in the rock record, 127
Hummocky crossbedding
    characteristics of, 276
    sequences of, 276
    in rock record, 298, 301, 303
Hummocky megaripples, origin of, 267

Inner shelf, 271, 289
Interdistributaries
    at delta margins, 211, 214
    on delta plains, 208
Interdunes
    processes in, 61
    sediments in, 62
    types of, 61
Intermittent streams
    characteristics of, 40
    controls on sequence development, 43
    cyclic deposition in, 43
    facies of, 42
    processes on, 41
    sequences in the rock record, 45
Internal waves, 331
Intraslope basins, 344
    origin, 326
Intrinsic stress defined, 18
Irish Valley member, 155

Japan fore-arc continental slope, 344
Jura Quartzite shelf sediments, 294

Kissinger Sandstone alluvial sequence, 113
Konigsfjord Formation submarine fan deposits, 380
Kootenay Formation shelf sediments, 301, 313

La Jolla Fan (submarine), 377
La Ventana Tongue barrier island deposits, 192
Labrador Sea abyssal plain, 384
Lagoons
    characteristics of, 173
    flood tidal deltas in, 174
    river deltas in, 173
    sediments in, 173
    tidal creeks in, 173
    washover fans in, 175
Lakes, arid (see Playas and saline lakes)
Lakes, humid
    characteristics, 119
    controls on sequence evolution, 123
    cyclic sedimentation, 123
    facies distribution, 121
    occurrence, 119
    processes, 119
    proglacial, 122
    sequences in the rock record, 124
    temperate, 120
    tropical, 122
Laminar flow, 5
Laney Member of Green River Formation, 125
Laurentian Fan (submarine), 403
Lavaca Bay estuarine deposits, 234
Law of Superposition, 22
Levees
    on delta plains, 208
    in meandering rivers, 93
    on submarine fans, 368
    in submarine trenches, 393
Lift force, 6
Lockatong Formation lake deposits, 123
Longitudinal bars in braided channels, 94, 97
Longshore bars, 160
Longshore currents, 160
Longshore troughs, 160
Low stand fan, 290
Low tidal flat, 141
Low-sinuosity meandering rivers, 90
Lower shoreface, 158
    on barriers, 169
Lyons Formation
    dunes, 67
    playas, 52

Macroprocesses, defined, 2
Manning equation, 7
Marnoso-areanacea Formation abyssal plain
    deposits, 388

Mass movements
  classification, 329, 337
  mechanisms of, 329
Matilija Sandstone
  submarine fan deposits, 378
  submarine ramp deposits, 378
Meander cutoff, 93
Meandering rivers
  characteristics of, 88
  controls on sequence evolution, 97
  facies of, 89
  facies sequences in, 93
  point bar in, 89
Medial facies
  on arid alluvial fans, 32, 35
  on humid alluvial fans, 78
  in intermittent streams, 45
  on submarine fans, 370
Melvaer Breccia, 83
Mesoprocesses, defined, 2
Microprocesses, defined, 2
Mid-channel islands in meandering rivers, 90
Mid-fan facies on humid alluvial fans, 78
Mid-tidal flat, 140
Middle Atlantic shelf, 268
Milankovich cycles, 132
Millerton Formation estuarine deposits, 243
Mississippi Delta, effect on slope sedimenta-
  tion, 342, 348
Mixed-sediment submarine fans, 371
Mont Saint-Michel, tidal flats on, 145
Monterey Fan (submarine), 376
Monterey Formation
  cyclic sedimentation in, 355
  slope deposits of, 352
Moroccan Paleozoic deltaic sequences, 228
Mt. Pleasant fluvial system, 225
Mud line, 334
Mud-rich submarine fans, 365, 373
Muddy coastlines
  evolution of depositional sequences, 153
  occurrence, 151
  processes along, 151
  sequences in the rock record, 155
Muddy Sandstone strandplain deposits, 197
Mudflats in playas, 49

Narrow Cape Formation shelf sediments, 318
Navy Fan (submarine), 376
Nepheloid layers, 333

New Red Sandstone
  alluvial fans, 39
  intermittent streams, 46
Niger Delta, architecture of, 214, 219
North American continental rise, 406
North Sea, continental shelf, 269, 285

Ocean basins
  characteristics of, 365
  components, 366
Oceanographer Canyon (submarine), 334
Old Red Sandstone
  alluvial fans, 38
  arid system, 73
Orcadian Basin lake deposits, 123
Oregon continental slope, 344
Oregon-Washington shelf
  processes on, 270
  sediment delivery to, 270
  sediment distribution on, 286
Outer shelf, 272, 289
Overbank facies
  on arid alluvial fans, 36
  on humid alluvial fans, 78
  on submarine fans, 369
  in submarine trenches, 391, 394
Oxnard-Ventura coastal deposits, 165
Oxygen minimum layer, 334

Padre Island barrier deposits, 187
Pamlico Sound, origin of, 216
Parkman Sandstone strandplain deposits, 197
Passive margins, 327
Pendleton Bay/Lagoon System, 245
Peninsula Formation shelf sediments, 301, 320
Physical mixing on shelf substrates, 283
Pigeon Point Formation slope deposits, 356
Plane bed, 7, 11
Playas and saline lakes
  characteristics of, 48
  controls on sequence development, 50
  cyclic deposition in, 50
  defined, 48
  equilibrium with alluvial fans, 52
  facies associations of, 50
  facies of, 49
  processes in, 48
Pleistocene alluvial sequences, 109
Pleistocene terraces
  of South Carolina, 201
  of south Georgia, 164

Point bars in meandering rivers, 90
Port Hacking estuarine deposits, 240
Prodelta
    facies in, 210
    processes in, 212
Progressive sorting on shelves, 284
Proximal facies
    on abyssal plains, 385
    on arid aluvial fans, 35
    in ergs, 65
    on humid alluvial fans, 78
    in intermittent streams, 45
    in submarine trenches, 394

Radial submarine fans, 365, 371
Rangitata Geosyncline
    slope deposits, 353
    submarine trench deposits, 399
Reynold's number, 5
Rhythmic time defined, 20
Ridge and runnel, 161
Rio Dell Formation slope deposits, 353
Rip channels, 160
Rip currents, 158
Ripples
    current, 7
    eolian, 9
    oscillatory, 11
Rockdale deltaic system, 225
Rogue Canyon, 348
Rotliegendes arid system, 69

Saltation, 9
San Marcos Strandplain/Bay System, 246
San Nicholas Basin, 386
Sand flats in braided channels, 97
Sand patches, 266
Sand ribbons, 266
Sand ridges on continental shelves, 263
Sand tongues, 142
Sand waves
    along tide-dominated coasts, 143
    on continental shelves, 257, 263, 294
    origin, 7
Sand-rich submarine fans, 365, 371
Scroll bars in meandering rivers, 90
Sea level changes, effects of
    on barrier islands, 178, 185
    on continental shelves, 268, 279, 289
    on continental slopes, 340, 360
    on continental rises, 404
    on deltas, 217

on estuarine coasts, 233
on submarine canyons, 347
on submarine trenches, 394
on tidal coasts, 145
Sediment entrainment, 6
Sediment mass movement, 12
Sheetfloods on alluvial fans, 35
Shoreface bypassing, 261
Shublik Formation shelf sediments, 303, 318
Shumagin Formation trench deposits, 399
Sigsbee Abyssal Plain, 386
Silver Abyssal Plain, 385
Sinai Peninsula intermittent streams, 45
Siwalik Group alluvial sequences, 114
Slide zones, 337
Slides, mechanics of, 12
Slumps, mechanics of, 12
Sohm Abyssal Plain, 403
Solund Basin alluvial fans, 83
Sörlandet Sandstone, 83
Southern Pyrenees alluvial sequences, 109
Southwest Africa continental slope sequence,
    342
Splays in meandering rivers, 93
St. Peter Sandstone coastal deposits, 167
Steady time defined, 19
Stephanian coalfield alluvial fans, 85
Strandplains
    characteristics of, 192, 194
    controls on sequence evolution, 196
    evolution from barrier islands, 189
    origin of, 194, 196
    sequences in the rock record, 196
Stratigraphy, defined, 1
Stream flow, 5
Stream flow on arid alluvial fans, 34
Stream power, 6
Submarine canyons
    controls on sequence evolution, 350, 359
    effects of climate, 360
    effects of sea level fluctuation, 347
    hemipelagic sedimentaton in, 350
    origin, 324
    relation to sources, 349
    sediment movement in, 324
    sediments in, 334
    sedimentation in, 347, 349, 352, 358
    tidal currents in, 332
    turbidity flows in, 337
    wave-forced currents in, 332
Submarine channels, in trenches, 393, 394

Submarine fans
  on continental rises, 403
  controls on fan growth, 376
  controls on sequence evolution, 371
  effects of sea level variations, 376
  facies of, 369
  progradational sequences, 373, 378
  retrogradational sequences, 374, 380
  in rock record, 377
  shape of, 365
  subdivisions of, 366
  in trenches, 393, 397
Submarine ramps, 369, 373
Submarine trenches
  controls on sequence evolution, 394
  defined, 390
  effects of convergence, 394, 398
  morphology of, 391, 393
  sediment assemblages in, 391
  sediment transport in, 391, 393
  sequences in the rock record, 399
Sudmoor Point Sandstone alluvial sequence,
  105
Supercritical flow, 6
Suprafan, lobes, 368
Supratidal zone, 140
Surface creep, 9
Surinam Coast chenier plain, 152
Suspension, 8

Tanafjord Group shelf sediments, 295
Te Anau Assemblage trench deposits, 399
Texas-Louisiana chenier plain, 153
Threshold stress defined, 16
Tidal bundles, 274
Tidal channels
  in Cohansey Sand, 190
  effects on facies preservation, 141
  in estuaries, 233, 238
  in Fall River Formation, 147
  on Niger Delta, 216
Tidal creeks
  effects on barrier evolution, 183, 189
  on Kiawah Island, 190
  in lagoons, 173
Tidal currents
  in Cretaceous Seaway, 308
  in Devonian Catskill Sea, 308
  in epeiric seas, 307
  generation on shelves, 259, 306
  in submarine canyons, 342

Tidal flats
  on delta plains, 209
  in estuaries, 233
  on tide-dominated coasts, 138
Tidal inlets
  characteristics of, 175
  effect on barrier island evolution, 185, 189,
    190
  in estuaries, 233
  processes in, 177
  sequences in, 177
Tide-dominated coasts
  characteristics of, 138
  evolution of depositional sequences, 144
  facies of, 140
  occurrence, 138
  sequences in the rock record, 147
Tides, origin of, 138
Time
  cyclic, 19
  graded, 18
  steady, 18
Torrey Canyon slope deposits, 358
Tranquil flow, 5
Transform margins, 329
Transition zones defined, 26
Transverse bars
  in braided channels, 95
  in meandering rivers, 90
Trentishoe Formation intermittent streams, 46
Turbidity currents
  mechanics of, 13
  products of, 14
  on submarine fans, 368, 376
Turbulent flow, 5

Uinta Basin humid system sequences, 127
Uniformity defined, 15
Upper shoreface, 160
  on barriers, 169

Vaeroy Conglomerate, 83
Vail model of sequence evolution, 289
Van Horn Sandstone alluvial fan, 81

Walther's principle, 23
Washover fans, 175
Water waves, mechanics of, 10
Wave base defined, 10

Wave-dominated coasts
  barred coasts, 160
  controls on depositional sequences, 164
  facies of, 160
  high wave energy, 165
  low wave energy, 164
  non-barred coasts, 161
  occurrence, 157
  processes along, 158
  sequences in the rock record, 166
  systems on, 169

Wave-induced sediment movement, 10
Waves
  internal, 331
  properties of, 157
  transformation in shallow water, 158
Wilcox Group
  coastal zone deposits, 245
  deltaic deposits, 225
Wilkins Peak Member saline lakes, 53
Willapa Bay estuarine deposits, 234
Wind flows, sand transport, 65